Annals of Mathematics Studies

Number 122

HARMONIC ANALYSIS IN PHASE SPACE

by

Gerald B. Folland

PRINCETON UNIVERSITY PRESS

PRINCETON, NEW JERSEY

1989

Printed in the United States of America by

Princeton University Press, 41 William Street,

Princeton, New Jersey 08540

The Annals of Mathematics Studies are edited by William Browder,

Robert Langlands, John Milnor, and Elias M. Stein

Corresponding editors:

Stefan Hildebrandt, H. Blaine Lawson, Louis Nirenberg, and David Vogan

Clothbound editions of Princeton University Press books are

printed on acid-free paper, and binding materials are chosen for

strength and durability. Paperbacks, while satisfactory for

personal collections, are not usually suitable for library rebinding

Library of Congress Cataloging in Publication Data

Folland, G. B.

    Harmonic analysis in phase space / by Gerald B. Folland

    p. cm.- (The Annals of mathematics studies ; 122)

    Bibliography: p.

    Includes index.

    1. Phase space (Statistical physics) 2. Harmonic analysis.

I. Title. II. Series: Annals of mathematics studies ; no. 122.

QC174.85.P48F65 1989   530.1'3   88-26693

ISBN 0-691-08527-7 (alk. paper)

ISBN 0-691-08528-5 (pbk.)

9 8 7 6 5 4 3 2

CONTENTS

PREFACE

The phrase "harmonic analysis in phase space" is a concise if somewhat inadequate name for the area of analysis on $\mathbf{R}^n$ that involves the Heisenberg group, quantization, the Weyl operational calculus, the metaplectic representation, wave packets, and related concepts: it is meant to suggest analysis *on* the configuration space $\mathbf{R}^n$ done by working *in* the phase space $\mathbf{R}^n \times \mathbf{R}^n$. The ideas that fall under this rubric have originated in several different fields—Fourier analysis, partial differential equations, mathematical physics, representation theory, and number theory, among others. As a result, although these ideas are individually well known to workers in such fields, their close kinship and the cross-fertilization they can provide have often been insufficiently appreciated. One of the principal objectives of this monograph is to give a coherent account of this material, comprising not just an efficient tour of the major avenues but also an exploration of some picturesque byways.

Here is a brief guide to the main features of the book. Readers should begin by perusing the Prologue and perhaps refreshing their knowledge about Gaussian integrals by glancing at Appendix A.

Chapter 1 is devoted to the description of the representations of the Heisenberg group and various integral transforms and special functions associated to them, with motivation from physics. The material in the first eight sections is the foundation for all that follows, although readers who wish to proceed quickly to pseudodifferential operators can skip Sections 1.5–1.7.

The main point of Chapter 2 is the development of the Weyl calculus of pseudodifferential operators. As a tool for studying differential equations, the Weyl calculus is essentially equivalent to the standard Kohn-Nirenberg calculus—in fact, this equivalence is the principal result of Section 2.2—but it is somewhat more elegant and more natural from the point of view of harmonic analysis. Its close connection with the Heisenberg group yields some insights which are useful in the proofs of the Calderón-Vaillancourt $(0,0)$ estimate and the sharp Gårding inequality in Sections 2.5 and 2.6 and in the arguments of Section 3.1. Since my aim is to provide a reasonably accessible introduction rather than to develop a general theory (in contrast to Hörmander [70]), I mainly restrict attention to the standard symbol classes $S^m_{\rho,\delta}$. Moreover, I assume that the relevant estimates on symbols and their derivatives hold uniformly on all of $\mathbf{R}^n$ rather than on compact sets. This simplification makes the theory cleaner without restricting its generality in an essential way, as the study of localized symbols can generally be reduced to the study of global ones by standard tricks involving cutoff functions.

Chapter 3 grew out of my attempt to understand the Córdoba-Fefferman paper [33] on wave packet transforms in the context of the Weyl calculus. What has resulted, in Sections 3.1 and 3.2, is a new approach to their results which shows that, for some purposes, their Gaussian wave packets can be replaced by arbitrary (nonzero) even Schwartz class functions.

Chapter 4 is devoted to the metaplectic representation. It is more comprehensive than most other accounts in the literature, but it is still only an introduction. An exhaustive discussion of the many facets and applications of this beautiful representation and its siblings and children would require a book by itself.

Finally, Chapter 5 is my own retelling of some recent work of R. Howe [76], whom I wish to thank for permission to include his results. It ties together a number of strands from the previous chapters and provides, in my opinion, a satisfying conclusion to the book.

One problem with writing a book like this is deciding what background to expect of the readers. Basically, I take for granted a knowledge of real analysis, Fourier analysis, and basic functional analysis such as can be found in the first eight chapters of my text [50]; on this foundation, plus a few additional facts from functional analysis and Lie theory that are needed here and there, the book is pretty much self-contained. However, the material in it impinges on a number of subjects, including partial differential equations, spectral theory and the analysis of self-adjoint operators, Hamiltonian mechanics, quantum mechanics, Lie groups, and representation theory. Readers who are acquainted with these subjects will find their appreciation enhanced thereby; those who are not will find a few places where the path is hard to follow, but they are urged to forge resolutely ahead.

Another problem is deciding where to stop. I have allowed the selection of topics treated in detail to be governed by personal taste, while providing an extensive list of references to related material. These references, together with the references *they* contain, should be enough to keep anyone busy for quite a while. Nonetheless, many readers will undoubtedly ask at some point or other, "Why didn't you mention topic X or the work of Y?" In some cases I can plead the necessity of keeping the scope of the book within reasonable bounds, but in others the answer must be—as Samuel Johnson said when asked why he had given an erroneous definition in his Dictionary—"Ignorance, madam, sheer ignorance."

The seeds in my mind from which this book grew began to germinate after I attended the conference on Harmonic Analysis and Schrödinger Equations at the University of Colorado in March 1986. The book took shape during my sabbatical in the following academic year. In particular, I gave a series of lectures at the Bangalore Centre of the Indian Statistical Institute in the winter of 1987 that constituted a sort of first draft of Chapters 1, 2, and 4. I am grateful to the Indian Statistical Institute, the University of New South Wales, the Australian National University, and especially my friends on their faculties, for providing me with extremely congenial environments in which to work during the first half of 1987.

Gerald B. Folland

Seattle, Washington
June, 1988

HARMONIC ANALYSIS IN PHASE SPACE

PROLOGUE.
SOME MATTERS OF NOTATION

Readers are urged to examine this material before proceeding further.

Integrals. The integral of a function f on $\mathbf{R}^n$ or $\mathbf{C}^n$ with respect to Lebesgue measure will be denoted by $\int f(\xi)\,d\xi$ where ξ is any convenient dummy variable, be it real or complex. Thus, if z is a (scalar) complex variable, dz denotes the element of area on $\mathbf{C}$ and not the holomorphic differential used in defining contour integrals.

Functions and Distributions. We use the standard notation of [50] for function spaces on $\mathbf{R}^n$ or $\mathbf{C}^n$. For example, $L^p(\mathbf{R}^n)$ is the L^p space with respect to Lebesgue measure, $\|f\|_p$ is the L^p norm of f, and $C_c^\infty(\mathbf{R}^n)$ is the space of C^∞ functions with compact support. The inner product on L^2 is denoted by

$$(1) \qquad \langle f, g \rangle = \int f(x)\overline{g(x)}\,dx.$$

Inner products on other Hilbert spaces will also be denoted by $\langle \cdot, \cdot \rangle$ or by $\langle \cdot, \cdot \rangle_*$ where $*$ is a subscript to label the Hilbert space in question.

$\mathcal{S}(\mathbf{R}^n)$ and $\mathcal{S}'(\mathbf{R}^n)$ are the Schwartz spaces of rapidly decreasing smooth functions and tempered distributions, respectively. "Convergence in $\mathcal{S}'$" always means convergence in the weak $*$ topology on $\mathcal{S}'$. The pairing between $\mathcal{S}$ and $\mathcal{S}'$ will be denoted either by integrals or by pointed brackets, in a manner consistent with equation (1). Thus, if $f \in \mathcal{S}'$ and $\phi \in \mathcal{S}$ we write

$$\int \phi(x)f(x)\,dx = \lim \int \phi(x)f_j(x)\,dx,$$

$$\langle \phi, f \rangle = \overline{\langle f, \phi \rangle} = \lim \int \phi(x)\overline{f_j(x)}\,dx,$$

where $\{f_j\}$ is any sequence of smooth functions that converges in $\mathcal{S}'$ to f. In general, we shall be quite cavalier about writing distributions as if they were functions, especially under integral signs. δ will denote the Dirac distribution defined by $\langle \delta, f \rangle = f(0)$.

Of fundamental importance for us are the operators X_j and D_j on distributions on $\mathbf{R}^n$ defined by

$$(2) \qquad (X_j f)(x) = x_j f(x), \qquad D_j f = \frac{1}{2\pi i}\frac{\partial f}{\partial x_j}.$$

We generally regard X_j and D_j as continuous operators on $\mathcal{S}$ or $\mathcal{S}'$. When we regard them as unbounded operators on L^2, their domains are the obvious maximal ones: the domain of X_j is the set of $f \in L^2$ such that $X_j f \in L^2$, and likewise for D_j.

Matrices and Vectors. We denote by $M_n(\mathbf{R})$ and $M_n(\mathbf{C})$ the space of $n \times n$ matrices over $\mathbf{R}$ and $\mathbf{C}$, respectively. We identify linear endomorphisms of $\mathbf{R}^n$ and $\mathbf{C}^n$ with their matrices with respect to the canonical basis, and hence think of elements of $M_n(\mathbf{R})$ and $M_n(\mathbf{C})$ as either matrices or linear maps, according to context. The transpose and Hermitian adjoint of a matrix A are denoted by $A^\dagger$ and A^*; for real matrices, when these two notions coincide, we generally use the notation A^*. We employ the standard notation for the classical groups of invertible matrices: $GL(n, \mathbf{R})$, $U(n)$, $SO(n)$, etc. The $n \times n$ identity matrix is denoted by I_n when precision is needed, but more often simply by I. For powers of determinants, we use the convention that is common for trig functions: $\det^\alpha A = (\det A)^\alpha$.

Except in a few instances where clarity demands otherwise, we denote the dot product of two vectors in $\mathbf{R}^n$ or $\mathbf{C}^n$ by simple juxtaposition:

$$xy = \sum_1^n x_j y_j \qquad (x, y \in \mathbf{R}^n \text{ or } \mathbf{C}^n).$$

Thus, the Hermitian inner product of $z, w \in \mathbf{C}^n$ is $z\overline{w}$. We also set

$$x^2 = xx = \sum x_j^2 \qquad (x \in \mathbf{R}^n \text{ or } \mathbf{C}^n),$$
$$|z|^2 = z\overline{z} = \sum |z_j|^2 \qquad (z \in \mathbf{C}^n).$$

When linear mappings intervene in such products, we shall generally take care to write them between the two vectors, thus:

$$xAy = yA^\dagger x = x \cdot Ay = \sum x_j A_{jk} y_k \qquad (x, y \in \mathbf{C}^n, \ A \in M_n(\mathbf{C})).$$

The notation xAy can be regarded as a shorthand for either the matrix product $x^\dagger A y$ (where x and y are regarded as column vectors) or the physicists' bra–ket notation $\langle \overline{x} | A | y \rangle$. These "dotless products" may look a bit peculiar at first, but they are usually very efficient.

One other bit of notation for vectors will be frequently used in connection with pseudodifferential operators: if $\xi \in \mathbf{R}^n$,

$$\langle \xi \rangle = (1 + \xi^2)^{1/2}.$$

The Fourier Transform. In this book the Fourier transform and its inverse are defined by

$$\mathcal{F}f(\xi) = \widehat{f}(\xi) = \int e^{-2\pi i x \xi} f(x)\, dx,$$

$$\mathcal{F}^{-1}f(x) = \int e^{2\pi i x \xi} f(\xi)\, d\xi,$$

for $f \in \mathcal{S}(\mathbf{R}^n)$. (Note the "dotless products," as discussed above, in the exponents.) $\mathcal{F}$ and $\mathcal{F}^{-1}$, of course, extend uniquely to linear automorphisms of $\mathcal{S}'(\mathbf{R}^n)$. The placement of the 2π's in the exponent is uncommon in partial differential equations but almost mandatory in harmonic analysis, for it is the only way, short of renormalizing Lebesgue measure, to make $\mathcal{F}$ both an isometry on L^2 and an algebra homomorphism on L^1:

$$\|\widehat{f}\|_2 = \|f\|_2 \qquad \text{and} \qquad (f * g)\widehat{\ } = \widehat{f}\widehat{g},$$

where

$$f * g(x) = \int f(x - y)g(y)\, dy = \int f(y)g(x - y)\, dy.$$

From the physical point of view, this convention regarding the 2π's amounts to setting Planck's constant h, rather than the more common $\hbar = h/2\pi$, equal to 1. It is the reason for the 2π in the definition of D_j (formula (2) above). It also has the effect that our definition of Hermite functions is not quite the standard one; see Section 1.7.

Incidentally, the Fourier inversion formula

$$\iint e^{2\pi i(u-v)\xi} f(v)\, dv\, d\xi = f(u)$$

can be expressed neatly in the language of distributions as

(3) $$\int e^{2\pi i x \xi} d\xi = \delta(x).$$

Sometimes the most perspicuous way of evaluating an iterated integral involving exponentials is to pretend that the integral (3) is absolutely convergent and interchange the order of integration. This trick is used several times in the text; in each instance the reader may verify that it is an application of the Fourier inversion theorem.

Phase Space. There seems to be no system of terminology for the various objects associated with phase space that is consistent with itself as well as with the traditions of classical mechanics and differential equations and that leads to the most elegant formulas in all situations. The system used in this book was not adopted without considerable thought, but it sometimes leads to formulations that readers (including the author) may find discordant. The following paragraphs are in the nature of an apology for this state of affairs.

In the first place, although the functorially correct definition of phase space is $(\mathbf{R}^n)^* \times \mathbf{R}^n$ (or, in some contexts, the cotangent bundle of $\mathbf{R}^n$), I have abandoned any attempt to distinguish between $\mathbf{R}^n$ and $(\mathbf{R}^n)^*$. Maintaining this distinction seems to be more trouble than it is worth, especially when (as frequently happens) we have to consider both $(\mathbf{R}^n)^* \times \mathbf{R}^n$ and *its* dual space. So, in this book, phase space is just $\mathbf{R}^n \times \mathbf{R}^n$, or $\mathbf{R}^{2n}$ for short.

Next, there is the question of what to call points in phase space, or the coordinate functions on phase space. In classical mechanics the usual choice is (p, q), where $p \in \mathbf{R}^n$ denotes momentum and $q \in \mathbf{R}^n$ denotes position. On the other hand, in the literature of partial differential equations the same variables are usually denoted by ξ and x. I have found it convenient to employ both of these sets of labels: (ξ, x) for one copy of $\mathbf{R}^{2n}$ on which the symbols of pseudodifferential operators live, and (p, q) for another copy of $\mathbf{R}^{2n}$ (actually, its dual) on which their Fourier transforms live. The resulting usage of the letters p and q is sometimes, but not always, consistent with their interpretation as momentum and position.

I chose the ordering (p, q) in order to make certain formulas involving the Heisenberg group come out naturally (essentially, to avoid making Planck's constant negative). To prevent massive confusion, I was then forced to order the dual variables as (ξ, x) rather than (x, ξ). Consequently, in this book pseudodifferential operators are written as $\sigma(D, X)$, in flagrant disregard of the custom of writing them as $\sigma(X, D)$ or $\sigma(x, D)$. This, however, may serve the useful purpose of reminding readers conversant with pseudodifferential operators that $\sigma(D, X)$ is defined here by the Weyl calculus instead of the Kohn-Nirenberg calculus.

There are two canonical symplectic forms on $\mathbf{R}^n \times \mathbf{R}^n$, differing from each other by a factor of -1. One must simply make a choice; the symplectic form used here is denoted by square brackets and defined by

$$[(p, q), (p', q')] = pq' - qp' = \sum_1^n (p_j q_j' - q_j p_j'),$$

or

$$[X, Y] = X \mathcal{J} Y \qquad \text{where} \qquad \mathcal{J} = \begin{pmatrix} 0 & I \\ -I & 0 \end{pmatrix}.$$

The point that I found most troublesome is the question of whether to use the Euclidean Fourier transform or the symplectic Fourier transform on $\mathbf{R}^{2n}$,

$$\mathcal{F}f(\xi, x) = \iint e^{-2\pi i(p\xi + qx)} f(p, q)\, dp\, dq = \iint e^{-2\pi i(p,q)\cdot(\xi,x)} f(p, q)\, dp\, dq$$

or

$$\mathcal{F}_{\text{symp}} f(\xi, x) = \iint e^{2\pi i(px - q\xi)} f(p, q)\, dp\, dq = \iint e^{2\pi i[(p,q),\,(\xi,x)]} f(p, q)\, dp\, dq,$$

and correspondingly, whether to parametrize the Schrödinger representation by

$$\rho(p, q) = e^{2\pi i(pD + qX)} \qquad \text{or} \qquad \rho'(p, q) = e^{2\pi i(pX - qD)},$$

where

$$e^{2\pi i(pD + qX)} f(x) = e^{\pi i pq + 2\pi i qx} f(x + p),$$
$$e^{2\pi i(pX - qD)} f(x) = e^{-\pi i pq + 2\pi i px} f(x - q).$$

Since the symplectic structure of $\mathbf{R}^{2n}$ is of fundamental importance, the symplectic Fourier transform is in some ways more appropriate. Moreover, the operator $e^{2\pi i(pX - qD)}$ is geometrically more natural than $e^{2\pi i(pD + qX)}$, because it transforms wave packets whose mean momentum and position are a and b to wave packets whose mean momentum and position are $a + p$ and $b + q$, rather than $a + q$ and $b - p$. However, I was persuaded to use the Euclidean Fourier transform and the operators $e^{2\pi i(pD + qX)}$ by the following three considerations.

1. I find the symplectic Fourier transform confusing to use in performing specific calculations.

2. Occasionally we need to view $\mathbf{R}^{2n}$ both as the phase space of $\mathbf{R}^n$ and as a configuration space in its own right (i.e., as $\mathbf{R}^{n'}$ where n' happens to be $2n$), and consistency then demands the use of the Euclidean Fourier transform.

3. The parametrization $e^{2\pi i(pX - qD)}$ leads to some unsightly factors of $-i$ in the correspondence between the Schrödinger and Fock models.

This dilemma is in any event not of earthshaking importance, because the symplectic Fourier transform is simply the composition of the Euclidean Fourier transform with the map $(\xi, x) \to (x, -\xi)$, which belongs to both $SO(2n)$ and $Sp(n, \mathbf{R})$; and the operators $e^{2\pi i(pD + qX)}$ and $e^{2\pi i(pX - qD)}$ are intertwined by the Fourier transform on $\mathbf{R}^n$:

$$e^{2\pi i(pX - qD)} = \mathcal{F} e^{2\pi i(pD + qX)} \mathcal{F}^{-1}.$$

It is therefore a simple matter to translate formulas from one scheme to the other.

Index of Symbols. Here is a list of symbols frequently used in this book. Membership in it is generally limited to symbols whose use extends over more than one section of the text.

CHAPTER 1.
THE HEISENBERG GROUP
AND ITS REPRESENTATIONS

The Heisenberg group and its Lie algebra were born long before they were christened.

The Heisenberg Lie algebra is so named because its structure equations are Heisenberg's canonical commutation relations in quantum mechanics. These relations, however, are merely the quantized version of the Poisson bracket relations for canonical coordinates in Hamiltonian mechanics, and the importance of the latter was recognized as long ago as 1843 in Jacobi's lectures on dynamics ([82, Vorlesung 35]). Moreover, the Heisenberg group and its discrete variants have long played an implicit role in the theory of theta functions and related parts of analysis and number theory. But the names "Heisenberg group" and "Heisenberg algebra" did not come into common usage until the 1970's, and only since that time has the Heisenberg group received the recognition it deserves.

In an abstract sense, the Heisenberg group has only one locally faithful irreducible unitary representation. More precisely, up to unitary equivalence it has a one-parameter family of such representations, all of which are related to one another via automorphisms of the Heisenberg group. There are, however, several quite different "natural" ways of realizing these representations concretely in particular Hilbert spaces, and the really interesting part of the representation theory of the Heisenberg group consists of studying these realizations, the relations among them, and the integral transforms and special functions derived from them, in detail. It is to this task that this chapter is largely devoted.

1. Background from Physics

Much of the material in this monograph is motivated or illuminated by ideas coming from physics. The relevant physics is on a very basic level: the classical and quantum kinematics of a single particle moving in n-dimensional space. (We reassure the reader, perhaps unnecessarily, that the case $n = 3$ is not the only physically meaningful one: for example, if $n = 3k$, a "particle

in n-space" can be a mathematical model for k particles in 3-space.) In this
section we provide a very brief review of the classical and quantum pictures and
the relationship between them. For further information, the following are good
references: for classical mechanics, Abraham–Marsden [1], Arnold [6], Goldstein
[58]; for quantum mechanics, Landau–Lifshitz [94], Mackey [98], Messiah [103].

Hamiltonian Mechanics. According to Newton's second law, the motion of a particle is governed by a second-order ordinary differential equation
involving the forces acting on the particle, so that once these forces are known,
the motion of the particle is completely determined by its position and velocity at a particular time. In other words, position and velocity give a complete
specification of the "state" of the particle. However, in the Hamiltonian description it is found to be preferable to replace velocity by momentum ($=$ mass $\times$
velocity). Therefore, we shall take as the state space the so-called **phase space**
$\mathbf{R}^{2n}$ with coordinates

$$(p, q) = (p_1, ..., p_n, q_1, ..., q_n)$$

where p is the momentum vector of the particle and q is its position vector.
(From a coordinate-invariant point of view, one should regard momentum as
a cotangent vector and $\mathbf{R}^{2n}$ as the cotangent bundle of $\mathbf{R}^n$.) The physical
observables are real-valued functions on phase space. Thus every observable
is a function of position and momentum—another version of the fact that the
latter quantities specify the state of the system.

The time evolution of the system, and its various symmetries, are given by
certain transformations of phase space. The characteristic feature of Hamiltonian mechanics is that these transformations are not arbitrary diffeomorphisms,
but rather share the property of leaving invariant the differential form*

$$\Omega = \sum_1^n dp_j \wedge dq_j.$$

Ω is a translation-invariant bilinear form on tangent vectors; if we identify the
tangent space to $\mathbf{R}^{2n}$ at any point with $\mathbf{R}^{2n}$ itself, Ω becomes the standard
symplectic form on $\mathbf{R}^{2n}$, which we denote by brackets:

(1.1) $$[(p, q), (p', q')] = pq' - qp'.$$

---

* The formalism of differential geometry will be used for the next couple
of pages and then will disappear from the scene; the reader to whom it is
unfamiliar is asked to be indulgent during this interlude.

The diffeomorphisms of $\mathbf{R}^{2n}$ that preserve Ω are called **canonical transformations** or **symplectomorphisms**. The group of linear canonical transformations, i.e., the group of all $T \in GL(2n, \mathbf{R})$ such that

$$[T(p,q), T(p',q')] = [(p,q), (p',q')]$$

is the **symplectic group** $Sp(n, \mathbf{R})$.

Since the form Ω is nondegenerate, it provides an identification of tangent vectors with cotangent vectors. Actually, it provides two equally good ones, differing from each other by a minus sign; we shall choose the one that assigns to each tangent vector X at a point (p,q) the cotangent vector ω_X at (p,q) whose action as a linear form on tangent vectors at (p,q) is given by

$$\omega_X(Y) = \Omega(Y,X).$$

If f is a smooth observable (= function on $\mathbf{R}^{2n}$), the vector field associated to the one-form df under this correspondence is called the **Hamiltonian vector field** of f and is denoted by X_f. Explicitly, X_f is given by

$$\Omega(Y, X_f) = df(Y)$$

$$\text{or} \qquad X_f = \sum_1^n \left(\frac{\partial f}{\partial p_j} \frac{\partial}{\partial q_j} - \frac{\partial f}{\partial q_j} \frac{\partial}{\partial p_j} \right).$$

If f and g are smooth observables, their **Poisson bracket** $\{f, g\}$ is the observable defined by

(1.2) $$\{f,g\} = \Omega(X_f, X_g) = \sum \left(\frac{\partial f}{\partial p_j} \frac{\partial g}{\partial q_j} - \frac{\partial f}{\partial q_j} \frac{\partial g}{\partial p_j} \right).$$

It is easily verified that the Poisson bracket is skew-symmetric and satisfies the Jacobi identity

$$\{f, \{g, h\}\} + \{h, \{f, g\}\} + \{g, \{h, f\}\} = 0,$$

so that it makes the space of smooth observables into a Lie algebra. It is also easily verified that the map $f \to X_f$ is a Lie algebra homomorphism,

$$[X_f, X_g] = X_{\{f,g\}},$$

whose kernel consists of the constant functions. The Poisson bracket of the coordinate functions p_j, q_k are given by

(1.3) $$\{p_j, p_k\} = \{q_j, q_k\} = 0, \qquad \{p_j, q_k\} = \delta_{jk}.$$

Any system of coordinates p'_j, q'_k that satisfies (1.3) is called **canonical**, and the canonical transformations of $\mathbf{R}^{2n}$ are characterized by the fact that they map each canonical coordinate system into another one.

The "infinitesimal automorphisms" of the system are the vector fields X such that the flow generated by X consists of (local) canonical transformations. This holds precisely when the Lie derivative of Ω along X vanishes, i.e.,

$$i_X(d\Omega) + d(i_X\Omega) = 0$$

where, for any k-form ϕ, $i_X\phi$ is the contraction of X with ϕ:

$$i_X\phi(Y_1, ..., Y_{k-1}) = \phi(X, Y_1, ..., Y_{k-1}).$$

But $d\Omega = 0$ and $i_X\Omega = -\omega_X$ in the terminology introduced above, so X is an infinitesimal automorphism iff ω_X is closed. Since $\mathbf{R}^{2n}$ is simply connected, all closed one-forms are exact, so the closedness of ω_X means that X is a Hamiltonian vector field. In short, the map $f \to X_f$ establishes a one-to-one correspondence between smooth observables modulo constants and infinitesimal automorphisms. The latter comprise (in a rough sense) the "Lie algebra" of the group of canonical transformations, and the Poisson bracket is the pullback of this Lie algebra structure to the observables. If H is a smooth observable, the canonical transformations generated by X_H are obtained by integrating Hamilton's equations:

$$\frac{dp_j}{dt} = -\frac{\partial H}{\partial q_j}, \qquad \frac{dq_j}{dt} = \frac{\partial H}{\partial p_j}.$$

Quantum Mechanics. In classical mechanics, once the state of the system is specified, the value of every observable is completely determined. In the quantum world this is no longer true: in a given state, observables have only probability distributions of values, which may be, but usually are not, concentrated at a single point. The mathematical setup is as follows. The state space for a quantum system is a projective Hilbert space $\mathbf{P}\mathcal{H}$, that is, the set of all (complex) lines through the origin in a Hilbert space $\mathcal{H}$. Normally we think of states as being given by unit vectors in $\mathcal{H}$, with the understanding that two unit vectors define the same state when they are scalar multiples of one another. The observables are projection-valued Borel measures on $\mathbf{R}$, that is, mappings Π from the Borel sets in $\mathbf{R}$ to the orthogonal projections on $\mathcal{H}$ such that $\Pi(\mathbf{R}) = I$ and if $E_1, E_2, ...$ are disjoint Borel sets,

(1.4) $\Pi(E_j)\Pi(E_k) = 0 \quad \text{for} \quad j \neq k, \qquad \sum_j \Pi(E_j) = \Pi(\bigcup_j E_j).$

If $\mathcal{R}(E)$ is the range of $\Pi(E)$, (1.4) is equivalent to

$$\mathcal{R}(E_j) \perp \mathcal{R}(E_k) \quad \text{for} \quad j \neq k, \qquad \bigoplus_j \mathcal{R}(E_j) = \mathcal{R}\left(\bigcup_j E_j\right).$$

If Π is such a measure and $u \in \mathcal{H}$ is a unit vector, then $E \to \langle \Pi(E)u, u \rangle$ is an ordinary probability measure on $\mathbf{R}$, and this is the probability distribution of the observable Π in the state u.

By the spectral theorem, projection-valued measures Π are in one-to-one correspondence with self-adjoint operators A on $\mathcal{H}$ (cf. Appendix B):

$$A = \int \lambda \, d\Pi(\lambda), \qquad \Pi(E) = \chi_E(A).$$

Thus one can, and generally does, think of observables as self-adjoint operators. We note that if A is a self-adjoint operator and u is a unit vector in the domain of A, the mean or expectation of the observable A in the state u is

$$\int \lambda \langle d\Pi(\lambda)u, u \rangle = \langle Au, u \rangle.$$

The probability distribution of A in the state u is concentrated at a single point λ precisely when u is an eigenvector of a with eigenvalue λ. Moreover, if the spectrum of A is purely discrete, so that A has an orthonormal eigenbasis $\{e_j\}$ with eigenvalues $\{\lambda_j\}$, the probability distribution of A in any state is given by

$$(1.5) \qquad E \longrightarrow \sum_{\lambda_j \in E} |\langle u, e_j \rangle|^2.$$

For the system we are interested in, a particle moving in n-space, the Hilbert space $\mathcal{H}$ is taken to be $L^2(\mathbf{R}^n)$. If $f \in L^2(\mathbf{R}^n)$ is a unit vector, $|f|^2$ is interpreted as the probability density of the position of the particle in the state f; that is, the probability that the particle will be found in a set $B \subset \mathbf{R}^n$ is $\int_B |f|^2$. From this, we can easily identify the self-adjoint operators $Q_1, ..., Q_n$ corresponding to the classical coordinate functions $q_1, ..., q_n$. Namely, if $E \subset \mathbf{R}$, the probability that the jth coordinate x_j of the particle will lie in E is

$$(1.6) \qquad \int_{x_j \in E} |f(x)|^2 \, dx.$$

Thus the projection-valued measure Π_j for this observable is given by

$$\Pi_j(E) = \text{multiplication by the characteristic function of } \{x: x_j \in E\},$$

and it follows easily that the operator

$$Q_j = \int \lambda \, d\Pi_j(\lambda)$$

is multiplication by the jth coordinate function, which we generally denote by X_j:

(1.7) $$Q_j f(x) = X_j f(x) = x_j f(x),$$

defined on the domain of all $f \in L^2$ such that $x_j f \in L^2$.

We observe that there are no states $f \in L^2$ for which the observables Q_j have definite values: the simultaneous eigenfunctions of the Q_j's are the delta functions $\delta_{x_0}(x) = \delta(x - x_0)$. These, however, can be considered as a set of "idealized states" that form a "continuous orthonormal basis":

$$\langle \delta_{x_1}, \delta_{x_2} \rangle = \int \delta(x - x_1)\delta(x - x_2) \, dx = \delta(x_1 - x_2),$$

$$f = \int f(x)\delta_x \, dx = \int \langle f, \delta_x \rangle \delta_x \, dx,$$

all integrals being interpreted in the sense of distributions. The formula (1.6) for the distribution of Q_j is then the analogue of (1.5).

What about momentum? According to the principles of wave mechanics (cf. Messiah [103]), the eigenfunctions for momentum are the plane waves $e_\xi(x) = e^{2\pi i x \xi}$: the momentum of e_ξ is $h\xi$, where h is Planck's constant. Like the delta functions, the e_ξ's are not in L^2 but form a "continuous orthonormal basis":

$$\langle e_{\xi_1}, e_{\xi_2} \rangle = \int e^{2\pi i (\xi_1 - \xi_2)x} \, dx = \delta(\xi_1 - \xi_2),$$

$$f = \int \widehat{f}(\xi) e_\xi \, d\xi = \int \langle f, e_\xi \rangle e_\xi \, d\xi.$$

Here $\widehat{f}$ is the Fourier transform of f, and these equations are restatements of the Fourier inversion formula. By analogy with (1.6) and (1.7), we deduce that the probability in the state f that the jth component of the momentum will lie in E is

$$\int_{h\xi_j \in E} |\widehat{f}(\xi)|^2 \, d\xi$$

and hence that the self-adjoint operator P_j corresponding to the classical observable p_j is given by

$$(P_j f)\widehat{}(\xi) = h\xi_j \widehat{f}(\xi), \qquad \text{or} \qquad P_j = h\mathcal{F}^{-1} Q_j \mathcal{F}.$$

In other words,

$$(1.8) \qquad P_j = \frac{h}{2\pi i} \frac{\partial}{\partial x_j} = hD_j.$$

The "automorphisms" of a quantum mechanical system are the bijections on the state space $\mathbf{P}\mathcal{H}$ that preserve the projective inner product $|\langle u, v \rangle|^2$. (Here u and v are unit vectors in $\mathcal{H}$, but $|\langle u, v \rangle|^2$ depends only on the lines containing u and v.) By a theorem of Wigner (see Bargmann [12]), these are precisely the maps of projective space induced by unitary or anti-unitary operators on $\mathcal{H}$. For our purposes we may ignore the anti-unitary operators and regard the automorphism group as the unitary group on $\mathcal{H}$ modulo scalar multiples of the identity. The "infinitesimal automorphisms" are the generators of one-parameter groups of unitary operators, and by Stone's theorem (cf. Nelson [114], Reed–Simon [122]) these are precisely the skew-adjoint operators. If $B = -B^*$, the unitary group generated by B is e^{tB}; this is projectively trivial precisely when B is an imaginary multiple of I. If we disregard the (very substantial) difficulties involved in algebraically manipulating unbounded operators, we have a natural Lie algebra structure on the set of infinitesimal automorphisms, given by the commutator

$$[B_1, B_2] = B_1 B_2 - B_2 B_1.$$

Just as in classical mechanics, there is a correspondence between observables, modulo scalar multiples of I, and infinitesimal automorphisms, for we can convert any self-adjoint operator into a skew-adjoint one by multiplying by an imaginary constant. We shall find it convenient to take this constant to be $2\pi i$. Thus, the one-parameter unitary group associated to an observable A is $e^{2\pi i t A}$, and the induced (formal) Lie algebra structure on the set of observables is

$$(A_1, A_2) \longrightarrow \frac{1}{2\pi i}[2\pi i A_1, 2\pi i A_2] = 2\pi i[A_1, A_2].$$

(Remark: the "physically correct" choice of constant is not $2\pi i$ but $-2\pi i/h$. This is irrelevant for our purposes; it would result in the relabeling of some parameters when we describe certain unitary representations, but it does not ultimately affect the quantization procedures discussed in Chapter 2.) In particular, the basic observables Q_j and P_j satisfy the **canonical commutation relations**

$$(1.9) \qquad [P_j, P_k] = [Q_j, Q_k] = 0, \qquad [P_j, Q_k] = \frac{h \delta_{jk}}{2\pi i} I.$$

Quantization. By the "quantization problem" we shall mean the problem of setting up a correspondence $f \to A_f$ between classical and quantum observables, i.e., between functions on $\mathbf{R}^{2n}$ and self-adjoint operators on $L^2(\mathbf{R}^n)$, such that the properties of the classical observables are reflected as much as possible in their quantum counterparts in a way consistent with the probabilistic interpretation of quantum observables. Since this discussion is intended only to provide motivation, we shall ignore all technical difficulties associated with unbounded operators. On the formal level, then, a quantization procedure $f \to A_f$ ideally should have the following properties.

(i) The quantum counterparts of the position and momentum coordinates q_j and p_j should be the operators Q_j and P_j defined by (1.7) and (1.8). Moreover, if f is a constant function c, the probability that $f = c$ is one no matter which state the system is in, whence it follows that the quantum counterpart of f must be the operator cI. Thus:

$$(1.10) \qquad A_{q_j} = Q_j, \qquad A_{p_j} = P_j, \qquad A_c = cI.$$

(ii) If f, g are classical observables, the expectation of A_{f+g} in any state should be the sum of the expectations of A_f and A_g, that is,

$$\langle A_{f+g} u, u \rangle = \langle A_f u, u \rangle + \langle A_g u, u \rangle.$$

But if A is a self-adjoint operator, the diagonal matrix elements $\langle Au, u \rangle$ determine all matrix elements $\langle Au, v \rangle$, and hence the operator A, by polarization. Therefore,

$$(1.11) \qquad A_{f+g} = A_f + A_g.$$

(iii) Suppose $\phi: \mathbf{R} \to \mathbf{R}$ is a Borel function. If $E \subset \mathbf{R}$ and $\mathcal{Q}$ is a probabilistically determined quantity, the probability that $\phi(\mathcal{Q}) \in E$ is the probability that $\mathcal{Q} \in \phi^{-1}(E)$. Thus, if f is a classical observable and $A_f = \int \lambda \, d\Pi_f(\lambda)$, the spectral projections for $A_{\phi \circ f}$ should be $\Pi_{\phi \circ f}(E) = \Pi_f(\phi^{-1}(E))$. But these are the spectral projections for the operator $\phi(A_f)$ defined by the spectral functional calculus (cf. Appendix B). Thus:

$$(1.12) \qquad A_{\phi \circ f} = \phi(A_f).$$

(iv) A much weaker requirement than (1.12) is that $A_{cf} = cA_f$ (c constant) and $A_{f^2} = (A_f)^2$, and this together with (1.11) implies that the quantum counterpart of a product fg should be the Jordan product of A_f and A_g. Indeed,

$$(A_f + A_g)^2 = (A_{f+g})^2 = A_{(f+g)^2} = (A_f)^2 + 2A_{fg} + (A_g)^2,$$

so that

$$(1.13) \qquad A_{fg} = \tfrac{1}{2}(A_f A_g + A_g A_f).$$

(v) Finally, there should be a correspondence between the Lie algebra structures of classical and quantum observables: $[A_f, A_g]$ should be a constant multiple of $A_{\{f,g\}}$. In view of (1.9) and (1.10), the constant must be $h/2\pi i$:

$$(1.14) \qquad [A_f, A_g] = \frac{h}{2\pi i} A_{\{f,g\}}.$$

Now, how much of this can be accomplished? We shall insist on the basic position-momentum correspondence (1.10), the additivity (1.11), and the very special case $A_{cf} = cA_f$ ($c \in \mathbf{R}$) of (1.12): in other words, we require $f \to A_f$ to be a *linear* map satisfying (1.10). However, there is no such map that satisfies either (1.13) or (1.14) (or (1.12), which is stronger than (1.13)). That (1.13) is impossible can easily be seen as follows. Let $f(p,q) = p_1$ and $g(p,q) = q_1$. If (1.13) were true we would have

$$\tfrac{1}{4}(P_1 Q_1 + Q_1 P_1)^2 = (A_{fg})^2 = A_{f^2 g^2} = \tfrac{1}{2}(P_1^2 Q_1^2 + Q_1^2 P_1^2).$$

But a simple calculation shows that

$$\tfrac{1}{4}(P_1 Q_1 + Q_1 P_1)^2 = \frac{-h^2}{4\pi^2}\left[x_1^2 \frac{\partial^2}{\partial x_1^2} + 2x_1 \frac{\partial}{\partial x_1} + \frac{1}{4} \right],$$

$$\tfrac{1}{2}(P_1^2 Q_1^2 + Q_1^2 P_1^2) = \frac{-h^2}{4\pi^2}\left[x_1^2 \frac{\partial^2}{\partial x_1^2} + 2x_1 \frac{\partial}{\partial x_1} + 1 \right].$$

The proof that (1.14) cannot be satisfied is a bit more involved; we shall present it in Section 4.4 (Groenewold's theorem).

We should not be too surprised or disappointed at these negative results: life would be dull if things were so simple! We shall, however, keep (1.12), (1.13), and (1.14) in mind as guidelines, and in Chapter 2 we shall construct quantization procedures which satisfy them in an approximate sense for large classes of observables, the approximation being good when Planck's constant h is small. We shall also investigate the extent to which the (pointwise) boundedness or positivity of a classical observable can be reflected in the (operator-theoretic) boundedness or positivity of its quantum counterpart.

2. The Heisenberg Group

The Poisson bracket relations (1.3) for canonical coordinates in Hamiltonian mechanics and the commutation relations (1.9) for their quantum analogues are formally identical, and the abstract algebraic structure underlying them is the following. We consider $\mathbf{R}^{2n+1}$ with coordinates

$$(p_1, ..., p_n, q_1, ..., q_n, t) = (p, q, t),$$

and we define a Lie bracket on $\mathbf{R}^{2n+1}$ by

(1.15) $[(p, q, t), (p', q', t')] = (0, 0, pq' - qp') = (0, 0, [(p, q), (p', q')])$,

where the bracket on the right is the symplectic form on $\mathbf{R}^{2n}$. It is easily verified that the bracket (1.15) makes $\mathbf{R}^{2n+1}$ into a Lie algebra, called the **Heisenberg Lie algebra** and denoted by h_n. If $P_1, ..., P_n, Q_1, ..., Q_n, T$ is the standard basis for $\mathbf{R}^{2n+1}$, the Lie algebra structure is given by

(1.16) $[P_j, P_k] = [Q_j, Q_k] = [P_j, T] = [Q_j, T] = 0,$ $[P_j, Q_k] = \delta_{jk} T.$

Thus, (1.3) and (1.9) say that in both classical and quantum mechanics, the momentum, position, and constant observables span a Lie algebra isomorphic to h_n.

In order to identify the Lie group corresponding to h_n, it is convenient to use a matrix representation. Given $(p, q, t) \in \mathbf{R}^{2n+1}$, we define the matrix $m(p, q, t) \in M_{n+2}(\mathbf{R})$ by

$$
m(p, q, t) = \begin{pmatrix}
0 & p_1 & \cdots & p_n & t \\
0 & 0 & \cdots & 0 & q_1 \\
\vdots & \vdots & \ddots & \vdots & \vdots \\
0 & 0 & \cdots & 0 & q_n \\
0 & 0 & \cdots & 0 & 0
\end{pmatrix}
$$

(where all entries are zero except on the first row and last column). Moreover, we define

$$M(p, q, t) = I + m(p, q, t).$$

It is easily verified that

(1.17) $m(p, q, t)m(p', q', t') = m(0, 0, pq'),$

(1.18) $M(p, q, t)M(p', q', t') = M(p + p', q + q', t + t' + pq').$

From (1.17) it follows that

$$[m(p, q, t), m(p', q', t')] = m(0, 0, pq' - qp'),$$

where the bracket now denotes the commutator. Hence the correspondence $X \to m(X)$ is a Lie algebra isomorphism from h_n to $\{m(X) : X \in \mathbf{R}^{2n+1}\}$, and to obtain the corresponding Lie group we can simply apply the matrix exponential map. From (1.17) we have

$$m(p, q, t)^2 = m(0, 0, pq) \quad \text{and} \quad m(p, q, t)^k = 0 \quad \text{for} \quad k \geq 3,$$

so

(1.19) $e^{m(p,q,t)} = I + m(p,q,t) + \frac{1}{2}m(0,0,pq) = M(p,q,t+\frac{1}{2}pq)$.

Thus the exponential map is a bijection from $\{m(X) : X \in \mathbf{R}^{2n+1}\}$ to $\{M(X) : X \in \mathbf{R}^{2n+1}\}$, and the latter is a group with group law (1.18). We could take this to be the Lie group corresponding to h_n, but we prefer to use a slightly different model. It is easily verified that

$$\exp m(p,q,t)\exp m(p',q',t') = \exp m\big(p+p',\, q+q',\, t+t'+\tfrac{1}{2}(pq'-qp')\big).$$

Therefore, if we identify $X \in \mathbf{R}^{2n+1}$ with the matrix $e^{m(X)}$, we make $\mathbf{R}^{2n+1}$ into a group with group law

(1.20) $(p,q,t)(p',q',t') = \big(p+p',\, q+q',\, t+t'+\tfrac{1}{2}(pq'-qp')\big)$.

We call this group the **Heisenberg group** and denote it by $\mathbf{H}_n$. The exponential map from h_n to $\mathbf{H}_n$ is then merely the identity, and the inverse of the element (p,q,t) is simply $(-p,-q,-t)$.

Occasionally it is better to identify (p,q,t) with the matrix $M(p,q,t)$, which by (1.18) yields the group law

$$(p,q,t)(p',q',t') = (p+p',\, q+q',\, t+t'+pq').$$

We call $\mathbf{R}^{2n+1}$ with this group law the **polarized Heisenberg group** and denote it by $\mathbf{H}_n^{\mathrm{pol}}$. By (1.19), the map

$$(p,q,t) \longrightarrow (p,q,t+\tfrac{1}{2}pq)$$

is an isomorphism from $\mathbf{H}_n$ to $\mathbf{H}_n^{\mathrm{pol}}$, and it is also the exponential map from h_n to $\mathbf{H}_n^{\mathrm{pol}}$. The inverse of (p,q,t) in $\mathbf{H}_n^{\mathrm{pol}}$ is $(-p,-q,-t+pq)$.

We observe that

$$\mathcal{Z} = \{(0,0,t) : t \in \mathbf{R}\}$$

is the center, and also the commutator subgroup, of both $\mathbf{H}_n$ and $\mathbf{H}_n^{\mathrm{pol}}$. Moreover, Lebesgue measure on $\mathbf{R}^{2n+1}$ is a bi-invariant Haar measure on both $\mathbf{H}_n$ and $\mathbf{H}_n^{\mathrm{pol}}$.

The Automorphisms of the Heisenberg Group. We denote by $\mathrm{Aut}(\mathbf{H}_n)$ and $\mathrm{Aut}(\mathsf{h}_n)$ the automorphism groups of $\mathbf{H}_n$ and h_n (as a topological group and a Lie algebra, respectively). Since the underlying set of both $\mathbf{H}_n$ and h_n is $\mathbf{R}^{2n+1}$, $\mathrm{Aut}(\mathbf{H}_n)$ and $\mathrm{Aut}(\mathsf{h}_n)$ are both sets of mappings from $\mathbf{R}^{2n+1}$ to itself. In fact, they are equal. This is an instance of a general theorem about automorphisms of simply connected Lie groups and their Lie algebras, but we can give a simple direct proof.

(1.21) Proposition. $\mathrm{Aut}(\mathbf{H}_n) = \mathrm{Aut}(\mathfrak{h}_n)$.

Proof: By (1.15) and (1.20), the Heisenberg group and algebra structures on $\mathbf{R}^{2n+1}$ are related by

$$XY = X + Y + \tfrac{1}{2}[X, Y] \qquad (X, Y \in \mathbf{R}^{2n+1}).$$

From this it is clear that if $\alpha \in \mathrm{Aut}(\mathfrak{h}_n)$ then $\alpha \in \mathrm{Aut}(\mathbf{H}_n)$. On the other hand, suppose $\alpha \in \mathrm{Aut}(\mathbf{H}_n)$. If $[X, Y] = 0$ then

$$
\begin{aligned}
\alpha(X + Y) &= \alpha(XY) = \alpha(X)\alpha(Y) = \alpha(X) + \alpha(Y) + \tfrac{1}{2}[\alpha(X), \alpha(Y)] \\
&= \alpha(YX) = \alpha(Y)\alpha(X) = \alpha(X) + \alpha(Y) - \tfrac{1}{2}[\alpha(X), \alpha(Y)]
\end{aligned}
$$

and hence $\alpha(X + Y) = \alpha(X) + \alpha(Y)$. In particular, α is additive on every one-dimensional subspace, and it is continuous; hence it commutes with scalar multiplication. Therefore, if $X, Y \in \mathbf{R}^{2n+1}$ and $s \in \mathbf{R}$, we have

$$
\begin{aligned}
s\alpha\big(X + Y + \tfrac{1}{2}s[X, Y]\big) &= \alpha\big(sX + sY + \tfrac{1}{2}s^2[X, Y]\big) = \alpha\big((sX)(sY)\big) \\
&= \big(s\alpha(X)\big)\big(s\alpha(Y)\big) = s\alpha(X) + s\alpha(Y) + \tfrac{1}{2}s^2[\alpha(X), \alpha(Y)].
\end{aligned}
$$

If we divide through by s and let $s \to 0$ we obtain $\alpha(X + Y) = \alpha(X) + \alpha(Y)$. Thus α is linear; taking this into account, the above equation also shows that $\alpha([X, Y]) = [\alpha(X), \alpha(Y)]$. In short, $\alpha \in \mathrm{Aut}(\mathfrak{h}_n)$. ∎

We now identify the automorphisms of $\mathbf{H}_n$ explicitly. It is easy to write down several families of them:

(i) *Symplectic maps.* If $S \in Sp(n, \mathbf{R})$, the map

$$(p, q, t) \longrightarrow \big(S(p, q), t\big)$$

is clearly in $\mathrm{Aut}(\mathbf{H}_n)$. For the moment, we denote the group of such automorphisms of $\mathbf{H}_n$ by G_1.

(ii) *Inner Automorphisms.* It is easily checked that

$$(a, b, c)(p, q, t)(a, b, c)^{-1} = (p,\ q,\ t + aq - bp).$$

We denote the group of inner automorphisms of $\mathbf{H}_n$ by G_2.

(iii) *Dilations.* If $r > 0$, the map $\delta[r]$ defined by

$$\delta[r](p, q, t) = (rp, rq, r^2 t)$$

is obviously in $\mathrm{Aut}(\mathbf{H}_n)$; moreover, $\delta[rs] = \delta[r]\delta[s]$. We denote the group of dilations by G_3.

(iv) *Inversion.* The map i defined by

$$i(p, q, t) = (q, p, -t)$$

is in $\mathrm{Aut}(\mathbf{H}_n)$. We denote the two-element group consisting of i and the identity by G_4.

(1.22) Theorem. *With notation as above, every automorphism of* $\mathbf{H}_n$ *can be written uniquely as* $\alpha_1\alpha_2\alpha_3\alpha_4$ *with* $\alpha_j \in G_j$.

Proof: If $\alpha \in \mathrm{Aut}(\mathbf{H}_n)$, α maps the center $\mathcal{Z}$ to itself, and by Proposition (1.21), α is linear; hence α must be of the form

$$\alpha(p,q,t) = \big(T(p,q),\ ap + bq + st\big)$$

with $T \in GL(2n,\mathbf{R})$, $a,b \in \mathbf{R}^n$, and $s \in \mathbf{R} \setminus \{0\}$. By composing α with the inversion i if necessary, we can make $s > 0$; then by composing with the dilation $\delta[s^{-1/2}]$ we can make $s = 1$; finally by composing with a suitable inner automorphism we can make $a = b = 0$. What is left is a map of the form $(p,q,t) \to (S(p,q),t)$ where $S \in GL(2n,\mathbf{R})$, and clearly this is an automorphism of $\mathbf{H}_n$ iff $S \in Sp(n,\mathbf{R})$. ∎

3. The Schrödinger Representation

We recall that the quantum-mechanical position and momentum operators are $Q_j = X_j$ (multiplication by x_j) and $P_j = hD_j$ ($h/2\pi i$ times differentiation with respect to x_j). We may regard these operators as continuous operators on the Schwartz class $\mathcal{S}(\mathbf{R}^n)$. As such, they satisfy the commutation relations (1.9), and it follows that the map $d\rho_h$ from the Heisenberg algebra $\mathfrak{h}_n$ to the set of skew-Hermitian operators on $\mathcal{S}$ defined by

$$d\rho_h(p,q,t) = 2\pi i(hpD + qX + tI)$$

is a Lie algebra homomorphism. We wish to exponentiate this representation of $\mathfrak{h}_n$ to obtain a unitary representation of the Heisenberg group $\mathbf{H}_n$.

For the moment we take $h = 1$. The main point is to compute the operators $e^{2\pi i(pD+qX)}$. If $f \in L^2$, let

$$g(x,t) = [e^{2\pi it(pD+qX)}f](x).$$

g is the solution of the differential equation $\partial g/\partial t = 2\pi i(pD + qX)g$ subject to the initial condition $g(x,0) = f(x)$, that is,

$$\frac{\partial g}{\partial t} - \sum p_j \frac{\partial g}{\partial x_j} = 2\pi i qx g, \qquad g(x,0) = f(x).$$

The expression on the left is just the directional derivative of g along the vector $(-p,1)$, so if we set

$$x(t) = x - tp, \qquad G(t) = g\big(x(t),t\big),$$

we obtain

$$G'(t) = 2\pi i q(x - tp)G(t), \qquad G(0) = f(x).$$

This ordinary differential equation is easily solved:

$$g(x - tp, t) = G(t) = f(x)e^{2\pi i t q x - \pi i t^2 pq}.$$

Setting $t = 1$ and replacing x by $x + p$, we obtain the desired result:

$$(1.23) \qquad e^{2\pi i(pD+qX)} f(x) = e^{2\pi i qx + \pi i pq} f(x + p).$$

From this formula it is evident that $e^{2\pi i(pD+qX)}$ is a unitary operator on L^2, and it is easily checked that

$$(1.24) \qquad e^{2\pi i(pD+qX)} e^{2\pi i(rD+sX)} = e^{\pi i(ps-qr)} e^{2\pi i[(p+r)D+(q+s)X]}.$$

It follows therefore that the map ρ from $\mathbf{H}_n$ to the group of unitary operators on L^2 defined by

$$\rho(p, q, t) = e^{2\pi i(pD+qX+tI)} = e^{2\pi i t} e^{2\pi i(pD+qX)},$$

that is,

$$(1.25) \qquad \rho(p, q, t)f(x) = e^{2\pi i t + 2\pi i qx + \pi i pq} f(x + p),$$

is a unitary representation of $\mathbf{H}_n$. Moreover, the operators $\rho(p, q, t)$, besides being unitary on L^2, are continuous on $\mathcal{S}$ and extend to continuous operators on $\mathcal{S}'$; and we shall frequently regard them as such.

At this point we can put Planck's constant back in. Namely, we set

$$\rho_h(p, q, t) = \rho(hp, q, ht) = e^{2\pi i h t} e^{2\pi i(hpD+qX)},$$

or

$$(1.26) \qquad \rho_h(p, q, t)f(x) = e^{2\pi i h t + 2\pi i qx + \pi i hpq} f(x + hp).$$

Then for any real number h, ρ_h is a unitary representation of $\mathbf{H}_n$ on $L^2(\mathbf{R}^n)$, and the corresponding representation $d\rho_h$ of $\mathbf{h}_n$ is given by (1.23). Moreover, ρ_h and $\rho_{h'}$ are inequivalent for $h \neq h'$. (It suffices to observe that the central characters $e^{2\pi i h t}$ and $e^{2\pi i h' t}$ are inequivalent for $h \neq h'$.) We shall see in the next section that ρ_h is irreducible for $h \neq 0$.

We call ρ_h the **Schrödinger representation** of $\mathbf{H}_n$ with parameter h. Generally we shall take $h = 1$ and restrict attention to the representation

$\rho = \rho_1$; the generalization to other (nonzero) values of h is an easy exercise which we shall omit except when it leads to something of particular interest.

As we pointed out in the Prologue, in some ways it is more natural to replace ρ by the representation

$$(1.27) \qquad \rho'(p,q,t) = \rho(-q,p,t) = e^{2\pi i t}e^{2\pi i(pX-qD)},$$

in which the symplectic form rather than the Euclidean inner product is used to pair (p,q) with (D,X) in the exponent. Indeed, we have

$$e^{2\pi i(pX-qD)}f(x) = e^{2\pi i px - \pi i pq}f(x-q)$$
$$[e^{2\pi i(pX-qD)}f]\widehat{\;}(\xi) = e^{-2\pi i q\xi + \pi i pq}\widehat{f}(\xi - p).$$

Thus if the mean values of the position and momentum of f are x_0 and ξ_0, the mean values of the position and momentum of $e^{2\pi i(pX-qD)}f$ are $x_0 + q$ and $\xi_0 + p$. The operator $e^{2\pi i(pX-qD)}$ therefore represents a translation in position space by q and a translation in momentum space by p, which accords with the usual interpretation of p and q as momentum and position variables. However, ρ' also has its disadvantages, and we shall generally stick to ρ. In any case, it is easy to go from one representation to the other. On the Heisenberg group side, it is just a matter of composing with the automorphism $(p,q,t) \to (-q,p,t)$ of $\mathbf{H}_n$; and on the L^2 side, ρ and ρ' are intertwined by the Fourier transform, as one can easily check:

$$\mathcal{F}\rho(p,q,t)\mathcal{F}^{-1} = \rho'(p,q,t).$$

The kernel of ρ is $\{(0,0,k) : k \in \mathbf{Z}\}$. For some purposes it is better to throw away this kernel, so we define the **reduced Heisenberg group** $\mathbf{H}_n^{\text{red}}$ to be the quotient

$$\mathbf{H}_n^{\text{red}} = \mathbf{H}_n \,/\, \{(0,0,k) : k \in \mathbf{Z}\}.$$

We still write elements of $\mathbf{H}_n^{\text{red}}$ as (p,q,t), with the understanding that t is taken to be a real number mod 1, and we regard ρ as a representation of $\mathbf{H}_n^{\text{red}}$, which is now *faithful*. In fact, since the central variable t always acts in a simple-minded way, as multiplication by the scalar $e^{2\pi i t}$, it is often convenient to disregard it entirely; we therefore define

$$\rho(p,q) = \rho(p,q,0) = e^{2\pi i(pD+qX)}.$$

The Integrated Representation. The unitary representation ρ of $\mathbf{H}_n^{\text{red}}$ determines a representation of the convolution algebra $L^1(\mathbf{H}_n^{\text{red}})$, still denoted by ρ, in the usual way: if $\Phi \in L^1(\mathbf{H}_n^{\text{red}})$,

$$\rho(\Phi) = \int_{\mathbf{H}_n^{\text{red}}} \Phi(X)\rho(X)\,dX = \iiint \Phi(p,q,t)\rho(p,q,t)\,dp\,dq\,dt.$$

The integral here is a Bochner integral, and $\rho(\Phi)$ is an operator on $L^2(\mathbf{R}^n)$ satisfying $\|\rho(\Phi)\| \le \|\Phi\|_1$.

Given $\Phi \in L^1(\mathbf{H}_n^{\mathrm{red}})$, we can expand it in a Fourier series in the central variable t:

$$\Phi(p, q, t) = \sum_{-\infty}^{\infty} \Phi_k(p, q)e^{2\pi i k t}.$$

(This series can be interpreted, for example, as the limit in the L^1 norm of its Cesàro means.) Since $\rho(p, q, t) = e^{2\pi i t}\rho(p, q)$, we have

$$\rho(\Phi) = \sum_{-\infty}^{\infty} \iiint \Phi_k(p, q)\rho(p, q)e^{2\pi i(k+1)t}\, dp\, dq\, dt$$

$$= \iint \Phi_{-1}(p, q)\rho(p, q)\, dp\, dq.$$

Thus, the only part of Φ that contributes to $\rho(\Phi)$ is the (-1)th Fourier component Φ_{-1}, so we might as well consider ρ as a representation of $L^1(\mathbf{R}^{2n})$ (with a nonstandard convolution structure, to be discussed below) rather than of $L^1(\mathbf{H}_n^{\mathrm{red}})$. Accordingly, for $F \in L^1(\mathbf{R}^{2n})$ let us define

$$(1.28) \qquad \rho(F) = \iint F(p, q)\rho(p, q)\, dp\, dq = \iint F(p, q)e^{2\pi i(pD+qX)}\, dp\, dq.$$

($\rho(F)$ is sometimes called the "Weyl transform" of F, but this is historically inaccurate. In fact, the "Weyl transform" of F should be $\rho(\widehat{F})$, as we shall explain in Chapter 2.) The explicit formula for the operator $\rho(F)$ is as follows:

$$\rho(F)f(x) = \iint F(p, q)e^{2\pi i q x + \pi i p q}f(x + p)\, dp\, dq$$

$$= \iint F(y - x, q)e^{\pi i q(x+y)}f(y)\, dy\, dq.$$

In other words, $\rho(F)$ is an integral operator with kernel

$$(1.29) \qquad K_F(x, y) = \int F(y - x, q)e^{\pi i q(x+y)}\, dq$$

$$= (\mathcal{F}_2^{-1}F)\left(y - x, \frac{y + x}{2}\right),$$

where $\mathcal{F}_2$ denotes Fourier transformation in the second variable. From this we easily deduce:

(1.30) Theorem. *The map ρ from $L^1(\mathbf{R}^{2n})$ to the space of bounded operators on $L^2(\mathbf{R}^n)$, defined by (1.28), extends uniquely to a bijection from $\mathcal{S}'(\mathbf{R}^{2n})$ to the space of continuous linear maps from $\mathcal{S}(\mathbf{R}^n)$ to $\mathcal{S}'(\mathbf{R}^n)$. Moreover, ρ maps $L^2(\mathbf{R}^{2n})$ unitarily onto the space of Hilbert-Schmidt operators on $L^2(\mathbf{R}^n)$, and $\rho(F)$ is a compact operator on $L^2(\mathbf{R}^n)$ for all $F \in L^1(\mathbf{R}^{2n})$.*

Proof: The kernel K_F is obtained from F by partial Fourier transformation followed by an invertible and measure-preserving change of variable. These operations make sense when F is an arbitrary tempered distribution and define K_F as a tempered distribution. In this case the operation $f \to \int K_F(\cdot, y)f(y)\,dy$ defines a continuous linear map from $\mathcal{S}(\mathbf{R}^n)$ to $\mathcal{S}'(\mathbf{R}^n)$. Explicitly, if $f, g \in \mathcal{S}(\mathbf{R}^n)$,

$$\left\langle \int K_F(\cdot, y)f(y)\,dy, g \right\rangle = \langle K_F, g \otimes \overline{f} \rangle = \langle F, h \rangle$$

where $h \in \mathcal{S}(\mathbf{R}^{2n})$ is defined by

$$h(p, q) = \int e^{-2\pi i qx - \pi i pq}\overline{f(x + p)}g(x)\,dx.$$

By the Schwartz kernel theorem (see Treves [138]), every continuous linear map from $\mathcal{S}(\mathbf{R}^n)$ to $\mathcal{S}'(\mathbf{R}^n)$ is of this form. Moreover, the map $F \to K_F$ is clearly unitary on $L^2(\mathbf{R}^{2n})$, which shows that $\rho(L^2(\mathbf{R}^n))$ is the set of Hilbert-Schmidt operators. (For background on Hilbert-Schmidt operators, see Reed–Simon [122].) In particular, $\rho(F)$ is compact for $F \in L^1 \cap L^2$, and hence for all $F \in L^1$ since $\|\rho(F)\| \le \|F\|_1$ and the norm limit of compact operators is compact. ∎

Remark. For conditions for the operator $\rho(F)$ to be bounded on $L^p(\mathbf{R}^n)$, $1 < p < \infty$, see Mauceri [102].

For future reference, we record how the original representation $\rho(p, q)$ combines with the integrated representation $\rho(F)$. The following proposition is an easy consequence of the definitions.

Proposition (1.31). *If $F \in \mathcal{S}'(\mathbf{R}^{2n})$ and $a, b \in \mathbf{R}^n$, we have*

$$\rho(a, b)\rho(F) = \rho(G) \qquad \text{and} \qquad \rho(F)\rho(a, b) = \rho(H)$$

where

$$G(p, q) = e^{\pi i(bp - aq)}F(p - a, q - b) \qquad \text{and} \qquad H(p, q) = e^{\pi i(aq - bp)}F(p - a, q - b).$$

Twisted Convolution. We return for a moment to $L^1(\mathbf{H}_n^{\text{red}})$. This space is a Banach algebra under convolution,

$$\Phi * \Psi(X) = \int \Phi(Y)\Psi(Y^{-1}X)\,dY = \int \Phi(XY^{-1})\Psi(Y)\,dY,$$

and the representation ρ is an algebra homomorphism:

$$\rho(\Phi)\rho(\Psi) = \iint \Phi(X)\Psi(Y)\rho(XY)\,dX\,dY = \rho(\Phi * \Psi).$$

We wish to transfer this algebra structure to $L^1(\mathbf{R}^{2n})$. For $F \in L^1(\mathbf{R}^{2n})$ we have

$$\rho(F) = \rho(F^0) \qquad \text{where} \qquad F^0 \in L^1(\mathbf{H}_n^{\text{red}}), \quad F^0(p,q,t) = F(p,q)e^{-2\pi it},$$

and if $F, G \in L^1(\mathbf{R}^{2n})$,

$$F^0 * G^0(p,q,t)$$
$$= \int_0^1 \iint_{\mathbf{R}^{2n}} F(p',q')e^{-2\pi it'} G(p-p', q-q')e^{-2\pi i(t-t')+\pi i(p'q-q'p)}\,dp'\,dq'\,dt'$$
$$= e^{-2\pi it} \iint F(p',q')G(p-p', q-q')e^{\pi i(p'q-q'p)}\,dp'\,dq'.$$

That is,

$$F^0 * G^0 = (F \natural G)^0,$$

where

$$(1.32) \qquad F \natural G(p,q) = \iint F(p',q')G(p-p', q-q')e^{\pi i(p'q-q'p)}\,dp'\,dq'$$
$$= \iint F(p-p', q-q')G(p',q')e^{\pi i(pq'-qp')}\,dp'\,dq'.$$

We call $F \natural G$ the **twisted convolution** of F and G. Its definition is set up so that

$$\rho(F \natural G) = \rho(F)\rho(G).$$

Twisted convolution enjoys most of the properties of ordinary convolution on $\mathbf{R}^{2n}$ except that it is not commutative. Like ordinary convolution, it extends from L^1 to other L^p spaces and satisfies Young's inequality:

$$\|F \natural G\|_r \le \|F\|_p \|G\|_q \quad \text{when} \quad \frac{1}{p} + \frac{1}{q} = \frac{1}{r} + 1.$$

But with respect to L^p estimates, twisted convolution is even better than ordinary convolution:

(1.33) Proposition. *If $F, G \in L^2(\mathbf{R}^{2n})$ then $F \natural G \in L^2(\mathbf{R}^{2n})$ and $\|F \natural G\|_2 \leq \|F\|_2 \|G\|_2$.*

Proof: This follows from the fact, observed above, that the map $F \to K_F$ defined by (1.29) is an isometry on L^2:

$$\|F \natural G\|_2 = \|K_{F \natural G}\|_2 = \left\| \int K_F(x, y) K_G(y, z)\, dy \right\|_{L^2(x, z)}$$

$$\leq \|K_F\|_2 \|K_G\|_2 = \|F\|_2 \|G\|_2.$$

The second equality is a restatement of the fact that $\rho(F \natural G) = \rho(F)\rho(G)$, and the next estimate follows from the Schwarz inequality. ∎

One can obtain other L^p estimates for twisted convolution by interpolating between Proposition (1.33) and Young's inequality.

The Uncertainty Principle. The uncertainty principle in its general form states that if A and B are quantum observables (i.e., self-adjoint operators), the probability distributions of A and B cannot both be concentrated near single points in any state u such that $\langle (AB - BA)u, u \rangle \neq 0$. To make this precise, when μ is a probability measure on $\mathbf{R}$ we shall adopt the second moment of μ about $a \in \mathbf{R}$,

$$\left[\int (\lambda - a)^2\, d\mu(\lambda) \right]^{1/2},$$

as a measure of how much μ fails to be concentrated at a. When μ is the distribution of the observable $A = \int \lambda\, d\Pi(\lambda)$ in the state u, i.e., $\mu(E) = \langle \Pi(E)u, u \rangle$, we have

$$\left[\int (\lambda - a)^2\, d\mu(\lambda) \right]^{1/2} = \langle (A - a)^2 u, u \rangle^{1/2} = \|(A - a)u\|.$$

The general uncertainty principle can then be enunciated as follows.

(1.34) Theorem. *If A and B are self-adjoint operators on a Hilbert space $\mathcal{H}$, then*

$$\|(A - a)u\|\, \|(B - b)u\| \geq \tfrac{1}{2} |\langle (AB - BA)u, u \rangle|$$

for all $u \in \mathrm{Dom}(AB) \cap \mathrm{Dom}(BA)$ and all $a, b \in \mathbf{R}$. Equality holds precisely when $(A-a)u$ and $(B-b)u$ are purely imaginary scalar multiples of one another.

Proof: We have

$$\langle (AB - BA)u, u \rangle = \langle [(A - a)(B - b) - (B - b)(A - a)]u, u \rangle$$
$$= \langle (B - b)u, (A - a)u \rangle - \langle (A - a)u, (B - b)u \rangle$$
$$= 2i\, \mathrm{Im}\langle (B - b)u, (A - a)u \rangle$$

and hence

$$\langle (AB - BA)u, u \rangle \le 2 \left| \langle (B-b)u, (A-a)u \rangle \right| \le 2 \| (A-a)u \| \, \| (B-b)u \|.$$

The first inequality is an equality precisely when $\langle (B-b)u, (A-a)u \rangle$ is imaginary, and the second one is an equality precisely when $(A-a)u$ and $(B-b)u$ are linearly dependent. ∎

If we apply this result to the position and momentum operators X and D on $L^2(\mathbf{R})$, we obtain:

(1.35) Corollary. *If $u \in L^2(\mathbf{R})$ and $a, b \in \mathbf{R}$ we have*

(1.36) $$\| (X-a)u \|_2 \| (D-b)u \|_2 \ge \frac{1}{4\pi} \| u \|_2,$$

with equality if and only if

$$u(x) = ce^{2\pi i b x} e^{-\pi r(x-a)^2} \qquad \text{for some} \qquad c \in \mathbf{C}, \quad r > 0.$$

Proof: The inequality is valid by Theorem (1.34) since $[D, X] = (2\pi i)^{-1} I$. (The preceding proof works when $u \in \text{Dom}(DX) \cap \text{Dom}(XD)$, but the result remains valid for all $u \in L^2$, with the understanding that, for example, $\| (D-b)u \|_2 = \infty$ if $u \notin \text{Dom}(D)$. This may be established by an approximation argument which we leave to the reader.) Equality holds iff

$$u'(x) - 2\pi i b u(x) = 2\pi r(x-a)u(x)$$

for some real r, and the solutions of this differential equation are the Gaussians described above. ∎

Another interesting variant of the uncertainty principle is the following:

(1.37) Corollary. *If $u \in L^2(\mathbf{R})$, we have*

(1.38) $$\| Xu \|_2^2 + \| Du \|_2^2 \ge \frac{1}{2\pi} \| u \|_2^2,$$

with equality if and only if $u(x) = ce^{-\pi x^2}$.

Proof: (1.38) follows from (1.36) (with $a = b = 0$) together with the numerical inequality $\alpha \beta \le \frac{1}{2}(\alpha^2 + \beta^2)$ ($\alpha, \beta \ge 0$). Equality holds here iff $\alpha = \beta$, which forces $r = 1$ above. ∎

(1.36) and (1.38) are actually equivalent. To deduce (1.36) from (1.38), apply (1.38) to the functions $u_\alpha(x) = \alpha^{1/2} u(\alpha x)$, $\alpha > 0$. Since $\|X u_\alpha\|_2 = \alpha^{-1} \|X u\|_2$ and $\|D u_\alpha\|_2 = \alpha \|D u\|_2$, the result is

$$\alpha^{-2} \|X u\|_2^2 + \alpha^2 \|D u\|_2^2 \geq \frac{1}{2\pi} \|u\|_2^2.$$

Minimizing the left side over all $\alpha > 0$ yields (1.36) with $a = b = 0$. Applying the latter result to the function $v(x) = e^{-2\pi i b(x+a)} u(x+a)$, one obtains (1.36) in general.

These results generalize in the obvious way to n dimensions. Namely, if $u \in L^2(\mathbf{R}^n)$,

(1.39)
$$\|(X_j - a_j)u\|_2 \|(D_j - b_j)u\|_2 \geq \frac{1}{4\pi} \|u\|_2^2 \qquad (a_j, b_j \in \mathbf{R}),$$

(1.40)
$$\|X_j u\|_2^2 + \|D_j u\|_2^2 \geq \frac{1}{2\pi} \|u\|_2^2.$$

Equality holds in (1.39) [resp. (1.40)] for a fixed j iff

$$u(x) = v(x) \exp(2\pi i b_j x_j - \pi r(x_j - a_j)^2) \qquad [\text{resp.} \quad u(x) = v(x) \exp(-\pi x_j^2)\,]$$

where v is independent of x_j, and it holds in (1.39) [resp. (1.40)] for all j iff

$$u(x) = c \exp \sum_1^n (2\pi i b_j x_j - \pi r_j(x_j - a_j)^2) \qquad [\text{resp.} \quad u(x) = c \exp(-\pi x^2)\,].$$

There are a number of other versions of the uncertainty principle in the literature: see de Bruijn [37] and the papers [34], [35], [119], [120], and [121] of Cowling, Price, and Sitaram (in various combinations).

One of the recurring themes of this monograph is the beauty and importance of the Gaussian functions

$$f(x) = e^{x A x + b x + c}.$$

(Here A is an $n \times n$ complex matrix with $\operatorname{Re} A$ negative definite, $b \in \mathbf{C}^n$, and $c \in \mathbf{C}$.) We have just seen the first indication of their fundamental nature, in the fact that when A is real and diagonal they are precisely the extremal functions for the uncertainty inequalities. In fact, every Gaussian is an extremal for the uncertainty inequalities for some set of operators $\{D'_j, X'_j\}_{j=1}^n$ obtained from $\{D_j, X_j\}_{j=1}^n$ by a symplectic linear transformation. (We shall explain this in detail in Section 4.5.) We shall see that the Gaussians play a special role in a number of other contexts. For the moment, we merely point out two sobriquets that Gaussians have acquired from their scientific applications. In the one-dimensional case, the functions

$$f(t) = e^{2\pi i \omega t} e^{-\pi a (t-\tau)^2} \qquad (t, \omega, \tau \in \mathbf{R}, \quad a > 0)$$

are known as **Gabor functions**, after a paper of Gabor [53] in which their utility as simple components for building electrical signals was demonstrated. (See Section 3.4.) Also, the functions

$$f(x) = \rho(p,q)\left[2^{n/4} e^{-\pi x^2}\right] = e^{2\pi i q x + \pi i p q} e^{-\pi (x+p)^2},$$

obtained by translating the basic Gaussian $2^{n/4} e^{-\pi x^2}$ in phase space, are known in quantum physics, and especially quantum optics, as **coherent states.**

4. The Fourier-Wigner Transform

In this section we study the matrix coefficients of the representation ρ. If $f, g \in L^2(\mathbf{R}^n)$, the matrix coefficient of ρ at (f, g) is the function M on $\mathbf{H}_n$ (or $\mathbf{H}_n^{\mathrm{red}}$) defined by

$$M(p,q,t) = \langle \rho(p,q,t) f, g \rangle.$$

Clearly $M(p,q,t) = e^{2\pi i t} M(p,q,0)$, so the t dependence carries no information and can best be ignored. Accordingly, for $f, g \in L^2(\mathbf{R}^n)$, we define the function $V(f,g)$ on $\mathbf{R}^{2n}$ by

(1.41)
$$\begin{aligned} V(f,g)(p,q) &= \langle \rho(p,q) f, g \rangle = \langle e^{2\pi i (pD + qX)} f, g \rangle \\ &= \int e^{2\pi i q x + \pi i p q} f(x+p) \overline{g(x)} \, dx \\ &= \int e^{2\pi i q y} f(y + \tfrac{1}{2} p) \overline{g(y - \tfrac{1}{2} p)} \, dy. \end{aligned}$$

The map V has no standard name; we shall call it the **Fourier-Wigner transform**, for reasons that will become clear in Section 1.8. It is clear from the Schwarz inequality that $V(f,g)$ is always a bounded, continuous function on $\mathbf{R}^{2n}$ satisfying $\|V(f,g)\|_\infty \le \|f\|_2 \|g\|_2$.

V can be extended in an obvious way from a sesquilinear map defined on $L^2(\mathbf{R}^n) \times L^2(\mathbf{R}^n)$ to a linear map $\widetilde{V}$ defined on the tensor product $L^2(\mathbf{R}^n) \otimes L^2(\mathbf{R}^n)$, which is naturally isomorphic to $L^2(\mathbf{R}^{2n})$. Namely, if $F \in L^2(\mathbf{R}^{2n})$ we define

$$\widetilde{V}(F)(p,q) = \int e^{2\pi i q y} F(y + \tfrac{1}{2} p, y - \tfrac{1}{2} p) \, dy.$$

We then have $V(f,g) = \tilde{V}(f \otimes \bar{g})$, where $f \otimes \bar{g}(x,y) = f(x)\bar{g}(y)$. $\tilde{V}$ is the composition of the measure-preserving change of variables $(y,p) \to (y+\frac{1}{2}p, y - \frac{1}{2}p)$ with inverse Fourier transformation in the first variable. Therefore it is unitary on $L^2(\mathbf{R}^{2n})$, maps $\mathcal{S}(\mathbf{R}^{2n})$ onto itself, and extends to a continuous bijection of $\mathcal{S}'(\mathbf{R}^{2n})$ onto itself. Transferring these results back to V, we obtain the following:

(1.42) Proposition. V maps $\mathcal{S}(\mathbf{R}^n) \times \mathcal{S}(\mathbf{R}^n)$ into $\mathcal{S}(\mathbf{R}^{2n})$ and extends to a map from $\mathcal{S}'(\mathbf{R}^n) \times \mathcal{S}'(\mathbf{R}^n)$ into $\mathcal{S}'(\mathbf{R}^{2n})$. Moreover, V is "sesqui-unitary" on L^2; that is, for all $f_1, g_1, f_2, g_2 \in L^2(\mathbf{R}^n)$,

$$\langle V(f_1, g_1), V(f_2, g_2) \rangle = \langle f_1, f_2 \rangle \overline{\langle g_1, g_2 \rangle}.$$

In the language of representation theory, this proposition says that ρ is square-integrable (modulo the center). The irreducibility of ρ is an easy corollary:

(1.43) Proposition. The representation ρ_h is irreducible for any $h \in \mathbf{R}\backslash\{0\}$.

Proof: Suppose $\mathcal{M} \subset L^2(\mathbf{R}^n)$ is a nonzero closed invariant subspace and $f \neq 0 \in \mathcal{M}$. If $g \perp \mathcal{M}$ then $g \perp e^{2\pi i(hpD+qX)}f$ for all $p,q \in \mathbf{R}^n$; in other words, $V(f,g) = 0$. But this implies that $\|f\|_2\|g\|_2 = 0$, whence $g = 0$ and $\mathcal{M} = L^2(\mathbf{R}^n)$. ∎

Here is what happens to $V(f,g)$ when f and g are transformed by the operators $\rho(a,b)$.

(1.44) Proposition. For any $a,b,c,d \in \mathbf{R}^n$ we have

(a)
$$V\big(\rho(a,b)f, \rho(c,d)g\big)(p,q)$$
$$= e^{\pi i(dp+da+pb-cq-cb-qa)}V(f,g)(p+a-c, q+b-d).$$

In particular,

(b) $\qquad V\big(\rho(a,b)f, g\big)(p,q) = e^{\pi i(pb-qa)}V(f,g)(p+a, q+b)$

(c) $\qquad V\big(f, \rho(c,d)g\big)(p,q) = e^{\pi i(dp-cq)}V(f,g)(p-c, q-d)$

(d) $\qquad V\big(\rho(a,b)f, \rho(a,b)g\big)(p,q) = e^{2\pi i(pb-qa)}V(f,g)(p,q)$

Proof: We have

$$V\big(\rho(a,b)f, \rho(c,d)g\big)(p,q) = \langle \rho(-c,-d)\rho(p,q)\rho(a,b)f, g \rangle$$

and, in $\mathbf{H}_n$,

$$(-c,-d,0)(p,q,0)(a,b,0) = (p+a-c, q+b-d, \tfrac{1}{2}(dp+da+pb-cq-cb-qa)).$$

(a) follows easily from these equations, and (b), (c), and (d) are special cases of (a). ∎

The matrix elements of the integrated representation can also be expressed in terms of the Fourier-Wigner transform. Indeed, we have
(1.45)
$$\langle \rho(F)f, g \rangle = \iint F(p,q)\langle \rho(p,q)f, g \rangle \, dp \, dq = \iint F(p,q)V(f,g)(p,q) \, dp \, dq.$$

An interesting thing happens when we use the conjugate of a Fourier-Wigner transform as input for the representation ρ:

(1.46) Proposition. *If $\phi, \psi \in L^2(\mathbf{R}^n)$ and $\Phi = \overline{V(\phi, \psi)}$ then*

$$\rho(\Phi)f = \langle f, \phi \rangle \psi \quad \text{for} \quad f \in L^2(\mathbf{R}^n).$$

Proof: By (1.45) and Proposition (1.42), we have

$$\langle \rho(\Phi)f, g \rangle = \int \overline{V(\phi,\psi)}V(f,g) = \langle V(f,g), V(\phi,\psi) \rangle$$
$$= \langle f, \phi \rangle \overline{\langle g, \psi \rangle} = \langle f, \phi \rangle \langle \psi, g \rangle,$$

whence the result is immediate. ∎

In other words, the operators $\rho(\Phi)$ where $\overline{\Phi}$ is a Fourier-Wigner transform are precisely the operators on L^2 with one-dimensional range. This leads to a nice formula for the twisted convolution of Fourier-Wigner transforms:

(1.47) Proposition. $\overline{V(\phi_1, \psi_1)} \natural \overline{V(\phi_2, \psi_2)} = \langle \psi_2, \phi_1 \rangle \overline{V(\phi_2, \psi_1)}.$

Proof: Let $\Phi_j = \overline{V(\phi_j, \psi_j)}$ and $\Psi = \overline{V(\phi_2, \psi_1)}$. Then

$$\rho(\Phi_1)\rho(\Phi_2)f = \rho(\Phi_1)\langle f, \phi_2 \rangle \psi_2 = \langle f, \phi_2 \rangle \langle \psi_2, \phi_1 \rangle \psi_1 = \langle \psi_2, \phi_1 \rangle \rho(\Psi)f.$$

But $\rho(\Phi_1)\rho(\Phi_2) = \rho(\Phi_1 \natural \Phi_2)$ and ρ is faithful, so $\Phi_1 \natural \Phi_2 = \langle \psi_2, \phi_1 \rangle \Psi$. ∎

We conclude this discussion with some calculations of Fourier-Wigner transforms of Gaussians that we shall need later.

(1.48) Proposition. *Let*

$$\phi(x) = 2^{n/4}e^{-\pi x^2}, \qquad \Phi = V(\phi, \phi), \qquad \Phi^{ab} = V\big(\phi, \rho(a,b)\phi\big).$$

Then

(a) $\qquad\qquad \Phi(p,q) = e^{-(\pi/2)(p^2+q^2)}.$

(b) $\qquad\qquad \Phi^{ab}(p,q) = e^{\pi i(bp-aq)}e^{-(\pi/2)[(p-a)^2+(q-b)^2]}.$

(c) $\qquad\qquad \rho(\Phi)\rho(a,b)\rho(\Phi) = e^{-(\pi/2)(a^2+b^2)}\rho(\Phi).$

(d) $\qquad\qquad \Phi \natural \Phi^{ab} = e^{(-\pi/2)(a^2+b^2)}\Phi.$

Proof: (a) follows from the Fourier transform formula for Gaussians (Appendix A):

$$\Phi(p,q) = 2^{n/2} \int e^{2\pi i q y} e^{-\pi[(y+(p/2)]^2 - \pi[(y-(p/2)]^2} \, dy$$

$$= 2^{n/2} e^{(-\pi/2)p^2} \int e^{2\pi i q y} e^{-2\pi y^2} \, dy = e^{-(\pi/2)(p^2+q^2)}.$$

(b) follows from (a) and Proposition (1.44c). As for (c), by Proposition (1.46),

$$\rho(\Phi)\rho(p,q)\rho(\Phi)f = \rho(\Phi)\big[\langle f, \phi\rangle \rho(a,b)\phi\big]$$

$$= \langle f, \phi\rangle \langle \rho(a,b)\phi, \phi\rangle \phi = \langle f, \phi\rangle \Phi(a,b)\phi = e^{-(\pi/2)(a^2+b^2)}\rho(\Phi)f.$$

Finally, by Proposition (1.31) and (b) we have $\rho(a,b)\rho(\Phi) = \rho(\Phi^{ab})$, so by (c), $\rho(\Phi \natural \Phi^{ab}) = e^{-(\pi/2)(a^2+b^2)}\rho(\Phi)$. Since ρ is faithful, (d) follows. $\blacksquare$

Radar Ambiguity Functions. We shall give a brief account of how Fourier-Wigner transforms turn up in the theory of radar.

A radar apparatus transmits an electromagnetic signal that reflects off a target and returns to the apparatus. The signal may be represented by a complex function of time, $f(t)$. [Technically, $f(t) = u(t) + iv(t)$ where u is the amplitude of the physical signal and v is the Hilbert transform of u. The energy of the signal is $\frac{1}{2}\|f\|_2^2$.] We assume that the frequencies of f are concentrated around some (large) number ω, so that we can write $f(t) = f_0(t)e^{2\pi i \omega t}$ where $f_0(t)$ is slowly varying in comparison to $e^{2\pi i \omega t}$. Let r be the distance of the target from the apparatus and $v = dr/dt$ its radial velocity. We assume that the signal f is essentially limited to a time interval Δt which is large in comparison to the period ω^{-1} but small enough so that r and v may be considered as constant in this interval. The reflected signal then arrives back at the apparatus after a time delay $\tau = 2r/c$ (where c is the propagation speed of the signal) and with frequencies dilated by a factor $1 - (2v/c)$ because of the Doppler effect. Since the frequencies of the transmitted signal are mostly near ω, we may assume instead that the frequencies are shifted by the amount $-\phi = -2\omega v/c$. In short, when these approximations have been made, the returning signal is

$$f_{\tau\phi}(t) = f(t - \tau)e^{-2\pi i \phi t}.$$

Now suppose there are two targets that produce returning signals $f_{\tau_1\phi_1}$ and $f_{\tau_2\phi_2}$. If these signals are similar to each other there will be a difficulty in distinguishing the two targets, so we are concerned with the mean squared difference,

$$(1.49) \qquad \int |f_{\tau_1\phi_1} - f_{\tau_2\phi_2}|^2 \, dt = 2\int |f|^2 \, dt - 2\mathrm{Re}\langle f_{\tau_1\phi_1}, f_{\tau_2\phi_2}\rangle.$$

Only the second term on the right depends on the targets, and since $f(t) = f_0(t)e^{2\pi i\omega t}$ we have

$$\langle f_{\tau_1\phi_1}, f_{\tau_2\phi_2} \rangle = e^{2\pi i\omega(\tau_2-\tau_1)} \int f_0(t-\tau_1)\overline{f_0(t-\tau_2)}e^{2\pi i(\phi_2-\phi_1)t}\,dt.$$

The integral varies slowly with τ_1 and τ_2, but the exponential in front is rapidly oscillating; so if we want (1.49) to be large in a way that is stable under small perturbations of τ_1 and τ_2, $|\langle f_{\tau_1\phi_1}, f_{\tau_2\phi_2}\rangle|$ must be near zero. If we set $\tau = \tau_1 - \tau_2$ and $\phi = \phi_1 - \phi_2$, we have

$$\langle f_{\tau_1\phi_1}, f_{\tau_2\phi_2} \rangle = e^{-2\pi i\phi\tau_1} \int f(t)\overline{f(t+\tau)}e^{-2\pi i\phi t}\,dt.$$

Since only the absolute value is important, and since nothing essential is changed by switching the two targets, we could equally well consider

$$A_1(\tau,\phi) = \int f(t)\overline{f(t+\tau)}e^{-2\pi i\phi t}\,dt$$

$$\text{or} \quad A_2(\tau,\phi) = e^{-\pi i\phi\tau}A_1(\tau,\phi)$$

$$\text{or} \quad A_3(\tau,\phi) = A_2(-\tau,-\phi) = \int f(t)\overline{f(t-\tau)}e^{2\pi i\phi t - \pi i\phi\tau}\,dt,$$

and we have

$$A_3 = V(f,f).$$

A_1, A_2, and A_3 and the squares of their absolute values are all referred to in one place or another as the **ambiguity function** of the signal f. Whichever variant is used, the intuitive significance is that if $|A_j(\tau,\phi)|$ is large, two targets whose associated time and frequency shifts differ by τ and ϕ will be hard to distinguish. In this connection, we observe that by the Schwarz inequality,

$$|A_j(\tau,\phi)| \le A_j(0,0) = \|f\|_2^2,$$

and by Proposition (1.42),

$$\iint |A_j(\tau,\phi)|^2\,d\tau\,d\phi = \|f\|_2^4.$$

This last equation may be interpreted as saying that for a signal f of fixed energy $\frac{1}{2}\|f\|_2^2$, there is a fixed amount of ambiguity distributed over the (τ,ϕ) plane that cannot be eliminated. This is sometimes called "conservation of ambiguity" or the "radar uncertainty principle."

Ambiguity functions were introduced into radar theory by Woodward [157], and the connection with the Wigner transform (cf. Section 1.8) was noted by Klauder [90]. A detailed account of the use of ambiguity functions in radar design can be found in Cook–Bernfeld [31]. Recently there has been interest in analyzing ambiguity functions by explicit use of the connection with the Heisenberg group: see Auslander–Tolimieri [9] and Schempp [125].

5. The Stone–von Neumann Theorem

We have constructed a family $\{\rho_h : h \in \mathbf{R}\backslash\{0\}\}$ of irreducible unitary representations of $\mathbf{H}_n$. We now prove the classic theorem of Stone [133] and von Neumann [146], which says in effect that any irreducible unitary representation of $\mathbf{H}_n$ that is nontrivial on the center is equivalent to some ρ_h. Since the irreducible representations that are trivial on the center are easily described, as we shall see below, we shall obtain a complete classification of the irreducible unitary representations of $\mathbf{H}_n$.

Nowadays the Stone–von Neumann theorem is usually obtained as a corollary of the Mackey imprimitivity theorem. Here we present von Neumann's original proof, a pretty argument that does not deserve the obscurity into which it has fallen. It actually does more than classify the irreducible representations: it also shows that any primary representation of $\mathbf{H}_n$ is a direct sum of copies of an irreducible representation, and hence that $\mathbf{H}_n$ is a type I group.

(1.50) The Stone–von Neumann Theorem. *Let π be a unitary representation of $\mathbf{H}_n$ on a Hilbert space $\mathcal{H}$, such that $\pi(0,0,t) = e^{2\pi i h t}I$ for some $h \in \mathbf{R}\backslash\{0\}$. Then $\mathcal{H} = \bigoplus \mathcal{H}_\alpha$ where the $\mathcal{H}_\alpha$'s are mutually orthogonal subspaces of $\mathcal{H}$, each invariant under π, such that $\pi \,|\, \mathcal{H}_\alpha$ is unitarily equivalent to ρ_h for each α. In particular, if π is irreducible then π is equivalent to ρ_h.*

Proof: We present the proof for $h = 1$; the argument in general is exactly the same. The crucial point is to identify the elements of $\mathcal{H}$ that correspond to the Gaussian $e^{-\pi x^2}$ in the Schrödinger representation. Concerning the latter, we adopt the following notation:

$$\phi(x) = 2^{n/4}e^{-\pi x^2}, \qquad \phi^{ab}(x) = \rho(a,b)\phi(x) = 2^{n/4}e^{2\pi i b x + \pi i a b}e^{-\pi(x+a)^2},$$

$$\Phi = V(\phi,\phi), \qquad \Phi^{ab} = V(\phi,\phi^{ab}).$$

By Proposition (1.48), we then have

$$(1.51) \qquad \langle \phi^{pq}, \phi^{ab} \rangle = \Phi^{ab}(p,q) = e^{\pi i(bp - aq)}e^{-(\pi/2)[(p-a)^2 + (q-b)^2]},$$

$$(1.52) \qquad \qquad \Phi \,\natural\, \Phi^{ab} = e^{-(\pi/2)(a^2+b^2)}\Phi.$$

Returning to the representation π, we mimic some constructions that we made with ρ in Section 1.3. First we set $\pi(p,q) = \pi(p,q,0)$, and we have

$$(1.53) \quad \pi(p,q)\pi(r,s) = \pi\big(p+r,\, q+s,\, \tfrac{1}{2}(ps - qr)\big) = e^{\pi i(ps - qr)}\pi(p+r,\, q+s).$$

We consider the integrated version of π,

$$\pi(F) = \iint F(p,q)\pi(p,q)\,dp\,dq \qquad (F \in L^1(\mathbf{R}^{2n})),$$

and just as with ρ, we have

(1.54) $\pi(F)\pi(G) = \pi(F \natural G)$,

(1.55) $\pi(F)\pi(a, b) = \pi(G)$ where $G(p, q) = e^{\pi i(aq-bp)}F(p - a, q - b)$,

(1.56) $\pi(a, b)\pi(F) = \pi(H)$ where $H(p, q) = e^{\pi i(bp-aq)}F(p - a, q - b)$.

Moreover, π is faithful on $L^1(\mathbf{R}^{2n})$. Indeed, if $\pi(F) = 0$ then, by (1.55) and (1.56), for any $u, v \in \mathcal{H}$ and $a, b \in \mathbf{R}^n$,

$$0 = \langle \pi(a, b)\pi(F)\pi(-a, -b)u, v \rangle$$
$$= \iint e^{2\pi i(bp-aq)}F(p, q)\langle \pi(p, q)u, v \rangle \, dp \, dq.$$

Thus by the Fourier inversion theorem,

$$F(p, q)\langle \pi(p, q)u, v \rangle = 0 \qquad \text{for a.e. } (p, q),$$

and since u and v are arbitrary, $F = 0$ a.e.

Now let us take F to be the function Φ defined above. By (1.51), (1.52), (1.54), and (1.56),

(1.57) $\pi(\Phi)\pi(a, b)\pi(\Phi) = \pi(\Phi \natural \Phi^{ab}) = e^{-(\pi/2)(a^2+b^2)}\pi(\Phi)$.

In particular, taking $a = b = 0$ we obtain $\pi(\Phi)^2 = \pi(\Phi)$, and since Φ is even and real it is easily seen that $\pi(\Phi)$ is self-adjoint. Thus $\pi(\Phi)$ is an orthogonal projection which is nonzero since $\Phi \neq 0$ and π is faithful. Let $\mathcal{R}$ denote the range of $\pi(\Phi)$. If $u, v \in \mathcal{R}$ then $u = \pi(\Phi)u$ and $v = \pi(\Phi)v$, so by (1.53),

(1.58) $\langle \pi(p, q)u, \pi(r, s)v \rangle = \langle \pi(-r, -s)\pi(p, q)\pi(\Phi)u, \pi(\Phi)v \rangle$
$$= e^{\pi i(ps-qr)}\langle \pi(\Phi)\pi(p - r, q - s)\pi(\Phi)u, v \rangle$$
$$= e^{\pi i(ps-qr)}e^{-(\pi/2)[(p-r)^2+(q-s)^2]}\langle u, v \rangle.$$

Let $\{v_\alpha\}$ be an orthonormal basis for $\mathcal{R}$, and let $\mathcal{H}_\alpha$ be the closed linear span of $\{\pi(p, q)v_\alpha : p, q \in \mathbf{R}^n\}$. By (1.58), $\mathcal{H}_\alpha \perp \mathcal{H}_\beta$ for $\alpha \neq \beta$, and $\mathcal{H}_\alpha$ is invariant under π by definition. Hence $\mathcal{N} = (\bigoplus \mathcal{H}_\alpha)^\perp$ is also invariant under π, and we have $\pi(\Phi)|\mathcal{N} = 0$. But this implies that $\mathcal{N} = \{0\}$, for otherwise we could apply the above reasoning to $\pi|\mathcal{N}$ to conclude that $\pi(\Phi)|\mathcal{N}$ were a nonzero orthogonal projection.

We claim that $\pi|\mathcal{H}_\alpha$ is equivalent to ρ for all α. Indeed, fix an α and let $v^{pq} = \pi(p, q)v_\alpha$. Then by (1.52) and (1.58),

$$\langle v^{pq}, v^{rs} \rangle = \langle \phi^{pq}, \phi^{rs} \rangle \qquad \text{for all } p, q, r, s.$$

It follows that if $u = \sum a_{jk}v^{p_j q_k}$ and $f = \sum a_{jk}\phi^{p_j q_k}$ then $\|u\|_\mathcal{H} = \|f\|_2$, and in particular $u = 0$ iff $f = 0$. Therefore the correspondence $v^{pq} \to \phi^{pq}$ extends by linearity and continuity to a unitary map from $\mathcal{H}_\alpha$ to $L^2(\mathbf{R}^n)$ that intertwines $\pi|\mathcal{H}_\alpha$ and ρ. ∎

We can now give a complete classification of the irreducible unitary representations of $\mathbf{H}_n$. Suppose π is such a representation. By Schur's lemma (cf. Appendix B), π must map the center $\mathcal{Z}$ of $\mathbf{H}_n$ homomorphically into the group $\{cI : |c| = 1\}$, so $\pi(0,0,t) = e^{2\pi i h t}I$ for some $h \in \mathbf{R}$. If $h \neq 0$, the Stone–von Neumann theorem shows that π is equivalent to ρ_h. If $h = 0$, on the other hand, π factors through the quotient group $\mathbf{H}_n/\mathcal{Z}$, which is isomorphic to $\mathbf{R}^{2n}$. The irreducible representations of the latter are all one-dimensional (Schur's lemma again) and hence are just the homomorphisms from $\mathbf{R}^{2n}$ into the circle group, namely, $(p,q) \to e^{2\pi i(ap+bq)}$. We have therefore proved:

(1.59) Theorem. *Every irreducible unitary representation of* $\mathbf{H}_n$ *is unitarily equivalent to one and only one of the following representations:*
(a) ρ_h ($h \in \mathbf{R}\backslash\{0\}$), *acting on* $L^2(\mathbf{R}^n)$,
(b) $\sigma_{ab}(p,q,t) = e^{2\pi i(ap+bq)}$ ($a,b \in \mathbf{R}^n$), *acting on* $\mathbf{C}$.

The Group Fourier Transform. If G is a locally compact group, let $\widehat{G}$ denote a collection of irreducible unitary representations of G containing exactly one member of each equivalence class. If $\pi \in \widehat{G}$ we denote by $\mathcal{H}_\pi$ the Hilbert space on which π acts. Given $f \in L^1(G)$ and $\pi \in \widehat{G}$, we define the operator $\widehat{f}(\pi)$ on $\mathcal{H}_\pi$ by

$$\widehat{f}(\pi) = \int_G f(x)\pi(x)^* \, dx = \int_G f(x)\pi(x^{-1}) \, dx,$$

where dx denotes Haar measure. The map $f \to \widehat{f}$ is called the **group Fourier transform**. For a large class of groups G there exists a measure μ on $\widehat{G}$ (the "Plancherel measure") such that for all sufficiently nice functions f on G one has the Fourier inversion formula

$$(1.60) \qquad f(x) = \int_{\widehat{G}} \mathrm{tr}\big(\widehat{f}(\pi)\pi(x)\big) \, d\mu(\pi)$$

and the Plancherel formula

$$(1.61) \qquad \int_G |f(x)|^2 \, dx = \int_{\widehat{G}} \mathrm{tr}\big(\widehat{f}(\pi)^*\widehat{f}(\pi)\big) \, d\mu(\pi) = \int_{\widehat{G}} \|\widehat{f}(\pi)\|_{HS}^2 \, d\mu(\pi).$$

(Here tr denotes trace and $\|\cdot\|_{HS}$ denotes the Hilbert-Schmidt norm.)

We now compute the Plancherel measure for $\mathbf{H}_n$, using the parametrization of $\widehat{\mathbf{H}_n}$ given by Theorem (1.59). Our analysis will show that we need only consider the representations ρ_h; that is, the one-dimensional representations σ_{ab} form a set of Plancherel measure zero.

If $f \in L^1(\mathbf{H}_n)$, $h \in \mathbf{R}\backslash\{0\}$, and $\phi \in L^2(\mathbf{R}^n)$, $\widehat{f}(\rho_h)\phi$ is given by

$$\widehat{f}(\rho_h)\phi(x) = \iiint f(p,q,t)\rho_h(-p,-q,-t)\phi(x)\,dp\,dq\,dt$$

$$= \iiint f(p,q,t)e^{-2\pi i qx + \pi i h pq - 2\pi i h t}\phi(x-hp)\,dp\,dq\,dt$$

$$= |h|^{-n}\iiint f(h^{-1}(x-y),\,q,\,t)e^{-\pi i(y+x)q - 2\pi i h t}\phi(y)\,dy\,dq\,dt.$$

Thus, $\widehat{f}(\rho_h)$ is an integral operator with kernel

$$\text{(1.62)} \qquad K_f^h(x,y) = |h|^{-n}\iint f(h^{-1}(x-y),\,q,\,t)e^{-\pi i(y+x)q - 2\pi i h t}\,dq\,dt$$

$$= |h|^{-n}\mathcal{F}_{2,3}f\big(h^{-1}(x-y),\,\tfrac{1}{2}(x+y),\,h\big),$$

where $\mathcal{F}_{2,3}$ denotes Fourier transformation in the second and third variables. Moreover,

$$\widehat{f}(\rho_h)\rho_h(p,q,t) = \iiint f(p',q',t')\rho_h(-p',-q',-t')\rho_h(p,q,t)\,dp'\,dq'\,dt'$$

$$= \iiint f(p',q',t')\rho_h\big(p-p',\,q-q',\,t-t'-\tfrac{1}{2}(q'p-p'q)\big)\,dp'\,dq'\,dt'$$

$$= \widehat{g}(\rho_h)$$

where

$$g(p',q',t') = f(p-p',\,q-q',\,t-t')e^{\pi i h(p'q - q'p)}.$$

Hence, in view of (1.62), the integral kernel of $\widehat{f}(\rho_h)\rho_h(p,q,t)$ is

$$F(x,y)$$

$$= |h|^{-n}\iint f(p-h^{-1}(x-y),\,q-q',\,t-t')e^{\pi i[(x-y)q - q'p] - \pi i(y+x)q' - 2\pi i h t'}\,dq'\,dt'$$

$$= |h|^{-n}\iint f(p-h^{-1}(x-y),\,q',\,t')e^{\pi i[(x-y)q-(q-q')p] - \pi i(x+y)(q-q') - 2\pi i h(t-t')}\,dq'\,dt'.$$

If f is such that all the integrals converge nicely, then, we have

$$\text{tr}\big(\widehat{f}(\rho_h)\rho_h(p,q,t)\big) = \int F(x,x)\,dx$$

$$= |h|^{-n}\iiint f(p,q',t')e^{\pi i p(q-q') - 2\pi i x(q-q') - 2\pi i h(t-t')}\,dq'\,dt'\,dx$$

$$= |h|^{-n}\iint f(p,q',t')e^{\pi i p(q-q') - 2\pi i h(t-t')}\delta(q-q')\,dq'\,dt$$

$$= |h|^{-n}\int f(p,q,t')e^{-2\pi i h(t-t')}\,dt'.$$

But by the (ordinary) Fourier inversion formula,

$$f(p,q,t) = \iint f(p,q,t')e^{-2\pi i h(t-t')} \, dt' \, dh = \int \text{tr}(\widehat{f}(\rho_h)\rho_h(p,q,t))|h|^n \, dh.$$

Thus (1.60) holds if we define the Plancherel measure on $\widehat{\mathbf{H}_n}$ to be $|h|^n \, dh$ on the family $\{\rho_h\}$ and 0 on the family $\{\sigma_{ab}\}$. Moreover, by (1.62) and the (ordinary) Plancherel theorem,

$$\|\widehat{f}(\rho_h)\|_{HS}^2 = \int |K_f^h(x,y)|^2 \, dx \, dy$$

$$= |h|^{-2n} \iint |\mathcal{F}_{2,3} f(h^{-1}(x-y), \tfrac{1}{2}(x+y), h)|^2 \, dx \, dy$$

$$= |h|^{-n} \iint |\mathcal{F}_{2,3} f(p,z,h)|^2 \, dp \, dz$$

$$= |h|^{-n} \int |\mathcal{F}_3 f(p,q,h)|^2 \, dp \, dq,$$

so that (1.61) also holds:

$$\|f\|_2^2 = \int |h|^n \|\widehat{f}(\rho_h)\|_{HS}^2 \, dh.$$

There is much more that can be said about the group Fourier transform on $\mathbf{H}_n$; see Geller [55], [56], [57].

6. The Fock–Bargmann Representation

There is a particularly interesting realization of the infinite-dimensional irreducible unitary representations of $\mathbf{H}_n$ in a Hilbert space of entire functions. We shall carry out the analysis for the representation ρ and indicate at the end how to generalize to ρ_h.

Let

$$\phi_0(x) = 2^{n/4} e^{-\pi x^2}$$

be the standard Gaussian on $\mathbf{R}^n$. Since $\|\phi_0\|_2 = 1$, by Proposition (1.42) the map $f \to V(f, \phi_0)$ is an isometry from $L^2(\mathbf{R}^n)$ into $L^2(\mathbf{R}^{2n})$. Explicitly, we have

$$V(f,\phi_0)(p,q) = \langle f, \rho(-p,-q)\phi_0 \rangle$$

$$= 2^{n/4} \int f(x)e^{2\pi i qx - \pi i pq} e^{-\pi(x-p)^2} \, dx$$

$$= 2^{n/4} e^{-(\pi/2)(p^2+q^2)} \int f(x)e^{2\pi x(p+iq) - \pi x^2 - (\pi/2)(p+iq)^2} \, dx.$$

For $z \in \mathbf{C}^n$ let us define

$$Bf(z) = 2^{n/4} \int f(x) e^{2\pi x z - \pi x^2 - (\pi/2) z^2} \, dx.$$

Then we have

$$V(f, \phi_0)(p, q) = e^{-(\pi/2)|z|^2} Bf(z), \qquad \text{with} \quad z = p + iq.$$

Bf is called the **Bargmann transform** of f. For $f \in L^2$, the integral defining $Bf(z)$ plainly converges uniformly for z in any compact subset of $\mathbf{C}^n$, so that Bf is an entire analytic function on $\mathbf{C}^n$. Moreover, since the map $f \to V(f, \phi_0)$ is an isometry on L^2, B is an isometry from $L^2(\mathbf{R}^n)$ into $L^2(\mathbf{C}^n, e^{-\pi|z|^2} dz)$. (Here and in the sequel, dz denotes Lebesgue measure on $\mathbf{C}^n$.) Hence B is an isometry from $L^2(\mathbf{R}^n)$ into the **Fock space**

$$\mathcal{F}_n = \left\{ F : F \text{ is entire on } \mathbf{C}^n \text{ and } \|F\|_{\mathcal{F}}^2 = \int |F(z)|^2 e^{-\pi|z|^2} \, dz < \infty \right\}.$$

We shall show below that B maps $L^2(\mathbf{R}^n)$ onto $\mathcal{F}_n$, and also explain the connection between $\mathcal{F}_n$ and the physicists' Fock space. First, we investigate the properties of $\mathcal{F}_n$. We denote the scalar product on $\mathcal{F}_n$ by $\langle \, , \, \rangle_{\mathcal{F}}$.

(1.63) Theorem. *Let*

$$\zeta_\alpha(z) = \sqrt{\frac{\pi^{|\alpha|}}{\alpha!}} z^\alpha.$$

Then $\{\zeta_\alpha : |\alpha| \geq 0\}$ is an orthonormal basis for $\mathcal{F}_n$.

   Proof: Orthonormality is easily proved by integrating in polar coordinates:

$$\sqrt{\frac{\alpha! \beta!}{\pi^{|\alpha|+|\beta|}}} \langle \zeta_\alpha, \zeta_\beta \rangle_{\mathcal{F}} = \prod_1^n \int_{\mathbf{C}} z_j^{\alpha_j} \bar{z}_j^{\beta_j} e^{-\pi|z_j|^2} \, dz_j$$

$$= \prod_1^n \int_0^\infty \int_0^{2\pi} e^{i\theta(\alpha_j - \beta_j)} r^{\alpha_j + \beta_j + 1} e^{-\pi r^2} \, d\theta \, dr.$$

The θ-integral is zero unless $\beta = \alpha$, in which case we get

$$\frac{\alpha!}{\pi^{|\alpha|}} \|\zeta_\alpha\|_{\mathcal{F}}^2 = (2\pi)^n \prod_1^n \int_0^\infty r^{2\alpha_j + 1} e^{-\pi r^2} \, dr = \prod_1^n \int_0^\infty \left(\frac{s}{\pi}\right)^{\alpha_j} e^{-s} \, ds = \frac{\alpha!}{\pi^{|\alpha|}}.$$

This calculation shows more generally that if

$$B_R = \{z \in \mathbf{C}^n : |z| \leq R\} \qquad \text{and} \qquad c_{R,\alpha} = \pi^{-|\alpha|} \prod_1^n \int_0^R s^{\alpha_j} e^{-s} \, ds$$

then $\{c_{R,\alpha}^{-1/2} z^\alpha\}$ is an orthonormal set in $L^2(B_R, e^{-\pi|z|^2} dz)$. To prove completeness, then, suppose $F \in \mathcal{F}_n$, and let $\sum a_\alpha z^\alpha$ be the Taylor series of F about 0. For all $R > 0$ this series converges to F uniformly on B_R, hence in $L^2(B_R, e^{-\pi|z|^2} dz)$. From the preceding remark it follows that $c_{R,\alpha}^{-1/2} a_\alpha$ is the αth Fourier coefficient of F with respect to the set $\{c_{R,\alpha}^{-1/2} z^\alpha\}$,

$$a_\alpha = c_{R,\alpha}^{-1} \int_{B_R} F(z) \bar{z}^\alpha e^{-\pi|z|^2} dz,$$

and the Parseval formula holds:

$$\int_{B_R} |F(z)|^2 e^{-\pi|z|^2} dz = \sum c_{R,\alpha} |A_\alpha|^2.$$

Let $R \to \infty$: then $c_{R,\alpha} \to \pi^{-|\alpha|} \alpha!$, so these equations become

$$\|F\|_{\mathcal{F}}^2 = \sum |\langle f, \zeta_\alpha \rangle_{\mathcal{F}}|^2.$$

Therefore $\{\zeta_\alpha\}$ is a basis. ∎

(1.64) Corollary. *If $F \in \mathcal{F}_n$ then the Taylor series of F converges to F in the topology of $\mathcal{F}_n$.*

(1.65) Corollary. *If $F \in \mathcal{F}_n$ then $|F(z)| \le e^{(\pi/2)|z|^2} \|F\|_{\mathcal{F}}$ for all $z \in \mathbf{C}^n$.*

Proof: The preceding argument shows that the Fourier series of F with respect to the basis $\{\zeta_\alpha\}$ is the Taylor series of F. Thus, if $F = \sum a_\alpha \zeta_\alpha$, the Schwarz inequality yields

$$|F(z)| = \left| \sum a_\alpha (\pi^{|\alpha|}/\alpha!)^{1/2} z^\alpha \right|$$
$$\le \left(\sum |a_\alpha|^2 \right)^{1/2} \left(\sum (\pi^{|\alpha|}/\alpha!) |z|^{2\alpha} \right)^{1/2} = \|F\|_{\mathcal{F}} e^{(\pi/2)|z|^2}. \quad ∎$$

By Corollary (1.65), for each z the map $F \to F(z)$ is a bounded linear functional on $\mathcal{F}_n$, so there exists $E_z \in \mathcal{F}_n$ such that

$$F(z) = \langle F, E_z \rangle_{\mathcal{F}}.$$

It is easy to identify E_z; we have

(1.66)
$$E_z(w) = \sum \langle E_z, \zeta_\alpha \rangle_{\mathcal{F}} \zeta_\alpha(w) = \sum \overline{\zeta_\alpha(z)} \zeta_\alpha(w) = \sum (\pi^{|\alpha|} \bar{z}^\alpha w^\alpha / \alpha!)$$
$$= e^{\pi w \bar{z}}.$$

Put in other terms, the function $K(z,\overline{w}) = e^{\pi z \overline{w}}$ is the reproducing kernel for the space $\mathcal{F}_n$:

$$F(z) = \int e^{\pi z \overline{w}} F(w) e^{-\pi |w|^2} dw, \qquad \text{for} \quad F \in \mathcal{F}_n, \quad z \in \mathbf{C}^n.$$

We observe also that

(1.67) $$\|E_z\|_{\mathcal{F}}^2 = \sum \frac{\pi^{|\alpha|}}{\alpha!} |z^\alpha|^2 = e^{\pi |z|^2}.$$

An important consequence of the existence of a reproducing kernel is that every bounded operator on $\mathcal{F}_n$ can be written as an integral operator. More precisely, we have:

(1.68) Proposition. *If T is a bounded operator on $\mathcal{F}_n$, let $K_T(z,\overline{w}) = TE_w(z)$. Then K_T is an entire function on $\mathbf{C}^{2n}$ that satisfies*

(a) $\quad K_T(\cdot, w) \in \mathcal{F}_n$ *for all w and $K_T(z, \cdot) \in \mathcal{F}_n$ for all z,*

(b) $\quad |K_T(z,\overline{w})| \leq e^{(\pi/2)(|z|^2 + |w|^2)} \|T\|,$

(c) $\quad TF(z) = \int K_T(z,\overline{w}) F(w) e^{-\pi |w|^2} dw$ *for all $F \in \mathcal{F}_n$ and $z \in \mathbf{C}^n$.*

Proof: We have

$$TF(z) = \langle TF, E_z \rangle_{\mathcal{F}} = \langle F, T^* E_z \rangle_{\mathcal{F}} = \int \overline{T^*(E_z)(w)} F(w) e^{-\pi |w|^2} dw,$$

and

$$\overline{T^* E_z(w)} = \overline{\langle T^* E_z, E_w \rangle} = \langle T E_w, E_z \rangle = T E_w(z).$$

These formulas show that K_T is entire (since E_z depends antiholomorphically on z) and satisfies (a) and (c). As for (b), by (1.67),

$$|K_T(z,\overline{w})| \leq \|TE_w\|_{\mathcal{F}} \|E_z\|_{\mathcal{F}} \leq \|T\| \, \|E_w\|_{\mathcal{F}} \|E_z\|_{\mathcal{F}} = e^{(\pi/2)(|w|^2 + |z|^2)} \|T\|. \quad \blacksquare$$

In this connection the following observation is sometimes useful:

(1.69) Proposition. *An entire function $K(z,\overline{w})$ of z and $\overline{w}$ is uniquely determined by its restriction to the diagonal $z = w$.*

Proof: Let $u = \frac{1}{2}(z + \overline{w})$ and $v = -\frac{1}{2}i(z - \overline{w})$, so that $z = u + iv$ and $\overline{w} = u - iv$. Then $K(z,\overline{w}) = G(u,v)$ where G is entire. But G is determined (by Taylor's formula, say) by its values for u and v real, and u and v are real precisely when $z = w$. $\blacksquare$

(1.70) Corollary. *A bounded operator T on $\mathcal{F}_n$ is uniquely determined by the function* $K_T(z, \bar{z}) = \langle TE_z, E_z \rangle$.

We now return to consideration of the Heisenberg group. The representation ρ can be transfered via the Bargmann transform to a representation β of $\mathbf{H}_n$ on $B(L^2(\mathbf{R}^n))$ (which, as we shall shortly see, coincides with $\mathcal{F}_n$). To describe this representation, it will be convenient to identify the underlying manifold of $\mathbf{H}_n$ with $\mathbf{C}^n \times \mathbf{R}$:

$$(p, q, t) \longleftrightarrow (p + iq, t).$$

In this parametrization of $\mathbf{H}_n$ the group law is given by

$$(z, t)(z', t') = \left(z + z', t + t' + \tfrac{1}{2}\mathrm{Im}\,\bar{z}z' \right).$$

The transferred representation β is then defined by

$$\beta(p + iq, t)B = B\rho(p, q, t).$$

As with ρ, we set

$$\beta(w) = \beta(w, 0), \quad \text{i.e.,} \quad \beta(w, t) = e^{2\pi i t}\beta(w).$$

We proceed to calculate β. Let $z = p + iq$, $w = r + is$. Then for $f \in L^2(\mathbf{R}^n)$,

$$
\begin{aligned}
[\beta(w)Bf](z) &= [B\rho(r, s)f](z) \\
&= e^{(\pi/2)|z|^2} V(\rho(r, s)f, \phi_0)(p, q) \\
&= e^{(\pi/2)|z|^2} e^{\pi i(ps - qr)} V(f, \phi_0)(p + r, q + s) \qquad \text{[by Prop. (1.44b)]} \\
&= e^{(\pi/2)|z|^2} e^{-\pi i \mathrm{Im}\, z\bar{w}} e^{-(\pi/2)|z + w|^2} Bf(z + w) \\
&= e^{-(\pi/2)|w|^2 - \pi z\bar{w}} Bf(z + w).
\end{aligned}
$$

In other words,

(1.71) $$\beta(w, t)F(z) = e^{-(\pi/2)|w|^2 - \pi z\bar{w} + 2\pi i t} F(z + w).$$

At this point we observe that

$$B\phi_0(z) = 2^{n/2} e^{-(\pi/2)|z|^2} \int e^{-2\pi z x - 2\pi x^2} dx = 1 = E_0(z),$$

and hence, if $w = r + is$,

(1.72) $$B\big(\rho(r, s)\phi_0\big)(z) = \beta(w)(1)(z) = e^{-(\pi/2)|w|^2 - \pi z\bar{w}} = e^{-(\pi/2)|w|^2} E_{-w}(z).$$

Thus all the E_w's are in the range of B, and since $\langle F, E_w \rangle_{\mathcal{F}} = 0$ only when $F = 0$, it follows that $B(L^2(\mathbf{R}^n)) = \mathcal{F}_n$ as claimed.

Incidentally, in physicists' terminology, (1.72) says that the coherent states in the Fock model are just the functions E_z.

Next, we compute the infinitesimal representation of β, that is, the operators corresponding to X_j and D_j under B. Here again it will be more suitable to consider the complex linear combinations

$$(1.73) \qquad A_j = \sqrt{\pi} B(X_j + iD_j)B^{-1}, \qquad A_j^* = \sqrt{\pi} B(X_j - iD_j)B^{-1}.$$

We set

$$w = r + is, \qquad \frac{\partial}{\partial w_j} = \frac{1}{2}\left(\frac{\partial}{\partial r_j} - i\frac{\partial}{\partial s_j}\right), \qquad \frac{\partial}{\partial \overline{w}_j} = \frac{1}{2}\left(\frac{\partial}{\partial r_j} + i\frac{\partial}{\partial s_j}\right).$$

Since

$$X_j f = \frac{1}{2\pi i}\frac{\partial}{\partial s_j}\rho(r,s)f|_{r=s=0} \qquad D_j f = \frac{1}{2\pi i}\frac{\partial}{\partial r_j}\rho(r,s)f|_{r=s=0},$$

by (1.71) we have

$$(1.74) \qquad \begin{aligned} A_j F &= \frac{\sqrt{\pi}}{2\pi i}B\left(\frac{\partial}{\partial s_j} + i\frac{\partial}{\partial r_j}\right)\rho(r,s)B^{-1}F|_{r=s=0} \\ &= \frac{1}{\sqrt{\pi}}\frac{\partial}{\partial w_j}\beta(w)F|_{w=0} = \frac{1}{\sqrt{\pi}}\frac{\partial F}{\partial z_j}, \end{aligned}$$

$$(1.75) \qquad \begin{aligned} A_j^* F &= \frac{\sqrt{\pi}}{2\pi i}B\left(\frac{\partial}{\partial s_j} - i\frac{\partial}{\partial r_j}\right)\rho(r,s)B^{-1}F|_{r=s=0} \\ &= \frac{-1}{\sqrt{\pi}}\frac{\partial}{\partial \overline{w}_j}\beta(w)F|_{w=0} = \sqrt{\pi}z_j F. \end{aligned}$$

We leave it as an easy exercise for the reader to verify that the operators A_j and A_j^*, defined on the obvious domains

$$\mathrm{Dom}(A_j) = \{F \in \mathcal{F}_n : \partial F/\partial z_j \in \mathcal{F}_n\}, \qquad \mathrm{Dom}(A_j^*) = \{F \in \mathcal{F}_n : z_j F \in \mathcal{F}_n\},$$

are adjoints of each other. Moreover, they satisfy the commutation relations

$$(1.76) \qquad [A_j, A_k] = [A_j^*, A_k^*] = 0, \qquad [A_j, A_k^*] = \delta_{jk}I.$$

The operators A_j and A_j^* act very simply on the basis $\{\zeta_\alpha\}$. If we define 1_j to be the multi-index whose jth entry is 1 and whose other entries are 0, we clearly have

$$\frac{\partial z^\alpha}{\partial z_j} = \alpha_j z^{\alpha - 1_j}, \qquad z_j z^\alpha = z^{\alpha + 1_j},$$

and hence

(1.77) $$A_j \zeta_\alpha = \sqrt{\alpha_j}\, \zeta_{\alpha-1_j}, \qquad A_j^* \zeta_\alpha = \sqrt{\alpha_j + 1}\, \zeta_{\alpha+1_j},$$

where $\zeta_{\alpha-1_j} = 0$ if $\alpha_j = 0$. In particular,

(1.78) $$\zeta_\alpha = \frac{1}{\sqrt{\alpha!}}(A_1^*)^{\alpha_1}\cdots(A_n^*)^{\alpha_n}\zeta_0.$$

Incidentally, by expanding $F \in \mathcal{F}_n$ in terms of the basis $\{\zeta_\alpha\}$ and using the Parseval equation, one can easily derive the formula

$$\|z_j F\|_{\mathcal{F}}^2 = \|F\|_{\mathcal{F}}^2 + \|\partial F/\partial z_j\|_{\mathcal{F}}^2,$$

from which it follows that

$$\mathrm{Dom}(A_j) = \mathrm{Dom}(A_j^*).$$

Let us now compute the inverse Bargmann transform. Since B is unitary, for $F \in \mathcal{F}_n$ and $g \in L^2(\mathbf{R}^n)$, we have

$$\langle B^{-1}F, g\rangle = \langle F, Bg\rangle_{\mathcal{F}} = 2^{n/4}\iint F(z)\overline{g(x)}e^{2\pi x\bar{z} - \pi x^2 - (\pi/2)\bar{z}^2 - \pi|z|^2}\,dx\,dz,$$

and hence

$$B^{-1}F(x) = 2^{n/4}\int F(z)e^{2\pi x\bar{z} - \pi x^2 - (\pi/2)\bar{z}^2 - \pi|z|^2}\,dz,$$

provided that the integrals are absolutely convergent. This will be the case if $|F(z)| \le Ce^{\delta|z|^2}$ for some $\delta < \pi/2$, and in particular if F is a polynomial. For a general $F \in \mathcal{F}_n$ the integral giving $B^{-1}F(x)$ may not converge, but we can compute $B^{-1}F$ by applying it to the partial sums of the Taylor series of F and taking the limit of the resulting functions in the L^2 norm. With this understanding, we can reformulate the Bargmann transform and its inverse as follows. We define the **Bargmann kernel** $B(z, x)$ by

(1.79) $$B(z, x) = 2^{n/4}e^{2\pi xz - \pi x^2 - (\pi/2)z^2} \qquad (z \in \mathbf{C}^n, \quad x \in \mathbf{R}^n),$$

and we then have

(1.80) $$Bf(z) = \int B(z, x)f(x)\,dx, \qquad B^{-1}F(x) = \int B(\bar{z}, x)F(z)e^{-\pi|z|^2}\,dz.$$

For future reference, we exhibit the relationship between Hilbert-Schmidt operators on $L^2(\mathbf{R}^n)$ and on $\mathcal{F}_n$.

(1.81) Proposition. *Suppose $k \in L^2(\mathbf{R}^{2n})$, and let*

$$Tf(x) = \int k(x,y)f(y)\,dy, \qquad f \in L^2(\mathbf{R}^n).$$

Then

$$BTB^{-1}F(z) = \int K(z,\overline{w})F(w)e^{-\pi|w|^2}dw, \qquad F \in \mathcal{F}_n,$$

where K is the (2n-dimensional) Bargmann transform of k.

Proof: We first observe that the Bargmann kernels B_{2n} and B_n in dimensions $2n$ and n are related by

$$B_{2n}\big((z,w),\,(x,y)\big) = B_n(z,x)B_n(w,y),$$

so that, with $B = B_n$,

$$K(z,w) = \iint B(z,x)B(w,y)k(x,y)\,dx\,dy.$$

If F is a polynomial, we can apply formula (1.80) to write

$$BTB^{-1}F(z) = \iiint B(z,x)k(x,y)B(\overline{w},y)F(w)e^{-\pi|w|^2}dw\,dy\,dx.$$

One easily computes that $\|B(z,\cdot)\|_2 = e^{\pi|z|^2/2}$, so by the Schwarz inequality, the above integral is majorized by

$$\|B(z,\cdot)\|_2\|k\|_2 \int \|B(\overline{w},\cdot)\|_2 |F(w)|e^{-\pi|w|^2}dw$$

$$= e^{\pi|z|^2/2}\|k\|_2 \int |F(w)|e^{-\pi|w|^2/2}dw,$$

which is finite since F is a polynomial. Thus we can integrate in x and y first to obtain

$$BTB^{-1}F(z) = \int K(z,\overline{w})F(w)e^{-\pi|w|^2}dw,$$

and this formula remains valid for arbitrary $F \in \mathcal{F}_n$ by continuity. ∎

Finally, we indicate how to modify the construction of the Fock–Bargmann representation for values of Planck's constant other than 1. If $h > 0$, we define the Fock space to be

$$\mathcal{F}_n^h = \{\, F : F \text{ is entire on } \mathbf{C}^n \text{ and } h^n\!\int |F(z)|^2 e^{-\pi h|z|^2}dz < \infty \,\}$$

and the Bargmann transform $B_h : L^2(\mathbf{R}^n) \to \mathcal{F}_n^h$ to be

$$B_h f(z) = e^{(\pi h/2)|z|^2} \langle \rho_h(p, q)f, \ \phi_h \rangle$$

where

$$z = p + iq \quad \text{and} \quad \phi_h(x) = \left(\frac{2}{h}\right)^{n/4} e^{-(\pi/h)x^2},$$

in other words,

$$B_h f(z) = \left(\frac{2}{h}\right)^{n/4} \int f(x) e^{2\pi x z - (\pi/h)x^2 - (\pi h/2)z^2} \, dx.$$

Then the representation

$$\beta_h(w) = B_h \rho_h(r, s) B_h^{-1} \qquad (w = r + is)$$

is given by

$$\beta_h(w)F(z) = e^{-(\pi h/2)|w|^2 - \pi h z \overline{w}} F(z + w).$$

On the other hand, if $h < 0$, the Fock space $\mathcal{F}_n^h$ consists of antiholomorphic functions:

$$\mathcal{F}_n^h = \{ \, F \circ c : F \in \mathcal{F}_n^{|h|} \, \}, \qquad \text{where} \qquad c(z) = \overline{z}.$$

The Bargmann transform is

$$B_h f(\overline{z}) = e^{(\pi|h|/2)|z|^2} \langle \rho_h(p, q)f, \phi_{|h|} \rangle$$

where $\overline{z} = p - iq$ and $\phi_{|h|}$ is as above, in other words,

$$B_h f(\overline{z}) = \left(\frac{2}{|h|}\right)^{n/4} \int f(x) e^{-2\pi x \overline{z} + (\pi/h)x^2 + (\pi h/2)\overline{z}^2} \, dx.$$

The representation

$$\beta_h(w) = B_h \rho(r, s) B_h^{-1} \qquad (w = r + is)$$

is then given by

$$\beta_h(w)F(\overline{z}) = e^{(\pi h/2)|w|^2 + \pi h w \overline{z}} F(\overline{z} + \overline{w}).$$

Some Motivation and History. We begin by explicating the relationship between our space $\mathcal{F}_n$ and the Fock space of quantum mechanics. If $\mathcal{H}$ is any separable Hilbert space, the **Fock space over** $\mathcal{H}$ is the complete tensor algebra over $\mathcal{H}$:

$$\mathcal{F}(\mathcal{H}) = \bigoplus_0^\infty (\otimes^k \mathcal{H}),$$

where $\otimes^k \mathcal{H}$ is the kth tensor power of $\mathcal{H}$ for $k \geq 1$ and $\otimes^0 \mathcal{H} = \mathbf{C}$. The Hilbert space structure on $\mathcal{F}(\mathcal{H})$ may be described as follows: if $\{e_j\}$ is an orthonormal basis for $\mathcal{H}$, then $\{e_{j_1} \otimes \cdots \otimes e_{j_k}\}$ is an orthonormal basis for $\otimes^k \mathcal{H}$, and the union of all these, together with the basis $\{1\}$ for $\otimes^0 \mathcal{H}$, is an orthonormal basis for $\mathcal{F}(\mathcal{H})$.

If $\mathcal{H}$ represents the state space for a quantum particle, $\otimes^k(\mathcal{H})$ can be considered as the state space for a system of k particles of the same type, and $\mathcal{F}(\mathcal{H})$ the state space for a system in which any number of particles can occur. In practice, however, particles are either bosons or fermions, which means that the k-particle states must be either symmetric or antisymmetric under interchancge of two particles. (The antisymmetry in the case of fermions is precisely the Pauli exclusion principle.) In these two cases, $\mathcal{F}(\mathcal{H})$ should be replaced by the **boson Fock space** $\mathcal{F}_s(\mathcal{H})$ consisting of all symmetric tensors or the **fermion Fock space** $\mathcal{F}_a(\mathcal{H})$ consisting of all antisymmetric tensors.

Our concern here is with the boson Fock space. The symmetrizer S which projects $\mathcal{F}(\mathcal{H})$ onto $\mathcal{F}_s(\mathcal{H})$ is given on $\otimes^k \mathcal{H}$ by

$$S(u_1 \otimes \cdots \otimes u_k) = \frac{1}{k!} \sum_\sigma u_{\sigma(1)} \otimes \cdots \otimes u_{\sigma(k)}$$

where σ ranges over the group of permutations of k letters. It is easy to verify that if $\{e_j\}$ is an orthonormal basis for $\mathcal{H}$, then

$$\left\{ E_\alpha = \sqrt{\frac{k!}{\alpha!}} \, S(e_1^{\alpha_1} \otimes e_2^{\alpha_2} \otimes \cdots) : \quad \sum \alpha_j = k, \quad k = 0, 1, 2, \ldots \right\}$$

is an orthonormal basis for $\mathcal{F}_s(\mathcal{H})$, where the superscripts α_j denote tensor powers.

Now let $\mathcal{H} = (\mathbf{C}^n)^*$, and let $\{e_j\}_1^n$ be the standard coordinate functions on $\mathbf{C}^n$. The elements of $\otimes^k \mathcal{H}$ are then k-linear functionals on $\mathbf{C}^n$, which are in one-to-one correspondence with homogeneous polynomials of degree k on $\mathbf{C}^n$. In order to make the normalizations come out right, we introduce a factor of $\sqrt{\pi^k/k!}$ into this correspondence:

$$S(e_1^{\alpha_1} \otimes \cdots \otimes e_n^{\alpha_n}) \longleftrightarrow \sqrt{\frac{\pi^k}{k!}} z_1^{\alpha_1} \cdots z_n^{\alpha_n} = \sqrt{\frac{\pi^k}{k!}} z^\alpha, \quad |\alpha| = k,$$

or in other words,

$$E_\alpha \longleftrightarrow \zeta_\alpha.$$

This correspondence clearly defines a unitary map from $\mathcal{F}_s((\mathbf{C}^n)^*)$ onto our space $\mathcal{F}_n$.

There are analogues of our operators A_j and A_j^* on an arbitrary $\mathcal{F}_s(\mathcal{H})$. Namely, given bases $\{e_j\}$ and $\{E_\alpha\}$ for $\mathcal{H}$ and $\mathcal{F}_s(\mathcal{H})$ as above, one defines A_j and A_j^* on the basis $\{E_\alpha\}$ by

$$A_j E_\alpha = \sqrt{\alpha_j}\, E_{\alpha-1_j}, \qquad A_j^* E_\alpha = \sqrt{\alpha_j + 1}\, E_{\alpha+1_j}.$$

It is not hard to verify that

$$A_j^* \left(\sum_0^\infty v^{(k)} \right) = \sum_0^\infty \sqrt{k+1}\, S(e_j \otimes v^{(k)}), \qquad v^{(k)} \in \otimes^k \mathcal{H}.$$

More generally, for any $u \in \mathcal{H}$ one can define A_u^* by replacing e_j by u in this formula, and then define A_u to be the adjoint of A_u^*. A_u^* and A_u map the k-particle states into $(k+1)$-particle states and $(k-1)$-particle states, respectively; they are called **creation** and **annihilation** operators in quantum field theory.

The Fock spaces $\mathcal{F}(\mathcal{H})$, $\mathcal{F}_s(\mathcal{H})$, and $\mathcal{F}_a(\mathcal{H})$ were introduced by Fock [48]. It was also Fock [47] who first described (on the level of formal calculation) the use of $A = \partial/\partial z$, $A' = z$ to solve the commutator equation $[A, A'] = I$. The rigorous development of the representation of $\mathbf{H}_n$ on $\mathcal{F}_n$ and the intertwining operator B is due to Bargmann [11]; the same ideas also appear in work of Segal [126], [127], done independently at about the same time.

We pulled the Fock space and the Bargmann transform out of a hat by using the Fourier-Wigner transform. It is perhaps more enlightening to see the heuristic method by which Bargmann [11] derived them. To begin with, we observe that if P_j and Q_j are self-adjoint operators satisfying the canonical commutation relations (1.9) with $h = 1$, then the operators $A_j = \pi^{1/2}(Q_j + iP_j)$ and $A_j^* = \pi^{1/2}(Q_j - iP_j)$ satisfy the commutation relations (1.76), and conversely. The latter relations are also satisfied by the differential operators $\pi^{-1/2}\partial/\partial z_j$ and $\pi^{1/2} z_j$, so we are led to look for a Hilbert space of holomorphic functions on which these operators are adjoints of each other. As a candidate for such a space, we try the space $\mathcal{H}$ of entire functions in $L^2(\mathbf{C}^n, \omega(z, \bar{z})dz)$ where ω is a suitable positive weight function, and the condition we require is

$$\pi^{1/2} \int z_j F\overline{G}\omega\, dz = \pi^{-1/2} \int F \overline{\frac{\partial G}{\partial z_j}} \omega\, dz \qquad \text{for} \qquad F, G \in \mathcal{H}.$$

But if we integrate by parts, assuming that F, G, and ω are such that the boundary term vanishes, since F is holomorphic we obtain

$$\int F \overline{\frac{\partial G}{\partial z_j}} \omega\, dz = -\int \overline{G}\, \frac{\partial(F\omega)}{\partial z_j}\, dz = -\int F\overline{G}\, \frac{\partial \omega}{\partial z_j}\, dz,$$

so we must have $\partial \omega / \partial \bar{z}_j = -\pi z_j \omega$ for all j, or, with $z = x + iy$,

$$\frac{\partial \omega}{\partial x_j} + i \frac{\partial \omega}{\partial y_j} = -2\pi(x_j + iy_j)\omega.$$

Since ω is positive, this means that

$$\frac{\partial \omega}{\partial x_j} = -2\pi x_j \omega \quad \text{and} \quad \frac{\partial \omega}{\partial y_j} = -2\pi y_j \omega, \qquad \text{or} \quad \nabla_{x,y}(\log \omega) = (-2\pi x, -2\pi y),$$

so that $\log \omega = -\pi(x^2 + y^2) + C = -\pi |z|^2 + C$. We may choose $C = 0$; then $\omega = e^{-\pi|z|^2}$, so we obtain the space $\mathcal{F}_n$.

Next, we look for an operator

$$Bf(z) = \int f(x)B(z,x)\,dx$$

that maps $L^2(\mathbf{R}^n)$ onto $\mathcal{F}_n$ and intertwines the operators

$$\pi^{1/2}(X_j + iD_j) = \pi^{1/2}x_j + \frac{1}{2\pi^{1/2}}\frac{\partial}{\partial x_j} \quad \text{and} \quad \pi^{1/2}(X_j - iD_j) = \pi^{1/2}x_j - \frac{1}{2\pi^{1/2}}\frac{\partial}{\partial x_j}$$

with $\pi^{1/2}z_j$ and $\pi^{-1/2}\partial/\partial z_j$. Again, a formal integration by parts yields

$$\int [(X_j \pm iD_j)f(x)]\,B(z,x)\,dx = \int f(x)[(X_j \mp iD_j)B(z,x)]\,dx,$$

so the intertwining conditions will be satisfied if

$$\left(\pi x_j + \frac{1}{2}\frac{\partial}{\partial x_j}\right) B(z,x) = \pi z_j B(z,x), \qquad \left(\pi x_j - \frac{1}{2}\frac{\partial}{\partial x_j}\right) B(z,x) = \frac{\partial B}{\partial z_j}(z,x),$$

or

$$\frac{\partial B}{\partial x_j} = 2\pi(z_j - x_j)B, \qquad \frac{\partial B}{\partial z_j} = \pi x_j B - \frac{1}{2}\frac{\partial B}{\partial x_j} = \pi(2x_j - z_j)B.$$

Hence

$$\frac{\partial(\log B)}{\partial x_j} = 2\pi(z_j - x_j), \qquad \frac{\partial(\log B)}{\partial z_j} = \pi(2x_j - z_j),$$

and integrating these equations gives

$$\log B = 2\pi xz - \pi x^2 - \tfrac{1}{2}\pi z^2 + C.$$

The constant C is chosen to be $\log 2^{n/4}$ to make the transform unitary, and the derivation is thereby complete.

7. Hermite Functions

The monomials $\zeta_\alpha(z) = \sqrt{\pi^{|\alpha|}/\alpha!}\, z^\alpha$ obviously play a distinguished role among all orthonormal bases for the Fock space $\mathcal{F}_n$, and one might therefore suspect that the corresponding functions $B^{-1}\zeta_\alpha$ should also be important in $L^2(\mathbf{R}^n)$. This is indeed the case. We call $B^{-1}\zeta_\alpha$ the αth (normalized, n-dimensional) **Hermite function** and denote it by h_α.

To compute h_α we utilize the operators

$$Z_j = (X_j + iD_j) = \pi^{-1/2}B^{-1}A_jB, \qquad Z_j^* = (X_j - iD_j) = \pi^{-1/2}B^{-1}A_j^*B$$

and their products

$$Z^\alpha = Z_1^{\alpha_1}\cdots Z_n^{\alpha_n}, \qquad Z^{*\alpha} = Z_1^{*\alpha_1}\cdots Z_n^{*\alpha_n}.$$

We observe that

$$Z_j^* f(x) = x_j f(x) - \frac{1}{2\pi}\frac{\partial f}{\partial x_j} = \frac{-1}{2\pi}e^{\pi x^2}\frac{\partial}{\partial x_j}\left(e^{-\pi x^2}f(x)\right),$$

so that

$$Z^{*\alpha}f(x) = \left(\frac{-1}{2\pi}\right)^{|\alpha|}e^{\pi x^2}\left(\frac{\partial}{\partial x}\right)^\alpha\left(e^{-\pi x^2}f(x)\right).$$

We have already noted in (1.72) that

$$h_0(x) = (B^{-1}\zeta_0)(x) = (B^{-1}E_0)(x) = 2^{n/4}e^{-\pi x^2}.$$

Therefore, by (1.78),

(1.81)
$$\begin{aligned}
h_\alpha(x) &= \sqrt{\frac{1}{\alpha!}}(B^{-1}A^{*\alpha}\zeta_0)(x) = \sqrt{\frac{\pi^{|\alpha|}}{\alpha!}}(Z^{*\alpha}h_0)(x) \\
&= \frac{2^{n/4}}{\sqrt{\alpha!}}\left(\frac{-1}{2\sqrt{\pi}}\right)^{|\alpha|}e^{\pi x^2}\left(\frac{\partial}{\partial x}\right)^\alpha\left(e^{-2\pi x^2}\right).
\end{aligned}$$

In particular, taking $n = 1$, we obtain the one-dimensional Hermite functions

$$h_j(x) = \frac{2^{1/4}}{\sqrt{j!}}\left(\frac{-1}{2\sqrt{\pi}}\right)^j e^{\pi x^2}\frac{d^j}{dx^j}\left(e^{-2\pi x^2}\right).$$

These are not quite the same as the Hermite functions usually found in the literature, because they are built from $e^{-\pi x^2}$ rather than $e^{-x^2/2}$ and are normalized differently. The **classical Hermite functions** on $\mathbf{R}$ are defined by

$$\tilde{h}_j(x) = (-1)^j e^{x^2/2}\frac{d^j}{dx^j}e^{-x^2}.$$

It follows easily that

$$h_j(x) = \frac{2^{1/4}}{\sqrt{2^j j!}} \tilde{h}_j(\sqrt{2\pi}\, x).$$

Returning now to the n-dimensional case, it is easy to read off from the above calculations a number of basic properties of the Hermite functions:

(i) The n-dimensional Hermite functions are products of one-dimensional Hermite functions; namely,

$$h_\alpha(x) = h_{\alpha_1}(x_1) \cdots h_{\alpha_n}(x_n).$$

(ii) The function $H_\alpha(x) = e^{\pi x^2} h_\alpha(x)$ is a polynomial of degree $|\alpha|$, called the αth **Hermite polynomial**. We have

$$H_\alpha(x) = 2^{(n/4)+|\alpha|} \sqrt{\frac{\pi^{|\alpha|}}{\alpha!}}\, x^\alpha + (\text{terms of degree } < |\alpha|).$$

(iii) Every polynomial of degree $\leq k$ on $\mathbf{R}^n$ is a linear combination of Hermite polynomials of degree $\leq k$. This follows from the preceding formula by induction on k.

(iv) Since $[Z_j, Z_k^*] = \pi^{-1}\delta_{jk} I$, one finds by induction that

$$[Z_j, Z^{*\alpha}] = \pi^{-1}\alpha_j Z^{*(\alpha - 1_j)}.$$

(v) We have

(1.82) $$Z_j h_\alpha = \sqrt{\frac{\alpha_j}{\pi}}\, h_{\alpha - 1_j}, \qquad Z_j^* h_\alpha = \sqrt{\frac{\alpha_j + 1}{\pi}}\, h_{\alpha + 1_j}.$$

Of course this follows immediately from the corresponding property (1.78) of the ζ_α's, but it can also be verified directly by using (iv) and the formula $h_\alpha = \sqrt{\pi^{|\alpha|}/\alpha!}\, Z^{*\alpha} h_0$.

(vi) In dimension one we have

$$ZZ^* = (X + iD)(X - iD) = X^2 + D^2 + i[D, X] = D^2 + X^2 + (2\pi)^{-1} I.$$

The operator

$$2\pi(D^2 + X^2) = 2\pi x^2 - \frac{1}{2\pi}\frac{d^2}{dx^2}$$

is called the **Hermite operator** (adapted to our use of $e^{-\pi x^2}$ rather than $e^{-x^2/2}$; the usual Hermite operator is $x^2 - (d/dx)^2$). The one-dimensional Hermite functions are the eigenfunctions of this operator: by (1.82) we have

$$2\pi(D^2 + X^2)h_k = (2\pi ZZ^* - 1)h_k = 2\sqrt{\pi(k+1)}\, Z h_{k+1} - h_k = (2k+1)h_k.$$

In n dimensions, this equation together with (i) shows that h_α is an eigenfunction of the Hermite operators in each variable,

$$(1.83a) \qquad 2\pi(D_j^2 + X_j^2)h_\alpha = (2\alpha_j + 1)h_\alpha,$$

as well as of the n-dimensional Hermite operator $2\pi(D^2 + X^2)$, that is, $2\pi \sum_1^n (D_j^2 + X_j^2)$:

$$(1.83b) \qquad 2\pi(D^2 + X^2)h_\alpha = (2|\alpha| + n)h_\alpha.$$

(vii) $\{h_\alpha\}$ is an orthonormal basis for $L^2(\mathbf{R}^n)$. This follows from the corresponding property of the ζ_α's, but we can prove it directly from (1.81) as follows. We have

$$\langle h_\alpha, h_\beta \rangle = \sqrt{\frac{\pi^{|\alpha|+|\beta|}}{\alpha!\beta!}} \langle h_0, Z^\alpha Z^{*\beta} h_0 \rangle.$$

Since $Z_j h_0 = 0$, if $\alpha_j > \beta_j$ for any j then repeated application of (iv) shows that $\langle h_\alpha, h_\beta \rangle = 0$, while if $\alpha = \beta$ it shows that $\|h_\alpha\|_2^2 = \|h_0\|_2^2$, and the latter number is 1. As for completeness, if $g \in L^2$ and $\langle g, h_\alpha \rangle = 0$ for all α, then by (iii), $\langle g, P(x)e^{-\pi x^2} \rangle = 0$ for all polynomials P. But then

$$\int g(x)e^{-\pi x^2} e^{2\pi i x \xi}\, dx = \sum_0^\infty \int g(x)e^{-\pi x^2} \frac{(2\pi i x \xi)^j}{j!}\, dx = 0,$$

so by Fourier uniqueness, $g(x)e^{-\pi x^2} = 0$ a.e., and hence $g = 0$.

(viii) The Hermite functions are eigenfunctions of the Fourier transform $\mathcal{F}$. Indeed, since $\mathcal{F}X_j = -D_j\mathcal{F}$ and $\mathcal{F}D_j = X_j\mathcal{F}$, we have

$$\mathcal{F}Z_j^* = \mathcal{F}(X_j - iD_j) = (-D_j - iX_j)\mathcal{F} = -iZ_j^*\mathcal{F}.$$

Moreover, $\mathcal{F}h_0 = h_0$, so

$$(1.84) \qquad \mathcal{F}h_\alpha = \sqrt{\frac{\pi^{|\alpha|}}{\alpha!}}\, \mathcal{F}Z^{*\alpha}h_0 = (-i)^{|\alpha|}\sqrt{\frac{\pi^{|\alpha|}}{\alpha!}}\, Z^{*\alpha}h_0 = (-i)^{|\alpha|}h_\alpha.$$

We conclude this section by deriving two classical generating function identities for Hermite functions. The first one is more or less equivalent to the fact that the Bargmann transform maps the orthonormal basis $\{h_\alpha\}$ for $L^2(\mathbf{R}^n)$ to the orthonormal basis $\{\zeta_\alpha\}$ for $\mathcal{F}^n$; the second one is somewhat deeper. In order to obtain uniformity of convergence, we shall need the following lemma.

(1.85) Lemma. *There is a constant C, depending only on the dimension n, such that*

$$\|h_\alpha\|_\infty \le C(|\alpha|+1)^{n/2}.$$

Proof: In dimension 1 we have $\|h_j\|_2 = 1$ and, by (1.82),

$$\|Dh_j\|_2^2 = \frac{1}{2}\|(Z - Z^*)h_j\|_2^2 = \frac{1}{2\sqrt{\pi}}\|\sqrt{j}\,h_{j-1} - \sqrt{j+1}\,h_{j+1}\|_2^2 = \frac{1}{2}\sqrt{\frac{2j+1}{\pi}},$$

since $h_{j-1} \perp h_{j+1}$. Moreover, the Fourier inversion formula implies that $\|h_j\|_\infty \le \|\widehat{h}_j\|_1$, so by the Schwarz inequality and the Plancherel theorem,

$$\|h_j\|_\infty \le \int |\widehat{h}_j(\xi)|\,d\xi \le \left(\int (1+\xi^2)|\widehat{h}_j(\xi)|^2\,d\xi\right)^{1/2}\left(\int (1+\xi^2)^{-1}d\xi\right)^{1/2}$$
$$= (\|h_j\|_2^2 + \|Dh_j\|_2^2)^{1/2}\pi^{1/2}$$
$$= \tfrac{1}{2}\sqrt{2j+1+4\pi}.$$

The n-dimensional case now follows easily in view of (i) above. ∎

(1.86) Theorem. *We have*

$$\sum_{|\alpha|\ge 0} h_\alpha(x)\zeta_\alpha(z) = 2^{n/4}e^{2\pi xz - \pi x^2 - (\pi/2)z^2} = B(z,x),$$

where $B(z,x)$ is the Bargmann kernel (1.79). The series converges uniformly on $\mathbf{R}^n \times K$ for every compact $K \subset \mathbf{C}^n$, and also in $L^2(x)$ for each z.

Proof: The assertions about convergence follow from Lemma (1.85) and the fact that $\zeta_\alpha(z) = \sqrt{\pi^{|\alpha|}/\alpha!}\,z^\alpha$ tends to zero rapidly as $|\alpha| \to \infty$ for z in any compact set. To sum the series, observe that by (1.81),

$$h_\alpha(x) = \frac{2^{n/4}}{\sqrt{\alpha!}}\left(\frac{1}{2\sqrt{\pi}}\right)^{|\alpha|}e^{\pi x^2}\left(\frac{\partial}{\partial z}\right)^\alpha e^{-2\pi(x-z)^2}\big|_{z=0}.$$

Hence, by Taylor's theorem,

$$e^{-2\pi(x-z)^2} = \sum \frac{(2\sqrt{\pi})^{|\alpha|}\sqrt{\alpha!}}{2^{n/4}e^{\pi x^2}}h_\alpha(x)\frac{z^\alpha}{\alpha!} = 2^{-n/4}e^{-\pi x^2}\sum h_\alpha(x)\zeta_\alpha(2z),$$

and the desired result follows immediately on replacing z by $z/2$. ∎

(1.87) Mehler's Formula. *For* $x, y \in \mathbf{R}^n$ *and* $w \in \mathbf{C}$ *with* $|w| < 1$ *we have*

$$\sum_{|\alpha| \geq 0} w^{|\alpha|} h_\alpha(x) h_\alpha(y) = \left(\frac{2}{1 - w^2}\right)^{n/2} \exp\left[\frac{-\pi(1 + w^2)(x^2 + y^2) + 4\pi w x y}{1 - w^2}\right].$$

(Here $u = 2/(1 - w^2)$ lies in the right half plane, and the square root in $u^{n/2}$ is the branch that is positive for $u > 0$.) The series converges absolutely and uniformly on compact sets of $\mathbf{R}^{2n} \times \{|w| < 1\}$, and also in $L^2(y)$ for each x and w.

Proof: The assertions about convergence follow easily from Lemma (1.85) and the orthonormality of $\{h_\alpha\}$. To sum the series, we replace z by wz in Theorem (1.86):

$$\sum w^{|\alpha|} h_\alpha(x) \zeta_\alpha(z) = 2^{n/4} e^{2\pi w x z - \pi x^2 - (\pi/2) w^2 z^2}.$$

We apply the inverse Bargmann transform to both sides. On the one hand, for fixed w and x, the left side converges in the Fock space norm by Lemma (1.85), and its inverse Bargmann transform is clearly

$$B^{-1}\left(\sum w^{|\alpha|} h_\alpha(x) \zeta_\alpha\right)(y) = \sum w^{|\alpha|} h_\alpha(x) h_\alpha(y).$$

On the other hand, the right side satisfies

$$\left|2^{n/4} e^{2\pi w x z - \pi x^2 - (\pi/2) z^2}\right| \leq C e^{(\pi/2)(|w|^2 + \epsilon)|z|^2} \leq C e^{\delta |z|^2}$$

for some $\delta < \pi/2$, so we can apply (1.80) to see that its inverse Bargmann transform is

$$2^{n/2} \int e^{2\pi w x z - \pi x^2 - (\pi/2) w^2 z^2 + 2\pi y \bar{z} - \pi y^2 - (\pi/2) \bar{z}^2} e^{-\pi |z|^2} \, dz$$

$$= 2^{n/2} e^{-\pi(x^2 + y^2)} \int e^{(\pi/2)(-w^2 z^2 - \bar{z}^2 + 4 w x z + 4 y \bar{z})} e^{-\pi |z|^2} \, dz$$

By Theorem 3 of Appendix A, this last integral equals

$$(1 - w^2)^{-n/2} \exp\left[(-\pi)\frac{2 w^2 x^2 - 4 w x y + 2 w^2 y^2}{1 - w^2}\right],$$

and the result follows immediately. ∎

Several people have obtained extensions of Mehler's formula in which exponentials of more general quadratic functions are expanded in series of Hermite functions; see Louck [97] and the references given there.

8. The Wigner Transform

The **Wigner transform** of two functions f and g is the Fourier transform of their Fourier-Wigner transform:

$$W(f,g)(\xi, x) = \iint e^{-2\pi i(\xi p + xq)} V(f,g)(p, q)\, dp\, dq.$$

$W(f,g)$ was first introduced into the literature in the case $g = f$ by Wigner [156]; the general case seems to have been first studied by Moyal [108]. Since

$$(1.88) \qquad V(f,g)(p, q) = \int e^{2\pi i q y} f(y + \tfrac{1}{2}p)\overline{g(y - \tfrac{1}{2}p)}\, dy,$$

we have

$$W(f,g)(\xi, x) = \iiint e^{-2\pi i(\xi p + xq - yq)} f(y + \tfrac{1}{2}p)\overline{g(y - \tfrac{1}{2}p)}\, dy\, dp\, dq,$$

so by the Fourier inversion theorem,

$$(1.89) \qquad W(f,g)(\xi, x) = \int e^{-2\pi i \xi p} f(x + \tfrac{1}{2}p)\overline{g(y - \tfrac{1}{2}p)}\, dp$$

The expressions (1.88) and (1.89) are deceptively similar; it is hard to believe that they are always Fourier transforms of one another! In fact, a simple calculation shows that

$$(1.90) \qquad W(f,g)(\xi, x) = 2^n V(f, \tilde{g})(2x, -2\xi), \quad \text{where} \quad \tilde{g}(x) = g(-x).$$

Like V, the sesquilinear transform W can be regarded as the restriction to functions of the form $f(x)\overline{g(y)}$ of the linear transform

$$(1.91) \qquad \widetilde{W}F(\xi, x) = \int e^{-2\pi i \xi p} F(x + \tfrac{1}{2}p,\, x - \tfrac{1}{2}p)\, dp,$$

defined for functions F of $2n$ variables. $\widetilde{W}$ is the composition of the measure-preserving change of variables $(x, p) \to (x + \tfrac{1}{2}p,\, x - \tfrac{1}{2}p)$ with Fourier transformation in the second variable, so it preserves the classes $\mathcal{S}(\mathbf{R}^{2n})$ and $\mathcal{S}'(\mathbf{R}^{2n})$ and is unitary on $L^2(\mathbf{R}^{2n})$. Therefore:

(1.92) Proposition. W maps $\mathcal{S}(\mathbf{R}^n) \times \mathcal{S}(\mathbf{R}^n)$ into $\mathcal{S}(\mathbf{R}^{2n})$ and extends to a map from $\mathcal{S}'(\mathbf{R}^n) \times \mathcal{S}'(\mathbf{R}^n)$ into $\mathcal{S}'(\mathbf{R}^{2n})$. Moreover, W maps $L^2(\mathbf{R}^n) \times L^2(\mathbf{R}^n)$ into $L^2(\mathbf{R}^{2n}) \cap C_0(\mathbf{R}^{2n})$ and satisfies

$$(1.93) \qquad \langle W(f_1, g_1), W(f_2, g_2) \rangle = \langle f_1, f_2 \rangle \overline{\langle g_1, g_2 \rangle}$$

and $\|W(f,g)\|_\infty \leq \|f\|_2 \|g\|_2$.

Proof: The assertions about $\mathcal{S}$ and $\mathcal{S}'$ and the unitarity relation (1.93) are consequences of the properties of $\widetilde{W}$. The estimate on $\|W(f,g)\|_\infty$ comes from the Schwarz inequality, and the fact that $W(f,g) \in C_0$ for $f, g \in L^2$ then follows since $\mathcal{S}$ is dense in L^2. ∎

(1.93) is often called **Moyal's identity**.

We summarize the basic transformation properties of $W(f, g)$ in the following proposition. The verifications of these formulas are all easy exercises which we leave to the reader.

(1.94) Proposition. *For $t \in \mathbf{R} \setminus \{0\}$, let $f^t(x) = |t|^{n/2} f(tx)$. Then:*

(a) $\quad W(f^t, g^t)(\xi, x) = W(f, g)(t^{-1}\xi, tx)$.

(b) $\quad W\big(\rho(a, b)f,\ \rho(c, d)g\big)(\xi, x)$
$$= e^{\pi i(bc - ad) + 2\pi i[(a-c)\xi + (b-d)x]} W(f, g)\big(\xi - \tfrac{1}{2}(b + d),\ x + \tfrac{1}{2}(a + c)\big).$$

(c) $\quad W(\widehat{f}, \widehat{g})(\xi, x) = W(f, g)(x, -\xi)$.

(d) $\quad W(g, f) = \overline{W(f, g)}$.

As a special case of Proposition (1.94b), we have

(1.95a) $\qquad W\big(\rho(a, b)f,\ \rho(a, b)g\big)(\xi, x) = W(f, g)(\xi - b,\ x + a),$

or, what is sometimes more convenient, with $\rho'(a, b) = \rho(-b, a)$ as in (1.27),

(1.95b) $\qquad W\big(\rho'(a, b)f,\ \rho'(a, b)g\big)(\xi, x) = W(f, g)(\xi - a,\ x - b).$

The function $W(f, g)$ is of greatest intrinsic interest in the case $g = f$. In this case we shall write

$$W(f, f) = Wf$$

and call Wf the **Wigner distribution** of f. Wf was proposed by Wigner [156] as a substitute for the nonexistent joint probability distribution of momentum and position in the quantum state f. The motivation is as follows. Since the uncertainty principle imposes a limit on the precision with which momentum and position can be determined in the state f ($f \in L^2$, $\|f\|_2 = 1$), it does not make sense to speak of a joint probability distribution for these observables. However, if such a distribution existed, with density $\sigma(\xi, x)$, then the inverse Fourier transform of σ,

$$\iint e^{2\pi i(p\xi + qx)} \sigma(\xi, x)\, d\xi\, dx,$$

would be the expected value of the function $e^{2\pi i(p\xi + qx)}$ with respect to σ. The latter quantity has a natural and consistent interpretation in quantum mechanics, namely, the expected value of $e^{2\pi i(pD + qX)}$ in the state f:

$$\langle e^{2\pi i(pD + qX)} f,\ f \rangle = \langle \rho(p, q)f,\ f \rangle = V(f, f),$$

so we are led to take $\sigma = V(f, f)\widehat{} = Wf$.

Wf is usually not a genuine probability density, because it may assume negative values. (Indeed, if f is odd we have $Wf(0,0) = -\|f\|_2^2$.) It is, however, always real, by Proposition (1.94d), and in some sense it tries very hard to be a joint density for momentum and position. The following results provide supporting evidence for this heuristic assertion.

In the first place, Wf has the right marginal distributions: if we integrate out either position or momentum, we get the probability distribution for the other one.

(1.96) Proposition. *We have*

$$\int Wf(\xi, x)\, dx = |\widehat{f}(\xi)|^2 \qquad \text{and} \qquad \int Wf(\xi, x)\, d\xi = |f(x)|^2.$$

Proof: Letting $u = x + \frac{1}{2}p$ and $v = x - \frac{1}{2}p$, we have

$$\int Wf(\xi, x)\, dx = \iint e^{-2\pi i \xi p} f(x + \tfrac{1}{2}p)\overline{f(x - \tfrac{1}{2}p)}\, dp\, dx$$

$$= \iint e^{-2\pi i \xi u} f(u)\overline{e^{-2\pi i \xi v} f(v)}\, du\, dv$$

$$= \widehat{f}(\xi)\overline{\widehat{f}(\xi)}.$$

This proves the first assertion, and the second one follows from the Fourier inversion formula:

$$\int Wf(\xi, x)\, d\xi = \iint e^{-2\pi i \xi p} f(x + \tfrac{1}{2}p)\overline{f(x - \tfrac{1}{2}p)}\, dp\, d\xi$$

$$= \int \delta(p) f(x + \tfrac{1}{2}p)\overline{f(x - \tfrac{1}{2}p)}\, dp = |f(x)|^2. \quad \blacksquare$$

Remark. We have been a bit sloppy here. An examination of these calculations shows that they are rigorously correct if $f \in L^1$ (which guarantees that $Wf(\xi, \cdot) \in L^1$) and $\widehat{f} \in L^1$ (which guarantees that $Wf(\cdot, x) \in L^1$). If f is merely in L^2, the integrals $\int Wf(\xi, x)dx$ and $\int Wf(\xi, x)d\xi$ need not be absolutely convergent, but the above formulas remain valid if they are suitably interpreted, a task which we leave to the reader.

As a corollary, under suitable hypotheses to ensure convergence, we obtain

$$\iint x_j Wf(\xi, x)\, d\xi\, dx = \int x_j |f(x)|^2\, dx = \langle X_j f, f \rangle,$$

$$\iint \xi_j Wf(\xi, x)\, d\xi\, dx = \int \xi_j |\widehat{f}(\xi)|^2\, d\xi = \langle D_j f, f \rangle.$$

This implies that the center of mass of Wf is $(\bar{\xi}, \bar{x})$ where $\bar{\xi}$ and $\bar{x}$ are the centers of mass of $|\widehat{f}|^2$ and $|f|^2$ respectively.

The next result shows that if the position or momentum spectrum of f is limited, there is a corresponding limitation on the support of Wf. In what follows, "supp(f)" means the smallest closed set outside of which $f = 0$ a.e., and we may assume that $f = 0$ everywhere outside supp(f).

(1.97) Proposition. *Let π_1 and π_2 be the projections from $\mathbf{R}^n \times \mathbf{R}^n$ onto the first and second factors, and for $E \subset \mathbf{R}^n$ let $H(E)$ denote the closed convex hull of E. Then*

$$\pi_1\big(\mathrm{supp}(Wf)\big) \subset H\big(\mathrm{supp}(\widehat{f})\big) \quad \text{and} \quad \pi_2\big(\mathrm{supp}(Wf)\big) \subset H\big(\mathrm{supp}(f)\big).$$

Proof: From (1.89), $Wf(\xi, x) = 0$ unless there is some p for which $x + \frac{1}{2}p$ and $x - \frac{1}{2}p$ are in supp(f); in this case x, being halfway between these points, is in $H\big(\mathrm{supp}(f)\big)$. This proves the second assertion, and the first one follows in the same way since $Wf(\xi, x) = W\widehat{f}(-x, \xi)$ by Proposition (1.94c). ∎

In view of these results, the Wigner distribution Wf can be viewed as a sort of portrait of the quantum state f in phase space. Another fact which supports this point of view is that

$$W(e^{2\pi i(aX - bD)}f)(\xi, x) = Wf(\xi - a,\ x - b),$$

so that momentum-position translations of f (cf. the discussion following (1.27)) correspond to ordinary translations of Wf. In this connection we should observe that Wf determines f up to a phase factor:

(1.98) Proposition. $Wf = Wg$ *if and only if* $f = cg$ *for some $c \in \mathbf{C}$ with* $|c| = 1$.

Proof: From formula (1.89) and the Fourier inversion theorem we see that $Wf = Wg$ if and only if

$$f(x + \tfrac{1}{2}p)\overline{f(x - \tfrac{1}{2}p)} = g(x + \tfrac{1}{2}p)\overline{g(x - \tfrac{1}{2}p)} \quad \text{for almost every } x, p,$$

in other words, $f(u)\overline{f(v)} = g(u)\overline{g(v)}$ for almost every u, v. The assertion is now obvious. ∎

A more classical interpretation is also available. Take $n = 1$, and let $f(t)$ represent the amplitude of a vibration—say, a sound wave—at time t. Then the Fourier representation $f(t) = \int e^{2\pi i \omega t} \widehat{f}(\omega)d\omega$ tells how f is synthesized from waves of definite frequencies, and Wf gives a picture of f in time-frequency space. This is rather like what is done in music. f might represent a musical composition, but composers almost never try to describe either f or $\widehat{f}$ directly;

rather, they make a "time-frequency plot" of f by writing notes on musical staves. For this reason, de Bruijn [37] has dubbed Wf the "musical score" of f.

For any f in L^2, by the Schwarz inequality we have $|Wf(\xi, x)| \leq \|f\|_2^2$, and hence

$$\left| \iint_E Wf(\xi, x) \, d\xi \, dx \right| \leq \|f\|_2^2 \cdot \text{meas}(E).$$

On the other hand, by Proposition (1.96),

$$\iint_{\mathbf{R}^{2n}} Wf(\xi, x) \, d\xi \, dx = \|f\|_2^2.$$

Hence the mass of Wf cannot almost all be concentrated in a set E in phase space unless $\text{meas}(E) \geq 1$. This is a form of the uncertainty principle; some quantitative versions of the uncertainty principle for Wf can be found in de Bruijn [37].

Let us return to the question of the positivity of Wf. Although Wf may assume negative values, it tends to be positive "on the average": it is easy to make Wf positive by convolving it with a suitable function G.

(1.99) Proposition. *Suppose $G \in (L^1 + L^2)(\mathbf{R}^{2n})$ satisfies $\int (Wf)G \geq 0$ for all $f \in L^2(\mathbf{R}^n)$. Then $Wf * G \geq 0$ pointwise for all $f \in L^2(\mathbf{R}^n)$. In particular, this is the case if $G = Wg$ for some $g \in L^2(\mathbf{R}^n)$; in fact,*

$$Wf * Wg(\xi, x) = |V(\widetilde{f}, g)(-x, \xi)|^2$$

where $\widetilde{f}(x) = f(-x)$.

Proof: We observe that by (1.95) and Proposition (1.94a) (with $t = -1$),

$$Wf(\xi - \eta, \, x - y) = W\widetilde{f}(\eta - \xi, \, y - x) = W\big(\rho(-x, \xi)\widetilde{f}\big)(\eta, y)$$

and hence

$$Wf * G(\xi, x) = \iint W\big(\rho(-x, \xi)\widetilde{f}\big)(\eta, y)G(\eta, y) \, d\eta \, dy.$$

This proves the first assertion. If $G = Wg$ then G is real, so by Moyal's identity (1.93) the above equation gives

$$Wf * Wg(\xi, x) = \langle W\big(\rho(-x, \xi)\widetilde{f}\big), \, Wg \rangle$$
$$= |\langle \rho(-x, \xi)\widetilde{f}, \, g \rangle|^2 = |V(\widetilde{f}, g)(-x, \xi)|^2. \quad \blacksquare$$

Other examples of G's satisfying the condition $\int (Wf)G \geq 0$ for all f may be found in Janssen [85].

To understand the meaning of Proposition (1.99), let us consider a specific example. For $a > 0$, let

$$\phi_a^0(x) = (2a)^{n/4} e^{-\pi a x^2} \quad \text{and} \quad \Phi_a = W\phi_a^0.$$

A simple calculation shows that

$$\Phi_a(\xi, x) = 2^n e^{-2\pi(ax^2 + a^{-1}\xi^2)} = \phi_a(x)\phi_{1/a}(\xi) \quad \text{where} \quad \phi_a(x) = (2a)^{n/2} e^{-2\pi x^2},$$

so Proposition (1.99) shows that $Wf * \Phi_a \geq 0$ for all $f \in L^2$ and $a > 0$. Now, ϕ_a is a Gaussian of total mass 1 whose central peak has width roughly $\|X_j \phi_a\|_2 = 1/\sqrt{4\pi a}$, so convolving Wf by Φ_a more or less amounts to averaging Wf over balls of radius $1/\sqrt{4\pi a}$ in x and over balls of radius $\sqrt{a/4\pi}$ in ξ. The uncertainty principle does not allow position and momentum to be measured simultaneously with complete precision, but it allows position to be measured with an error ϵ and momentum to be measured with an error δ provided that $\epsilon\delta \geq (4\pi)^{-1}$. So it should be possible for the averages of position and momentum over balls of radius $\epsilon = 1/\sqrt{4\pi a}$ and $\delta = \sqrt{a/4\pi}$ to have a joint distribution—and this is $Wf * \Phi_a$.

More generally, if we average over larger balls (that is, convolve with more spread-out Gaussians) we get something strictly positive, while averaging over smaller balls doesn't work, as the following result of de Bruijn [37] shows:

(1.100) Proposition. *Let*

$$\Phi_{a,b}(\xi, x) = 2^n (ab)^{-n/2} \exp(-2\pi) \left(\frac{\xi^2}{a} + \frac{x^2}{b} \right), \qquad a, b > 0,$$

and suppose $f \in L^2(\mathbf{R}^n)$. If $ab = 1$ then $Wf * \Phi_{a,b} \geq 0$. If $ab > 1$ then $Wf * \Phi_{a,b} > 0$. If $ab < 1$ then $Wf * \Phi_{a,b}$ may be negative.

Proof: We have proved the first assertion above. If $ab > 1$, pick $c < a$ and $d < b$ with $cd = 1$. By the semigroup property of Gaussians (easily verified by taking Fourier transforms),

$$Wf * \Phi_{a,b} = (Wf * \Phi_{c,d}) * \Phi_{a-c,b-d}.$$

But $Wf * \Phi_{c,d} \geq 0$ and $\Phi_{a-c,b-d} > 0$, so the second assertion follows. As for the last one, we leave it as an exercise for the reader to verify that if $f(x) = x e^{-\pi x^2}$ on $\mathbf{R}$, or more generally $f(x) = x_1 e^{-\pi x^2}$ on $\mathbf{R}^n$, then $Wf * \Phi_{a,b}(0,0) < 0$ if $ab < 1$. ∎

There remains the question of when Wf is itself nonnegative. We have seen that this is the case when $f(x) = e^{-\pi a x^2}$. More generally, if

(1.101) $f(x) = e^{-xAx+bx+c}$ where $A \in GL(n, \mathbf{C})$, $b \in \mathbf{C}^n$, $c \in \mathbf{C}$,
and $\operatorname{Re} A$ is positive definite,

then $Wf > 0$. It is not hard to prove this by a brute force calculation of Wf, and the reader is welcome to do so. We shall return to this point in Section 4.5, when we shall have the machinery to make it utterly transparent. The remarkable thing, however, is that the functions (1.101) are the *only* ones with nonnegative Wigner distributions.

(1.102) Theorem. (Hudson [77]) *If $f \neq 0 \in L^2(\mathbf{R}^n)$ and $Wf \geq 0$ then f is of the form (1.101).*

Proof: For $z \in \mathbf{C}^n$ let $\psi_z(x) = e^{-\pi x^2 - 2\pi i z x}$. It is easily checked that

$$W\psi_z(\xi, x) = 2^{n/2} \exp\left(-2\pi x^2 - 2\pi(\xi + \operatorname{Re} z)^2 + 4\pi(\operatorname{Im} z)x\right) > 0,$$

so that if $Wf \geq 0$, (1.93) yields

$$|\langle \psi_z, f \rangle|^2 = \int (W\psi_z)(Wf) > 0 \quad \text{for all } z.$$

Therefore, if we set

$$G(z) = \langle \psi_z, f \rangle = \int \overline{f(x)} e^{-\pi x^2 - 2\pi i z x} dx,$$

G is a nonvanishing entire function of z. Moreover,

$$\|\psi_z\|_2^2 = \int e^{-2\pi x^2 + 4\pi(\operatorname{Im} z)x} dx = 2^{n/2} e^{2\pi(\operatorname{Im} z)^2},$$

so

$$|G(z)| \leq \|f\|_2 \|\phi_z\|_2 \leq C e^{\pi |z|^2}.$$

We claim that $G(z)$ is therefore of the form $e^{zAz+bz+c}$. Granted this, we must have $\operatorname{Re}(A)$ negative definite since $G|\mathbf{R}^n = (\overline{f}e^{-\pi x^2})^\frown \in L^2$. But then $\overline{f}(x)e^{-\pi x^2} = \mathcal{F}^{-1}(G|\mathbf{R}^n)(x)$ is of the form (1.101), and hence so is f.

As for the claim: this is a reasonably well known fact, at least in dimension 1, but here is a direct proof. Since G is entire and nonvanishing, the function

$$H(z) = \log G(0) + \int_{[0,z]} \frac{G'(t)}{G(t)} dt$$

(where $[0, z]$ is the line segment from 0 to z) is entire, and $G = e^H$. The bound on G means that

(1.103)
$$\operatorname{Re} H(z) \leq C' + \pi |z|^2.$$

Let $H(z) = \sum a_\alpha z^\alpha$. Given $r > 0$, we set

$$z(\theta_1, \ldots, \theta_n) = z(\theta) = (re^{i\theta_1}, \ldots, re^{i\theta_n}), \qquad 0 \leq \theta_j \leq 2\pi,$$

so that

$$\operatorname{Re} H(z(\theta)) = \tfrac{1}{2} \sum (a_\alpha r^{|\alpha|} e^{i\alpha\theta} + \bar{a}_\alpha r^{|\alpha|} e^{-i\alpha\theta}).$$

The right side is a Fourier series, so

$$a_\alpha r^{|\alpha|} = \frac{2}{(2\pi)^n} \int_{[0,2\pi]^n} \big[\operatorname{Re} H(z(\theta))\big] e^{-i\alpha\theta} \, d\theta \quad \text{for} \quad \alpha \neq 0,$$

$$\operatorname{Re} a_0 = \frac{1}{(2\pi)^n} \int_{[0,2\pi]^n} \big[\operatorname{Re} H(z(\theta))\big] \, d\theta,$$

and hence

$$|a_\alpha| r^{|\alpha|} + 2\operatorname{Re} a_0 \leq \frac{2}{(2\pi)^n} \int_{[0,2\pi]^n} \big[|\operatorname{Re} H(z(\theta))| + \operatorname{Re} H(z(\theta))\big] \, d\theta$$

$$= \frac{4}{(2\pi)^n} \int_{[0,2\pi]^n} \max\big(\operatorname{Re} H(z(\theta)), 0\big) \, d\theta$$

$$\leq 4(C' + \pi r^2),$$

by (1.103). Letting $r \to \infty$ we conclude that a_α must vanish unless $|\alpha| \leq 2$, so H is a polynomial of degree ≤ 2. ∎

A generalization of Hudson's theorem, pertaining to Wigner distributions of more general (non-L^2) functions, can be found in Janssen [87].

9. The Laguerre Connection

In this section we calculate the Fourier-Wigner and Wigner transforms of the Hermite functions. The answers turn out to involve the Laguerre polynomials $L_k^{(j)}$, defined for nonnegative integers j and k by

$$L_k^{(j)}(x) = \sum_{m=0}^{k} \frac{(k+j)!}{(k-m)!\,(j+m)!} \frac{(-x)^m}{m!}.$$

(For each j, the polynomials $L_k^{(j)}$ are orthogonal on $(0, \infty)$ with respect to the measure $x^j e^{-x}\, dx$.)

It suffices to consider the one-dimensional case, since n-dimensional Hermite functions are products of one-dimensional Hermite functions, and the Fourier-Wigner and Wigner transforms preserve the product structure; that is,

$$V(h_\alpha, h_\beta)(p, q) = \prod_{j=1}^{n} V(h_{\alpha_j}, h_{\beta_j})(p_j, q_j),$$

and similarly for W.

(1.104) Theorem. *Suppose $p, q \in \mathbf{R}$ and $w = p + iq$. Then*

$$V(h_j, h_k)(p, q) = \begin{cases} \sqrt{\dfrac{k!}{j!}}\, e^{-(\pi/2)|w|^2}(\sqrt{\pi}\, w)^{j-k} L_k^{(j-k)}(\pi|w|^2) & \text{for } j \geq k, \\[2ex] \sqrt{\dfrac{k!}{j!}}\, e^{-(\pi/2)|w|^2}(-\sqrt{\pi}\, w)^{k-j} L_j^{(k-j)}(\pi|w|^2) & \text{for } j \leq k. \end{cases}$$

In particular,

$$V(h_j, h_j)(p, q) = e^{-(\pi/2)|w|^2} L_j^{(0)}(\pi|w|^2).$$

Proof: The simplest method is to perform the calculations in Fock space. We have

$$V(h_j, h_k)(p, q) = \langle \rho(p, q)h_j, h_k \rangle = \langle \beta(w)\zeta_j, \zeta_k \rangle_{\mathcal{F}}$$

$$= \sqrt{\frac{\pi^{j+k}}{j! k!}} \int e^{-\pi z \overline{w} - (\pi/2)|w|^2}(z + w)^j \overline{z}^k e^{-\pi|z|^2}\, dz$$

$$= \sqrt{\frac{\pi^{j+k}}{j! k!}}\, e^{-(\pi/2)|w|^2} \sum_{m=0}^{j} \frac{j!\, w^{j-m}}{m!\,(j-m)!} \int z^m \overline{z}^k e^{-\pi z \overline{w} - \pi|z|^2}\, dz$$

$$= \sqrt{\frac{\pi^{j+k}}{j! k!}}\, e^{-(\pi/2)|w|^2} \sum_{m=0}^{j} \frac{j!\, w^{j-m}}{m!\,(j-m)!}(-\pi)^{-m}\left(\frac{\partial}{\partial \overline{w}}\right)^m \int \overline{z}^k e^{-\pi z \overline{w} - \pi|z|^2}\, dz.$$

But

$$\int \overline{z}^k e^{-\pi z \overline{w} - \pi|z|^2}\, dz = \langle E_{-w}, z^k \rangle_{\mathcal{F}} = \overline{\langle z^k, E_{-w} \rangle_{\mathcal{F}}} = (-\overline{w})^k.$$

Hence the sum in the last formula for $V(h_j, h_k)$ becomes

$$\sum_{m=0}^{\min(j,k)} \frac{j!\, w^{j-m}}{m!\,(j-m)!}(-1)^{k-m}\pi^{-m}\frac{k!\, \overline{w}^{k-m}}{(k-m)!}.$$

If $j \geq k$ we make the substitution $m \to k - m$, obtaining

$$V(h_j, h_k)(p, q) = \sqrt{\frac{k!}{j!}} e^{-(\pi/2)|w|^2} \sum_{m=0}^{k} \frac{j! (\sqrt{\pi} w)^{j-k} (-\pi|w|^2)^m}{(k-m)! (j+m)! m!}$$

$$= \sqrt{\frac{k!}{j!}} e^{-(\pi/2)|w|^2} (\sqrt{\pi} w)^{j-k} L_j^{(j-k)}(\pi|w|^2),$$

while if $j \leq k$ we make the substitution $m \to j - m$ and obtain similarly

$$V(h_j, h_k)(p, q) = \sqrt{\frac{j!}{k!}} e^{-(\pi/2)|w|^2} (-\sqrt{\pi} w)^{k-j} L_j^{(k-j)}(\pi|w|^2). \quad \blacksquare$$

This result sems to have been first pointed out in the case $j = k$ by Klauder [90]. In the general case it was proved independently by Itzykson [81], Miller [106], and Vilenkin [145], and it has since been rederived by Peetre [116], Howe [74], and possibly others. Let us examine some consequences of it for the Laguerre functions

$$l_j(t) = e^{-t/2} L_j^{(0)}(t).$$

In the first place, the orthonormality of the l_j's in $L^2(0, \infty)$ follows immediately from the unitarity of V:

$$\int_0^\infty l_j(t) l_k(t) \, dt = \int_0^\infty \int_0^{2\pi} l_j(\pi r^2) l_j(\pi r^2) r \, d\theta \, dr$$

$$= \int_{\mathbb{C}} l_j(\pi|w|^2) l_k(\pi|w|^2) \, dw = \langle V(h_j, h_j), V(h_k, h_k) \rangle = \delta_{jk}.$$

Secondly, if we set

$$\mathcal{L}_j(p, q) = l_j(\pi(p^2 + q^2)),$$

then, by Proposition (1.46), the operator $\rho(\mathcal{L}_j)$ is the orthogonal projection onto h_j. Hence, by (1.83), the spectral resolution of the Hermite operator $\pi(D^2 + X^2)$ is given by

$$\pi(D^2 + X^2) = \sum_0^\infty (j + \tfrac{1}{2}) \rho(\mathcal{L}_j).$$

More generally, suppose $\phi(t)$ is any measurable function on $(0, \infty)$ that is $O((1 + t)^N)$ for some N. Then we can expand ϕ in a series of Laguerre functions:

$$\phi = \sum_0^\infty c_j l_j, \qquad c_j = \int_0^\infty \phi(t) \lambda_j(t) \, dt.$$

(The series converges in L^2 if $\phi \in L^2$ or in a suitable weak sense for more general ϕ.) If

$$\Phi(p,q) = \phi\big(\pi(p^2 + q^2)\big),$$

we then have $\Phi = \sum c_j \mathcal{L}^j$ (the sum converging at least in $\mathcal{S}'(\mathbf{R}^2)$), and $\rho(\Phi) = \sum c_j \rho(\mathcal{L}_j)$. We can therefore summarize the situation as follows:

$$\rho\Big[\phi\big(\pi(p^2 + q^2)\big)\Big] = \tilde{\phi}\big(\pi(D^2 + X^2)\big),$$

where the right side denotes the ordinary functional calculus of self-adjoint operators and $\tilde{\phi}$ is the function on the spectrum of $\pi(D^2 + X^2)$ defined by

$$\tilde{\phi}(j + \tfrac{1}{2}) = \int_0^\infty \phi(t) l_j(t)\, dt.$$

This result is due to Peetre [116], who used it to give Fourier-analytic derivations of various properties of Laguerre functions. (See also Itzykson [81].)

Closely related to these results is a formula due to Geller [55] that expresses the group Fourier transform of radial functions on $\mathbf{H}_n$ (i.e., functions $f(p,q,t)$ that depend only on $p^2 + q^2$ and t) in terms of Laguerre transforms. The connection between Laguerre functions and Fourier analysis on $\mathbf{H}_n$ has been exploited in the study of various translation-invariant operators on $\mathbf{H}_n$ by de Michele and Mauceri [39], Jerison [88], Nachman [110], and Beals, Greiner, and Vauthier [20].

Next we compute $W(h_j, h_k)$. This can be done easily by applying equation (1.90) to our formula for $V(h_j, h_k)$, since the h_j's are all either odd or even. However, for the sake of variety, we shall present an independent derivation, following Janssen [85], that utilizes the Bargmann transform in a somewhat different way.

(1.105) Theorem. *Suppose* $\xi, x \in \mathbf{R}$ *and* $z = x + i\xi$. *Then*

$$W(h_j, h_k)(\xi, x) = \begin{cases} 2(-1)^k \sqrt{\frac{k!}{j!}}\, e^{-2\pi|z|^2} (2\sqrt{\pi}\, z)^{j-k} L_k^{(j-k)}(4\pi|z|^2) & \text{for } j \geq k, \\ 2(-1)^j \sqrt{\frac{j!}{k!}}\, e^{-2\pi|z|^2} (2\sqrt{\pi}\, z)^{k-j} L_j^{(k-j)}(4\pi|z|^2) & \text{for } j \leq k. \end{cases}$$

In particular,

$$W h_j(\xi, x) = (-1)^j e^{-2\pi|z|^2} L_j^{(0)}(4\pi|z|^2).$$

Proof: For $u \in \mathbf{C}$, let

$$B_u(x) = B(u, x) = 2^{1/4} e^{2\pi u x - \pi x^2 - (\pi/2)u^2}.$$

Then by formula (1.89), $W(B_u, B_{\bar{v}})(\xi, x)$ equals

$$2^{1/2} e^{2\pi x(u+v) - (\pi/2)(u^2+v^2) - 2\pi x^2} \int e^{-2\pi i p[\xi + i(u-v)/2]} e^{(-\pi/2)p^2} dp$$

$$= 2 e^{2\pi x(u+v) - (\pi/2)(u^2+v^2) - 2\pi x^2 - 2\pi[\xi + i(u-v)/2]^2}$$

$$= 2 e^{-2\pi|z|^2} e^{2\pi(\bar{z}u + zv) - \pi uv}$$

$$= 2 e^{-2\pi|z|^2} \sum_{l,m,n=0}^{\infty} \frac{(-\pi)^l (2\pi\bar{z})^m (2\pi z)^n}{l!\, m!\, n!} u^{m+l} v^{n+l}$$

$$= 2 e^{-2\pi|z|^2} \sum_{j,k=0}^{\infty} u^j v^k \sum_{l=0}^{\min(j,k)} \frac{(-\pi)^l (2\pi\bar{z})^{j-l} (2\pi z)^{k-l}}{l!\, (j-l)!\, (k-l)!}.$$

On the other hand, by Theorem (1.86),

$$W(B_u, B_{\bar{v}}) = \sum_{j,k=0}^{\infty} \sqrt{\frac{\pi^{j+k}}{j!k!}}\, u^j v^k W(h_j, h_k).$$

Hence

$$W(h_j, h_k)(\xi, x) = 2 e^{-2\pi|z|^2} \sqrt{\frac{j!k!}{\pi^{j+k}}} \sum_{l=0}^{\min(j,k)} \frac{(-\pi)^l (2\pi\bar{z})^{j-l} (2\pi z)^{k-l}}{l!\, (j-l)!\, (k-l)!},$$

and the same manipulations as in the proof of Theorem (1.104) can be used to express this last quantity in terms of Laguerre polynomials. ∎

Remark. If we compare the formulas for $V(h_j, h_j)$ and $W h_j = W(h_j, h_j)$ in Theorems (1.104) and (1.105), we obtain the following result. If

$$F_j(w) = e^{-(\pi/2)|w|^2} L_j^{(0)}(\pi|w|^2), \qquad w \in \mathbf{C} \cong \mathbf{R}^2,$$

then

$$\widehat{F}_j(z) = 2(-1)^j F_j(2z).$$

If this equation is written out in polar coordinates, it reduces to the formula

$$\int_0^{\infty} e^{-r^2/2} L_j^{(0)}(r^2) J_0(rs) r\, dr = (-1)^j e^{-s^2/2} L_j^{(0)}(s^2)$$

(where J_0 is the Bessel function of order zero), which can be found in tables of Hankel transforms.

10. The Nilmanifold Representation

In this section we discuss yet another interesting way of realizing the irreducible representations of $\mathbf{H}_n$. We restrict attention to the representation ρ and leave the generalization to ρ_h, $h \neq 1$, to the reader.

Here it will be convenient to use the polarized form $\mathbf{H}_n^{\text{pol}}$ of the Heisenberg group. We recall that this is $\mathbf{R}^{2n+1}$ with the group law

$$(p,q,t)(p',q',t') = (p + p',\; q + q',\; t + t' + pq'),$$

and that the map $\alpha : \mathbf{H}_n^{\text{pol}} \to \mathbf{H}_n$ defined by

$$\alpha(p,q,t) = (p,\; q,\; t - \tfrac{1}{2}pq)$$

is an isomorphism. The representation ρ of $\mathbf{H}_n$ corresponds to the representation $\rho^{\text{pol}} = \rho \circ \alpha$ of $\mathbf{H}_n^{\text{pol}}$, given by

$$(1.106) \qquad \rho^{\text{pol}}(p,q,t)f(x) = e^{2\pi i(t+qx)}f(x+p) = e^{2\pi it}e^{2\pi iqX}e^{2\pi ipD}f(x).$$

Let Γ denote the subset of $\mathbf{H}_n^{\text{pol}}$ consisting of points whose coordinates are all integers:

$$\Gamma = \{(p,q,t) \in \mathbf{H}_n^{\text{pol}} : \quad p,q \in \mathbf{Z}^n \text{ and } t \in \mathbf{Z}\}.$$

Then Γ is a discrete subgroup of $\mathbf{H}_n^{\text{pol}}$, and the right coset space

$$M = \Gamma \backslash \mathbf{H}_n^{\text{pol}}$$

is a compact nilmanifold. It is easily verified that the half-open unit cube

$$Q^{2n+1} = [0,1)^{2n+1} \subset \mathbf{H}_n^{\text{pol}}$$

is a fundamental domain for Γ, that is, each right coset of Γ contains precisely one point of Q^{2n+1}. Hence M can be considered topologically as the closed unit cube $\overline{Q}^{2n+1}$ with certain pieces of its boundary identified with each other, a sort of "twisted torus." (In fact, M is a nontrivial circle bundle over the $2n$-torus.) Moreover, Haar measure on $\mathbf{H}_n^{\text{pol}}$—namely, Lebesgue measure—induces an invariant measure on M, so measure-theoretically we can think of M as the cube Q^{2n+1} with Lebesgue measure. We shall identify functions on M with Γ-invariant functions on $\mathbf{H}_n^{\text{pol}}$ or with functions on Q^{2n+1}, as the occasion warrants.

The action of $\mathbf{H}_n^{\text{pol}}$ on M by right translation determines the regular representation R of $\mathbf{H}_n^{\text{pol}}$ on $L^2(M)$:

$$R(X)f(\Gamma Y) = f(\Gamma Y X).$$

$L^2(M)$ breaks up into a sum of R-invariant subspaces $\mathcal{H}_j$ according to the action of the center of $\mathbf{H}_n^{\text{pol}}$:

$$\mathcal{H}_j = \{f \in L^2(M) : R(0,0,t)f = e^{2\pi ijt}f\}, \qquad j \in \mathbf{Z}.$$

In other words,

$$(1.107) \qquad\qquad f \in \mathcal{H}_j \iff f(p,q,t) = e^{2\pi ijt}f(p,q,0).$$

If we think of $f \in L^2(M)$ as a function on Q^{2n+1}, the expansion $f = \sum_{-\infty}^{\infty} f_j$, $f_j \in \mathcal{H}_j$, is just the Fourier series of f in the variable t:

$$f_j(p,q,t) = e^{2\pi ijt} \int_{Q^{2n+1}} f(p,q,\tau)e^{-2\pi ij\tau}\, d\tau.$$

By the Stone–von Neumann theorem, the restriction of the representation R to $\mathcal{H}_j$ is equivalent to a direct sum of copies of ρ_j. Our interest here is in the case $j = 1$, where, as we shall see, the restriction of R is irreducible.

Let us define a map T from functions on $\mathbf{R}^n$ to functions on $\mathbf{H}_n^{\text{pol}}$ as follows:

$$Tf(p,q,t) = \sum_{K \in \mathbf{Z}^n} f(p+K)e^{2\pi iKq}e^{2\pi it}.$$

If $f \in \mathcal{S}(\mathbf{R}^n)$, say, this series clearly converges nicely to a C^∞ function on $\mathbf{H}_n^{\text{pol}}$. Moreover, if $(a,b,j) \in \Gamma$,

$$\begin{aligned}
Tf((a,b,j)(p,q,t)) &= Tf(p+a,\, q+b,\, t+j+aq) \\
&= \sum f(p+a+K)e^{2\pi iK(q+b)}e^{2\pi i(t+j+aq)} \\
&= \sum f(p+a+K)e^{2\pi i(a+K)q}e^{2\pi it}.
\end{aligned}$$

But on relabeling the index K of summation as $K - a$, we see that this last sum is nothing but $Tf(p,q,t)$. Thus Tf is Γ-invariant, and we may (and do) regard it as a function on M. As such, Tf is in $\mathcal{H}_1$ by (1.107), and we have

$$\begin{aligned}
\|Tf\|^2_{L^2(M)} &= \int_{Q^{2n+1}} |Tf(p,q,t)|^2\, dp\, dq\, dt \\
&= \int_{[0,1)^{2n}} \left|\sum f(p+K)e^{2\pi iKq}\right|^2 dp\, dq \\
&= \sum \int_{[0,1)^n} |f(p+K)|^2\, dp \qquad \text{(by Parseval)} \\
&= \|f\|_2^2,
\end{aligned}$$

so T is an isometry from $L^2(\mathbf{R}^n)$ into $\mathcal{H}_1$. T is actually surjective onto $\mathcal{H}_1$, for if $g \in \mathcal{H}_1$ we can expand g in a Fourier series on Q^{2n+1}:

$$g(p,q,t) = e^{2\pi it} \sum_{J,K \in \mathbf{Z}^n} a_{JK} e^{2\pi i(Jp+Kq)}, \qquad (p,q,t) \in Q^{2n+1}.$$

We then have $g = Tf$ where f is the function defined piecemeal on $\mathbf{R}^n$ by

$$f(x+K) = \sum_J a_{JK} e^{2\pi iJx} \quad \text{for} \quad x \in [0,1)^n, \quad K \in \mathbf{Z}^n.$$

Finally, observe that

$$
\begin{aligned}
T\big[\rho^{\mathrm{pol}}(p,q,t)f\big](p',q',t') &= \sum e^{2\pi i[t+q(p'+K)]} f(p'+p+K) e^{2\pi iKq'} e^{2\pi it'} \\
&= \sum f(p'+p+K) e^{2\pi iK(q'+q)} e^{2\pi i(t'+t+p'q)} \\
&= \big[R(p,q,t)Tf\big](p',q',t').
\end{aligned}
$$

We have therefore proved:

(1.109) Theorem. *The transform T defined by (1.108) is a unitary map from $L^2(\mathbf{R}^n)$ to $\mathcal{H}_1$ which intertwines ρ^{pol} and $R|\mathcal{H}_1$. In particular, $R|\mathcal{H}_1$ is an irreducible unitary representation of $\mathbf{H}_n^{\mathrm{pol}}$ that is equivalent to ρ^{pol}.*

Remark. For $|j| \geq 2$, the representation $R|\mathcal{H}_j$ is not irreducible, but it is of finite multiplicity. See Auslander [7] or Brezin [26] for a detailed analysis of its structure. For $j = 0$, functions on $\mathcal{H}_0$ can be identified with functions on the $2n$-torus $\mathbf{R}^{2n}/\mathbf{Z}^{2n}$, and the irreducible subspaces are the one-dimensional spans of the functions $e^{2\pi i(Jp+Kq)}$, $J,K \in \mathbf{Z}^n$.

The central variable t enters the above calculations, as usual, in a rather trivial way, so we can provide an alternative description of the space $\mathcal{H}_1$ and the transform T that does not mention it. Namely, if $f \in \mathcal{H}_1$, f is determined according to (1.107) by the function

$$f_0(p,q) = f(p,q,0)$$

on $\mathbf{R}^{2n}$. (Here we regard f as a function on $\mathbf{H}_n^{\mathrm{pol}}$.) The Γ-invariance of f translates into the following quasi-periodicity property of f_0:

$$(1.110) \qquad f_0(p+a, q+b) = e^{-2\pi iaq} f_0(p,q) \quad \text{for} \quad a,b \in \mathbf{Z}^n.$$

A function satisfying (1.110) is competely determined by its values on the unit cube Q^{2n} in $\mathbf{R}^{2n}$, and its absolute value is actually periodic in all variables.

Moreover, the norm of f in $L^2(M)$ equals the norm of f_0 in $L^2(Q^{2n})$. Hence if we define

$$\mathcal{H}^1 = \left\{ f : f \text{ satisfies (1.110) and } \int_{Q^{2n}} |f(x)|^2 dx < \infty \right\},$$

the map $f \to f_0$ is unitary from $\mathcal{H}_1$ to $\mathcal{H}^1$. The corresponding map T_0 : $L^2(\mathbf{R}^n) \to \mathcal{H}^1$ is given by

$$(1.111) \qquad T_0 f(p,q) = (Tf)_0(p,q) = \sum_{K \in \mathbf{Z}^n} f(p+K)e^{2\pi i K q},$$

and the corresponding representation R_0 of $\mathbf{H}_n^{\text{pol}}$ on $\mathcal{H}^1$ is given by

$$(1.112) \qquad R_0(r,s,t)f(p,q) = f(p+r, q+s)e^{2\pi i(t+sp)}.$$

The transform T and its close relative T_0 are referred to in the literature as the **Weil-Brezin transform** (the name we shall adopt) or the **Zak transform.** T_0 was described by Weil [151, pp. 164-5], although it is implicitly present in much earlier works; and T was introduced later by Brezin [26]. Meanwhile, T_0 and the representation R_0 were discovered independently by Zak [158], [159], who found them useful for solving problems in solid state physics involving motion in a periodic potential and for studying other quantum phenomena in which periodic variables occur. Zak calls the representation R_0 (or rather its infinitesimal version) the **kq representation,** k and q being the standard names of the quasi-position and quasi-momentum variables in solid state physics.

We now discuss some interesting properties of the Weil-Brezin transform T_0 and the representation R_0.

(i) The infinitesimal representation dR_0 of $\mathbf{h}_n$ on $\mathcal{H}^1$ is given by

$$dR_0(r,s,t)f = \sum_{1}^{n} \left(r_j \frac{\partial f}{\partial p_j} + s_j \frac{\partial f}{\partial q_j} + 2\pi i s_j p_j f \right) + 2\pi i t f.$$

This follows easily from (1.112). A similar formula holds for the infinitesimal representation dR on $\mathcal{H}_1$, provided that one considers elements of $\mathcal{H}_1$ as Γ-invariant functions on $\mathbf{H}_n^{\text{pol}}$ rather than functions on M. (The coordinates p_j are not well defined on M. The reader may check that if $f \in \mathcal{H}_1$, the function $\partial f/\partial q_j + 2\pi i p_j f$ is Γ-invariant although $\partial f/\partial q_j$ and $2\pi i p_j f$ individually are not.)

(ii) The operators $\rho(a,0) = e^{2\pi i a D}$ and $\rho(0,b) = e^{2\pi i b X}$ on $L^2(\mathbf{R}^n)$ commute when a and b are both in $\mathbf{Z}^n$, and the Weil-Brezin transform provides the spectral resolution for this family of commuting normal operators. Indeed, if $a \in \mathbf{Z}^n$,

$$T_0(e^{2\pi i a D}f)(p,q) = \sum f(p+a+K)e^{2\pi i K q}$$
$$= \sum f(p+K)e^{2\pi i(K-a)q} = e^{-2\pi i a q}T_0 f(p,q),$$

and if $b \in \mathbf{Z}^n$,

$$T_0(e^{2\pi i b X}f)(p,q) = \sum f(p+K)e^{2\pi i b(p+K)}e^{2\pi i K q}$$
$$= e^{2\pi i b p}T_0 f(p,q).$$

Combining these results with the fact that $\rho(a,0)\rho(0,b) = e^{\pi i a b}\rho(a,b)$, we have

(1.113) $T_0\big(\rho(a,b)f\big)(p,q) = (-1)^{ab}e^{2\pi i(bp-aq)}T_0 f(p,q)$ for $a,b \in \mathbf{Z}^n$.

(iii) T_0 is not only an isometry from $L^2(\mathbf{R}^n)$ to $L^2(Q^{2n})$ but also a contraction from $L^p(\mathbf{R}^n)$ to $L^p(Q^{2n})$ for $1 \le p \le 2$. Indeed, for $p = 1$ we have

$$\|T_0 f\|_1 = \int_{Q^{2n}} \left| \sum f(p+K)e^{2\pi i K q} \right| dp\, dq$$
$$\le \int_{Q^{2n}} \sum |f(p+K)|\, dp = \|f\|_1,$$

and the case $1 < p < 2$ follows by interpolation.

(iv) The following slightly bizarre property of the space $\mathcal{H}^1$ was observed by Zak and Janssen [86].

(1.114) **Proposition.** *Every continuous function in $\mathcal{H}^1$ has zeros.*

Proof: Suppose $f \in \mathcal{H}^1$ is continuous. Since $\mathbf{R}^{2n}$ is simply connected, if f were nonvanishing we could write $f(p,q) = e^{2\pi i \psi(p,q)}$ for some continuous function ψ. The quasi-periodicity (1.110) implies that for every $a,b \in \mathbf{Z}^n$ there exists $K_{a,b} \in \mathbf{Z}$ such that

$$\psi(p+a, q+b) = \psi(p,q) - aq + K_{a,b}.$$

But this is self-contradictory: on the one hand,

$$\psi(a,b) = \psi(0,b) - ab + K_{a,0} = \psi(0,0) - ab + K_{a,0} + K_{0,b},$$

and on the other,

$$\psi(a,b) = \psi(a,0) + K_{0,b} = \psi(0,0) + K_{a,0} + K_{0,b}. \quad \blacksquare$$

(v) We now assume $n = 1$. Given $\tau \in \mathbf{C}$ with $\operatorname{Im} \tau > 0$, let

$$\phi_\tau(x) = e^{\pi i \tau x^2}.$$

Then $\phi_\tau \in L^2(\mathbf{R})$, and we have

$$T_0 \phi_\tau(u, v) = \sum_{k=-\infty}^{\infty} e^{2\pi i k v + \pi i \tau (u+k)^2} = e^{\pi i \tau u^2} \vartheta_3(z, q),$$

where

$$z = \pi(v + \tau u), \qquad q = e^{\pi i \tau}, \qquad \vartheta_3(z, q) = \sum_{-\infty}^{\infty} q^{k^2} e^{2ikz}.$$

ϑ_3 is one of the basic Jacobi theta functions, in the notation of Whittaker and Watson [154], the others being

$$\vartheta_1(z, q) = -i e^{iz + i\pi\tau/4} \vartheta_3(z + \tfrac{1}{2}\pi(\tau - 1), q),$$
$$\vartheta_2(z, q) = \vartheta_1(z + \tfrac{1}{2}\pi, q) = e^{iz + i\pi\tau/4} \vartheta_3(z + \tfrac{1}{2}\pi\tau, q),$$
$$\vartheta_4(z, q) = \vartheta_3(z - \tfrac{1}{2}\pi, q),$$

where $q = e^{\pi i \tau}$ throughout. Since

$$T_0\big(\rho^{\mathrm{pol}}(r, s)\phi_\tau\big)(u, v) = R_0(r, s)T_0\phi_\tau(u, v) = e^{2\pi i s u} T_0\phi_\tau(u + r, v + s),$$

these other theta functions can be obtained (up to factors involving only elementary exponential functions) as Weil-Brezin transforms of the functions $\rho^{\mathrm{pol}}(r, s)\phi_\tau$ where $s + \tau r$ is $\tfrac{1}{2}(\tau - 1)$, $\tfrac{1}{2}\tau$, or $\tfrac{1}{2}$.

These relations suggest that the Heisenberg group and the nilmanifold M should be of use in the study of theta functions. That is indeed the case, and this connection has been much exploited in recent years. See Auslander [7], Auslander–Tolimieri [8], Igusa [80], and Mumford [109].

11. Postscripts

The Heisenberg group plays a role in many other parts of analysis besides the subjects discussed in this monograph. In this section we provide a brief description of some of these areas and a few selected references, mostly expository works from whose bibliographies the reader can obtain a more complete guide to the literature. At the outset, let us mention the article of Howe [75], which surveys several aspects of the Heisenberg group that are considered here from a somewhat different point of view.

Several Complex Variables. As is well known, the unit disc in $\mathbf{C}$ can be mapped onto the upper half plane by a fractional linear transformation, and the boundary of the upper half plane (namely the real axis) can be identified with the group of horizontal translations of the plane. There is a similar situation in higher dimensions. We work in $\mathbf{C}^{n+1}$ and denote points in $\mathbf{C}^{n+1}$ by (ζ, τ) where $\zeta \in \mathbf{C}^n$ and $\tau \in \mathbf{C}$. The analogue of the unit disc is the unit ball,

$$B_{n+1} = \{(\zeta, \tau) \in \mathbf{C}^{n+1} : |\zeta|^2 + |\tau|^2 < 1\},$$

and the analogue of the upper half plane is the Siegel domain

$$D_{n+1} = \{(\zeta, \tau) \in \mathbf{C}^{n+1} : \operatorname{Im} \tau > |\zeta|^2\}.$$

It is easily checked that the fractional linear transformation

$$\phi(\zeta, \tau) = \left(\frac{\zeta}{i(\tau - 1)}, \frac{\tau + 1}{i(\tau - 1)} \right)$$

maps B_{n+1} onto D_{n+1}. Finally, the analogue of the horizontal translation group is the Heisenberg group $\mathbf{H}_n$, which acts on $\mathbf{C}^{n+1}$ by holomorphic affine transformations:

$$L_{(z,t)}(\zeta, \tau) = (\zeta + z, \tau - 4t + i|z|^2 - 2i\bar{z}\zeta).$$

Here we are using complex cooridnates $z = p + iq$ on $\mathbf{H}_n$ as we did in discussing the Bargmann transform. The reader may verify that L is indeed a left action,

$$L_{(z,t)}L_{(z',t')} = L_{(z,t)(z',t')},$$

and that the transformations $L_{(z,t)}$ map the domain D_{n+1} and its boundary ∂D_{n+1} onto themselves. The action on ∂D_{n+1} is simply transitive, so $\mathbf{H}_n$ can be identified with ∂D_{n+1} by the correspondence

(1.115) $(z, t) \longleftrightarrow L_{(z,t)}(0, 0) = (z, i|z|^2 - 4t).$

At this point it should be said that in the complex analysis literature it is customary to use a different parametrization of $\mathbf{H}_n$. Namely, one replaces the coordinate t by $-t/4$ so that $(z, t) \in \mathbf{H}_n$ becomes identified with $(z, t + i|z|^2) \in \partial D_{n+1}$.

Since the action L of $\mathbf{H}_n$ is holomorphic, the Cauchy-Riemann operators on D_{n+1} are invariant under it, and the induced complex of operators on ∂D_{n+1}, the so-called $\bar{\partial}_b$ complex, can actually be considered as a complex of left-invariant operators on $\mathbf{H}_n$ via the identification (1.115). One can then apply Fourier-analytic techniques on $\mathbf{H}_n$ to study these operators in detail.

The unit ball and the Siegel domain D_{n+1} are the simplest members of the class of strongly pseudoconvex domains, which is of fundamental importance in the theory of several complex variables. One can show that if $\Omega \subset \mathbf{C}^{n+1}$ is strongly pseudoconvex, for any $P \in \partial\Omega$ there is a holomorphic coordinate system with origin at P in which $\partial\Omega$ closely approximates ∂D_{n+1} near the origin. Using this fact, the analysis on the Heisenberg group can be transferred to $\partial\Omega$ to yield refined information about the $\bar{\partial}$ and $\bar{\partial}_b$ complexes on general strongly pseudoconvex domains. This program was initiated in Folland–Stein [51]; see also the survey articles of Folland [49], Stanton [130], and Beals-Fefferman-Grossman [15], and the references given there.

The identification of $\mathbf{H}_n$ with ∂D_{n+1} also leads to another derivation of the Fock-Bargmann representation, as was pointed out to the author by F. Ricci. Transfer Lebesgue measure on $\mathbf{H}_n$ to ∂D_{n+1}, and consider the subspace H^2 of $L^2(\partial D_{n+1})$ consisting of functions that are nontangential limits of holomorphic functions on D_{n+1}. If $F \in H^2$ and $h \in \mathbf{R}$, let

$$F_h(\zeta) = \int_{\mathrm{Im}\,\tau = |\zeta|^2} F(-\zeta, \tau)e^{-\pi i h \tau/2}\,d\tau = \int_{\mathbf{R}} F(-\zeta, s + i|\zeta|^2)e^{-\pi i h(s + i|\zeta|^2)/2}\,ds.$$

(We use $-\zeta$ instead of ζ just to make the formulas below turn out more neatly.) The first equation shows that $F_h(\zeta)$ is an entire function of ζ for each h, as the contour of integration can be deformed to be locally independent of ζ. The second one shows that $e^{-\pi h|\zeta|^2/2}F_h(\zeta)$ is, for each ζ, the Fourier transform of the function $s \to F(-\zeta, s + i|\zeta|^2)$, evaluated at $h/4$. Since the latter function extends holomorphically to the half plane $\mathrm{Im}\,s > 0$, one sees by Cauchy's theorem that $F_h(\zeta) = 0$ for $h < 0$. Moreover, by the Plancherel theorem,

$$\|F\|_2^2 = \frac{1}{4}\int_0^\infty \int_{\mathbf{C}^n} |F_h(\zeta)|^2 e^{-\pi h|\zeta|^2}\,d\zeta\,dh.$$

This formula exhibits H^2 as the direct integral of the Fock spaces $\mathcal{F}_n^h$ $(h > 0)$ defined in Section 1.6.

The Heisenberg group acts unitarily on H^2 by left translation:

$$U_{(z,t)}F(\zeta, \tau) = F\big(L_{(z,t)}^{-1}(\zeta, \tau)\big) = F(\zeta - z, \tau + 4t + i|z|^2 - 2i\bar{z}\zeta).$$

Applying the transform $F \to F_h$, we obtain

$$(U_{(z,t)}F)_h(\zeta) = \int F(-\zeta - z, s + i|\zeta|^2 + 4t + i|z|^2 + 2i\bar{z}\zeta)e^{-\pi i h(s + i|\zeta|^2)/2}\,ds$$

$$= \int F(-\zeta - z, s + 4t - 2\,\mathrm{Im}\,\bar{z}\zeta + i|\zeta + z|^2)e^{-\pi i h(s + i|\zeta|^2)/2}\,ds$$

$$= \int F(-\zeta - z, s + i|\zeta + z|^2)e^{-\pi i h(s - 4t + 2\,\mathrm{Im}\,\bar{z}\zeta + i|\zeta|^2)/2}\,ds$$

$$= e^{-\pi i h(-4t - 2i\bar{z}\zeta - i|z|^2)/2}\int F(-\zeta - z, s + i|\zeta + z|^2)e^{-\pi i h(s + i|\zeta + z|^2)/2}\,ds$$

$$= e^{2\pi i h t - \pi h \bar{z}\zeta - (\pi h/2)|z|^2}F_h(\zeta + z),$$

which is the Fock–Bargmann representation on $\mathbf{H}_n$ on $\mathcal{F}_n^h$. (For $h < 0$, one plays the same game with $\overline{H}^2$, the space of boundary values of antiholomorphic functions on D_{n+1}.)

Representation Theory. The Stone–von Neumann theorem provided inspiration for, and is a paradigmatic special case of, two of the fundamental results of modern representation theory: the Mackey imprimitivity theorem and the Kirillov classification of irreducible unitary representations of nilpotent Lie groups. See Mackey [99] for a lucid account of the path that leads from the Stone–von Neumann theorem to the imprimitivity theorem, and Mackey [100] for more information on the imprimitivity theorem and its applications. For the Kirillov theory and some of its extensions, see Moore [107] and Wallach [150].

Partial Differential Equations and Harmonic Analysis. In the past two decades a considerable amount of study has been devoted to partial differential operators constructed from non-commuting vector fields, in which the non-commutativity plays an essential role in determining the regularity properties of the operators. (One of the most important instances of this situation is the $\overline{\partial}_b$ complex on the boundary of a domain on $\mathbf{C}^{n+1}$, as discussed above.) The operators of this sort that can be most readily analyzed are left-invariant operators on graded nilpotent Lie groups (such as the Heisenberg group) that are homogeneous with respect to the natural dilations on these groups. In this setting one can develop non-Abelian, non-isotropic analogues of many of the tools of Euclidean harmonic analysis—singular integrals, Green's functions, various function spaces (Sobolev, Lipschitz, Hardy, etc.). These techniques, together with the representation theory of the groups, yield precise results for invariant differential operators, which can then be transferred to more general operators. Among the foundational papers in this subject are Folland–Stein [51], Rothschild–Stein [124], and Rockland [123]; see also Folland [49], Folland–Stein [52], Helffer–Nourrigat [68], Taylor [136], and Taylor [137].

In 1957 Hans Lewy shocked the world of analysis by producing the first example of a differential equation that is not locally solvable. Lewy's unsolvable operator is nothing but $(X + iY)$ on $\mathbf{H}_1$, to which he was led because of its connection with complex analysis. (It is essentially $\overline{\partial}_b$ on $\mathbf{H}_1$.) More recently, Greiner, Kohn, and Stein [60] have used the techniques mentioned above to give a complete characterization of the functions g for which $(X + iY)f = g$ is solvable.

Abstract Heisenberg Groups. There is an analogue of the Heisenberg group (or, more precisely, of the reduced, polarized Heisenberg group) associated to an arbitrary locally compact Abelian group G. If G is such a group, let T denote the group of complex numbers of modulus one and $\widehat{G}$ the group of continuous homomorphisms from G to T. ($\widehat{G}$ is a locally compact group with the compact-open topology.) The **Heisenberg group** of G is the locally compact group

$H(G)$ whose underlying space is $G \times \widehat{G} \times T$ and whose group law is

$$(g, \chi, z)(g', \chi', z') = \left(gg', \chi\chi', zz'\chi'(g)\right).$$

$H(G)$ has a family of Schrödinger representations ρ_j, indexed by a nonzero integer j, that act on $L^2(G)$ by

$$\rho_j(g, \chi, z)f(g') = z^j \chi(g')^j f(gg').$$

The analogue of the Stone–von Neumann theorem (again a special case of the Mackey imprimitivity theorem) is the fact that up to unitary equivalence, the ρ_j's exhaust all irreducible unitary representations of $H(G)$ that are nontrivial on the center. (The representations that are trivial on the center are just the characters of $G \times \widehat{G}$, lifted to $H(G)$.) The groups $H(G)$, in which G is an adèle group or the additive group of a vector space over a local field, have been applied to problems in number theory by Weil [151].

CHAPTER 2.
QUANTIZATION AND
PSEUDODIFFERENTIAL OPERATORS

As we explained in Section 1.1, the quantization problem is to associate to each function f on the phase space $\mathbf{R}^{2n}$ an operator A_f on $L^2(\mathbf{R}^n)$ such that the coordinate functions correspond to the operators D_j and X_j, and such that the properties of the functions f (ring and Lie algebra structures, positivity, boundedness, etc.) are reflected in some sense in the properties of the operators A_f. On the other hand, the central idea in the theory of pseudodifferential operators is to assign to each operator S of a suitable class a function σ_S on phase space, called the symbol of S, such that the symbols of D_j and X_j are the coordinate functions and such that the properties of S are in some sense reflected in the properties of σ_S. Evidently we have here two aspects of the same phenomenon, and in this chapter we shall develop both aspects simultaneously.

The first general quantization procedure, and in many ways still the most satisfactory one, was proposed by Weyl [153, Section IV.14] not long after the invention of quantum mechanics. Weyl's ideas were later elaborated in papers in the physics literature such as Groenewold [61], Moyal [108], and Pool [118]. Meanwhile, back in the nation of mathematics, it was discovered that one can assign to a singular integral operator on $\mathbf{R}^n$ a function, called its symbol, such that the product of the symbols corresponds to the composition of the operators modulo regular integral operators; see Mihlin [105]. The need for a more precise analysis of the error terms in the product formula led to the theory of pseudodifferential operators, initiated by Kohn and Nirenberg [92] and later refined and extended by a number of other authors. Kohn and Nirenberg observed that their symbolic calculus is related to the Weyl correspondence but made no use of the relationship. More recently, however, the ideas of pseudodifferential operators have been brought to bear on mathematical physics in such papers as Grossmann–Loupias–Stein [64] and Voros [148], [149]; the Weyl correspondence has been developed as a theory of pseudodifferential operators by Hörmander [70], [72]; and Howe [74] has emphasized and exploited the foundations of both the physical and differential-operator aspects of the theory in harmonic analysis.

In this chapter our principal aim is the development of the theory of pseudodifferential operators, at the level of the symbol classes $S^m_{\rho,\delta}$, by means of the Weyl correspondence. Our treatment is not meant to be comprehensive; for more complete accounts of pseudodifferential operators and their applications, the reader should consult Taylor [135], Treves [139], or Hörmander [72]. These books deal (mainly or exclusively) with the Kohn-Nirenberg calculus rather than the Weyl calculus, but it is easy to switch from one to the other in view of the results in Section 2.2.

As we proceed we shall see the extent to which the Weyl calculus is a successful quantization procedure, and the ideas from physics and harmonic analysis developed in Chapter 1 will sometimes provide motivation and illumination for our arguments concerning pseudodifferential operators. In the last section we discuss two other quantization methods, the Wick and anti-Wick correspondences, which are motivated by quantum field theory and are related to the Fock-Bargmann representation.

1. The Weyl Correspondence

Weyl's prescription for assigning an operator $\sigma(D, X)$ to a function $\sigma(\xi, x)$ amounts to postulating that the exponential function $e^{2\pi i(p\xi + qx)}$ $(p, q \in \mathbf{R}^n)$ should correspond to the operator $\rho(p, q) = e^{2\pi i(pD + qX)}$. Once this is granted, one can expand an "arbitrary" $\sigma(\xi, x)$ in terms of exponentials via the Fourier transform,

$$\sigma(\xi, x) = \iint \widehat{\sigma}(p, q)e^{2\pi i(p\xi + qx)}\, dp\, dq,$$

and one is led to the definition

$$(2.1) \qquad \sigma(D, X) = \iint \widehat{\sigma}(p, q)e^{2\pi i(pD + qX)}\, dp\, dq = \rho(\widehat{\sigma}).$$

Here the integral is an ordinary Bochner integral if $\widehat{\sigma} \in L^1$, but as we have already seen, $\rho(\widehat{\sigma})$ makes sense as an operator from $\mathcal{S}(\mathbf{R}^n)$ to $\mathcal{S}'(\mathbf{R}^n)$ whenever $\widehat{\sigma}$, and hence σ, is any tempered distribution. Indeed, by (1.29), $\rho(\widehat{\sigma})$ is the operator whose distribution kernel is

$$K_\sigma(x, y) = (\mathcal{F}_2^{-1}\widehat{\sigma})\big(y - x, \tfrac{1}{2}(y + x)\big),$$

where $\mathcal{F}_j$ denotes the Fourier transform in the jth variable. But this is just

$$(2.2). \quad K_\sigma(x, y) = (\mathcal{F}_1\sigma)\big(y - x, \tfrac{1}{2}(y + x)\big) = \int \sigma\big(\xi, \tfrac{1}{2}(x + y)\big)e^{2\pi i(x - y)\xi}\, d\xi.$$

In other words,

$$(2.3) \qquad \sigma(D, X)f(x) = \iint \sigma\big(\xi, \tfrac{1}{2}(x + y)\big)e^{2\pi i(x - y)\xi}f(y)\, dy\, d\xi,$$

where the integral is to be suitably interpreted, depending on the nature of f and σ. Just for the fun of it, let us give a direct formal derivation of (2.3) from (2.1):

$$
\begin{aligned}
\sigma(D, X) f(x) &= \iint \hat{\sigma}(p, q) \left(e^{2\pi i (pD + qX)} f \right)(x) \, dp \, dq \\
&= \iiiint \sigma(\xi, w) e^{-2\pi i (p\xi + qw)} e^{2\pi i qx + \pi i pq} f(x + p) \, d\xi \, dw \, dp \, dq \\
&= \iiint \sigma(\xi, w) e^{-2\pi i p\xi} f(x + p) \delta(x - w + \tfrac{1}{2} p) \, d\xi \, dw \, dp \\
&= \iint \sigma(\xi, x + \tfrac{1}{2} p) e^{-2\pi i p\xi} f(x + p) \, dp \, d\xi \\
&= \iint \sigma\left(\xi, \tfrac{1}{2}(x + y)\right) e^{-2\pi i (x - y)\xi} f(y) \, dy \, d\xi.
\end{aligned}
$$

Remark. The definition of $\sigma(D, X)$ depends only on the correspondence $\xi \to D$, $x \to X$ and not on the parametrization of the representation ρ. For example, if we had chosen to use $\rho'(p, q) = e^{2\pi i (pX - qD)}$ instead, we would have written

$$
\sigma(D, X) = \iint \tilde{\sigma}(p, q) e^{2\pi i (pX - qD)} \, dp \, dq
$$

where $\tilde{\sigma}$ is the symplectic Fourier transform of σ,

$$
\tilde{\sigma}(p, q) = \iint \sigma(\xi, x) e^{2\pi i (qx - p\xi)} \, d\xi \, dx.
$$

This formula for $\sigma(D, X)$ agrees with (2.1).

Just as with the integrated Schrödinger representation, the map $\sigma \to K_\sigma$ given by (2.2) is an isomorphism on $\mathcal{S}(\mathbf{R}^{2n})$, $\mathcal{S}'(\mathbf{R}^{2n})$, and $L^2(\mathbf{R}^{2n})$ (unitary, in the latter case). It therefore follows from the Schwartz kernel theorem (see [138]) that *every* continuous $S : \mathcal{S}(\mathbf{R}^n) \to \mathcal{S}'(\mathbf{R}^n)$ is $\sigma(D, X)$ for a unique $\sigma \in \mathcal{S}'(\mathbf{R}^{2n})$. In this situation, σ is called the **(Weyl) symbol** of S. Moreover, the correspondence $\sigma \to \sigma(D, X)$ is a unitary isomorphism from $L^2(\mathbf{R}^{2n})$ to the space of Hilbert-Schmidt operators on $L^2(\mathbf{R}^n)$. (This last result is due to Pool [118].)

We can use (2.2) to recover the symbol σ from the kernel K_σ. Indeed, setting $u = x - y$ and $v = \tfrac{1}{2}(x + y)$, we have

$$
\mathcal{F}_1^{-1} \sigma(u, v) = K_\sigma(v + \tfrac{1}{2} u, \, v - \tfrac{1}{2} u),
$$

or

(2.4) $\qquad \sigma(\xi, x) = \int e^{-2\pi i \xi u} K_\sigma(v + \tfrac{1}{2} u, \, v - \tfrac{1}{2} u) \, du = \widetilde{W} K_\sigma(\xi, v),$

where $\widetilde{W}$ is the extended Wigner transform defined by (1.91). This leads to the fundamental connection between the Weyl calculus and the Wigner transform:

(2.5) Proposition. If $\sigma \in S'(\mathbf{R}^{2n})$ and $f, g \in S(\mathbf{R}^n)$,

$$\langle \sigma(D, X)f, g \rangle = \iint \sigma(\xi, x) W(f, g)(\xi, x) \, d\xi \, dx.$$

Proof: Setting $h(x, y) = g(x)\overline{h(y)}$, since $\widetilde{W}$ is unitary we have

$$\langle \sigma(D, X)f, g \rangle = \langle K_\sigma h \rangle = \langle \widetilde{W}^{-1}\sigma, h \rangle = \langle \sigma, \widetilde{W}h \rangle = \langle \sigma, W(g, f) \rangle$$
$$= \iint \sigma(\xi, x) \overline{W(g, f)(\xi, x)} \, d\xi \, dx = \iint \sigma(\xi, x) W(f, g)(\xi, x) \, d\xi \, dx.$$

Or, to put it another way,

$$\langle \sigma(D, x)f, g \rangle = \iint \hat{\sigma}(p, q) \langle \rho(p, q)f, g \rangle \, dp \, dq$$
$$= \iint \hat{\sigma} \cdot V(f, g) = \iint \sigma \cdot V(f, g)^\smile = \iint \sigma \cdot W(f, g). \quad \blacksquare$$

One other basic property of the Weyl correspondence is an easy consequence of the definitions, namely, the formula for adjoints. To fix the terminology: if $S : S(\mathbf{R}^n) \to S'(\mathbf{R}^n)$ is a continuous linear map, its adjoint S^* and its transpose $S^\dagger$ are defined by

$$\int (S^*f)\overline{g} = \langle S^*f, g \rangle = \langle f, Sg \rangle = \int f(\overline{Sg}),$$
$$\int (S^\dagger f)g = \langle S^\dagger f, \overline{g} \rangle = \langle f, \overline{Sg} \rangle = \int f(Sg).$$

If $K(x, y)$ is the distribution kernel of S, the kernels of S^* and $S^\dagger$ are $\overline{K(y, x)}$ and $K(y, x)$. In view of (2.2), then, the reader may readily verify:

(2.6) Proposition. For any $\sigma \in S'(\mathbf{R}^{2n})$,

$$\sigma(D, X)^* = \overline{\sigma}(D, X) \qquad \text{and} \qquad \sigma(D, X)^\dagger = \sigma(-D, X).$$

Let us now see how the Weyl correspondence works in some specific cases. To begin with, suppose σ_0 and σ_1 are (reasonable) functions on $\mathbf{R}^n$; then since $\{X_1, \ldots, X_n\}$ and $\{D_1, \ldots, D_n\}$ are commuting families of self-adjoint operators, there is a canonical way to define $\sigma_0(X)$ and $\sigma_1(D)$, by the functional calculus given by the spectral theorem. Namely,

$$(2.7) \qquad \sigma_0(X)f(x) = \sigma_0(x)f(x), \qquad [\sigma_1(D)f]^\smallfrown(\xi) = \sigma_1(\xi)\hat{f}(\xi).$$

(2.8) Proposition. *Suppose* $\sigma \in \mathcal{S}'(\mathbf{R}^{2n})$ *is a function of* x *or* ξ *alone, that is,* $\sigma(\xi, x) = \sigma_0(x)$ *or* $\sigma(\xi, x) = \sigma_1(\xi)$. *Then* $\sigma(D, X)$ *equals* $\sigma_0(X)$ *or* $\sigma_1(D)$ *where the latter operators are defined by* (2.7).

Proof: If $\sigma(\xi, x) = \sigma_1(\xi)$ this is obvious:

$$\sigma(D, X)f(x) = \iint \sigma_1(\xi)e^{2\pi i(x-y)\xi}f(y)\,dy\,d\xi = \mathcal{F}^{-1}(\sigma\widehat{f})(x).$$

The case $\sigma(\xi, x) = \sigma_0(x)$ follows from the Fourier inversion formula:

$$\sigma(D, X)f(x) = \iint \sigma_0\big(\tfrac{1}{2}(x+y)\big)e^{2\pi i(x-y)\xi}f(y)\,dy\,d\xi$$

$$= \int \sigma_0\big(\tfrac{1}{2}(x+y)\big)f(y)\delta(x-y)\,dy = \sigma_0(x)f(x). \quad \blacksquare$$

One might wonder if the Weyl correspondence gives the same result as the spectral calculus in other situations where the latter applies. For example, if $\sigma(D, X)$ is self-adjoint on L^2 and $\phi : \mathbf{R} \to \mathbf{R}$, is it true that $\phi \circ \sigma(D, X) = \phi\big(\sigma(D, X)\big)$, where the operator on the right is defined by the spectral calculus? It is not hard to see that the answer is usually *no*. In fact, the counterexample presented in Section 1.1 shows that we do not even have $(\sigma^2)(D, X) = \sigma(D, X)^2$ in general. Nevertheless, Proposition (2.8) does have a substantial generalization, which we shall present shortly.

Next, let us see what $\sigma(D, X)$ looks like when σ is a polynomial in ξ,

(2.9) $$\sigma(\xi, x) = \sum_{|\alpha| \le k} a_\alpha(x)\xi^\alpha.$$

In this case we have

$$\sigma(D, X)f(x) = \sum_{|\alpha| \le k} \iint \xi^\alpha a_\alpha\big(\tfrac{1}{2}(x+y)\big)e^{2\pi i(x-y)\xi}f(y)\,dy\,d\xi$$

$$= \sum_{|\alpha| \le k} \iint a_\alpha\big(\tfrac{1}{2}(x+y)\big)(-1)^{|\alpha|}D_y^\alpha\big(e^{2\pi i(x-y)\xi}\big)f(y)\,dy\,d\xi$$

$$= \sum_{|\alpha| \le k} \iint e^{2\pi i(x-y)\xi}D_y^\alpha\Big[a_\alpha\big(\tfrac{1}{2}(x+y)\big)f(y)\Big]\,dy\,d\xi.$$

Thus, by the Fourier inversion formula,

$$\sigma(D, X)f(x) = \sum_{|\alpha| \le k} D_y^\alpha\Big[a_\alpha\big(\tfrac{1}{2}(x+y)\big)f(y)\Big]_{y=x}$$

(2.10)

$$= \sum_{|\alpha| \le k} \sum_{\beta \le \alpha} \frac{\alpha!}{\beta!\,(\alpha-\beta)!\,2^{|\alpha-\beta|}}D^{\alpha-\beta}a_\alpha(x)D^\beta f(x).$$

(Here "$\beta \le \alpha$" means that $\beta_j \le \alpha_j$ for all j.)

(2.11) Proposition. $\sigma(D, X)$ *is a differential operator of order k if and only if $\sigma(\xi, x)$ is a polynomial of degree k in ξ, and $\sigma(D, X)$ is a differential operator with polynomial coefficients if and only if $\sigma(\xi, x)$ is a polynomial in ξ and x. If σ is given by (2.9), then*

$$(2.12) \qquad \sigma(D, X) = \sum_{|\alpha|=k} a_\alpha(x) D^\alpha + \text{ terms of order } \leq k - 1.$$

Proof: That $\sigma(D, X)$ is a differential operator (with polynomial coefficients) when $\sigma(\xi, x)$ is a polynomial in ξ (and x) follows from (2.10), as does (2.12). On the other hand, suppose $P = \sum_{|\alpha| \leq k} b_\alpha(x) D^\alpha$. If we set $\sigma_0(\xi, x) = \sum_{|\alpha|=k} b_\alpha(x) \xi^\alpha$, $P - \sigma_0(D, X)$ will be a differential operator of order $\leq k - 1$, by (2.12). It follows by induction on k that $P = \sigma(D, X)$ where $\sigma(\xi, x)$ is a polynomial of degree k in ξ, and that $\sigma(\xi, x)$ will also be a polynomial in x if the coefficients b_α are polynomials. ∎

Covariance Properties. We now consider what happens to $\sigma(D, X)$ when σ is composed with a translation or canonical transformation of phase space. The behavior under translations is very simple:

(2.13) Proposition. *If $\sigma'(\xi, x) = \sigma(\xi - a, x - b)$ then*

$$\sigma'(D, X) = \rho(-b, a)\sigma(D, X)\rho(b, -a) = \rho(-b, a)\sigma(D, X)\rho(-b, a)^{-1}.$$

Proof: By (1.95) and Proposition (2.5), we have

$$\langle \sigma'(D, X)f, g \rangle = \iint \sigma(\xi - a, x - b) W(f, g)(\xi, x) \, d\xi \, dx$$

$$= \iint \sigma(\xi, x) W(f, g)(\xi + a, x + b) \, d\xi \, dx$$

$$= \iint \sigma(\xi, x) W\big(\rho(b, -a)f, \, \rho(b, -a)g\big)(\xi, x) \, d\xi \, dx$$

$$= \langle \sigma(D, X)\rho(b, -a)f, \, \rho(b, -a)g \rangle$$

$$= \langle \rho(-b, a)\sigma(D, X)\rho(b, -a)f, \, g \rangle. \qquad ∎$$

The behavior under linear canonical transformations is also simple, although in a more sophisticated way. To explain it, we need to quote the following result from Section 4.2. The reader who so wishes may skip ahead to Chapter 4 to see the proof, as it does not depend on the intervening material.

(2.14) Lemma. *For every $A \in Sp(n, \mathbf{R})$ there is an isomorphism $\mu(A) : S'(\mathbf{R}^n) \to S'(\mathbf{R}^n)$, whose restriction to $S(\mathbf{R}^n)$ is an isomorphism on $S(\mathbf{R}^n)$ and whose restriction to $L^2(\mathbf{R}^n)$ is unitary on $L^2(\mathbf{R}^n)$, such that*

$$\rho\big(A(p, q)\big) = \mu(A)\rho(p, q)\mu(A)^{-1} \quad \text{for all } p, q \in \mathbf{R}^n.$$

In Chapter 4 we shall produce explicit formulas for the operators $\mu(A)$, but for now all we need is their existence. Since $\mu(A)$ preserves S and S', we can compose it with $\sigma(D, X)$ on either side, and we have:

(2.15) Theorem. *For any $\sigma \in \mathcal{S}'(\mathbf{R}^{2n})$ and any $\mathcal{A} \in Sp(n, \mathbf{R})$,*

$$(\sigma \circ \mathcal{A})(D, X) = \mu(\mathcal{A}^*)\sigma(D, X)\mu(\mathcal{A}^*)^{-1},$$

where $\mathcal{A}^*$ is the transpose of $\mathcal{A}$ and $\mu(\mathcal{A}^*)$ is as in Lemma (2.14).

Proof: Since $\det \mathcal{A} = 1$, we have $(\sigma \circ \mathcal{A})\widehat{} = \widehat{\sigma} \circ \mathcal{A}^{*-1}$, so

$$(\sigma \circ \mathcal{A})(D, X)$$
$$= \iint (\widehat{\sigma} \circ \mathcal{A}^{*-1})(p, q)\rho(p, q)\, dp\, dq = \iint \widehat{\sigma}(p, q)(\rho \circ \mathcal{A}^*)(p, q)\, dp\, dq$$
$$= \iint \widehat{\sigma}(p, q)\mu(\mathcal{A}^*)\rho(p, q)\mu(\mathcal{A}^*)^{-1}dp\, dq = \mu(\mathcal{A}^*)\sigma(D, X)\mu(\mathcal{A}^*)^{-1}.$$

This calculation is rigorously valid if, say, $\sigma \in \mathcal{S}$; it then follows in general by the density of $\mathcal{S}$ in $\mathcal{S}'$. ∎

With this result in hand, we can give the promised generalization of Proposition (2.8). We shall give this result in a reasonably general form, although the polynomial growth hypotheses could be weakened a bit. We say that $\phi : \mathbf{R}^n \to \mathbf{R}$ is **polynomially bounded** if $|\phi(x)| \leq (1 + |x|)^M$ for some M.

(2.16) Lemma. *Suppose $\tau : \mathbf{R}^n \to \mathbf{R}$ is polynomially bounded, and for $f \in \mathcal{S}(\mathbf{R}^n)$ let $Tf(x) = \tau(x)f(x)$. Then T maps $\mathcal{S}$ into L^2, and T is essentially self-adjoint on the domain $\mathcal{S}$.*

Proof: The first assertion is obvious. As for the second, it is easily verified that T^* is given by $T^*f(x) = \tau(x)f(x)$ on the domain $D = \{f \in L^2 : \tau f \in L^2\}$. So one must check that if $f \in D$ then there is a sequence $\{f_n\}$ in $\mathcal{S}$ such that $f_n \to f$ and $\tau f_n \to \tau f$ in the L^2 norm, so that T^* is the closure of T. This we leave as an exercise for the reader. ∎

(2.17) Lemma. *Suppose $l_1, \ldots, l_k$ are independent linear functionals on $\mathbf{R}^{2n}$ whose Poisson brackets $\{l_i, l_j\}$ all vanish. Then $k \leq n$, and there exists $\mathcal{A} \in Sp(n, \mathbf{R})$ such that $l_j \circ \mathcal{A}(\xi, x) = x_j$.*

Proof: If we identify $(\mathbf{R}^{2n})^*$ with $\mathbf{R}^{2n}$ via the symplectic form, the Poisson bracket on $(\mathbf{R}^{2n})^*$ is the symplectic form itself. The desired assertion can therefore be restated as follows: if $v_1, \ldots, v_k$ are linearly independent elements of $\mathbf{R}^{2n}$ such that $[v_i, v_j] = 0$ for all i, j, then there exists a basis $\{e_j, f_j\}_{j=1}^n$ for $\mathbf{R}^{2n}$ such that

$$[e_i, e_j] = [f_i, f_j] = 0, \quad [e_i, f_j] = \delta_{ij}, \quad \text{and} \quad f_j = v_j \text{ for } j \leq k.$$

(The matrix $\mathcal{A}$ is then the one that takes the basis $\{e_j, f_j\}$ to the standard basis.) We prove the latter assertion by induction on n. First, observe that since the symplectic form is nondegenerate, there exists $e_1 \in \mathbf{R}^{2n}$ such that $[e_1, v_1] = 1$ and $[e_1, v_j] = 0$ for $j \geq 2$. If $n = 1$ the story is over; otherwise, let

$$W = \{w \in \mathbf{R}^{2n} : [w, e_1] = [w, v_1] = 0\}.$$

The restriction of the symplectic form to $W \times W$ is again a symplectic form (i.e., is nondegenerate), and W contains $v_2, \ldots, v_k$. The proof is therefore completed by applying the inductive hypothesis on W. ∎

(2.18) Theorem. *Suppose $l_1, \ldots, l_k$ are independent linear functionals on $\mathbf{R}^{2n}$ whose Poisson brackets $\{l_i, l_j\}$ all vanish. Suppose also that $\tau : \mathbf{R}^k \to \mathbf{R}$ is polynomially bounded, and let*

$$\sigma(\xi, x) = \tau\big(l_1(\xi, x), \ldots, l_k(\xi, x)\big).$$

Then:
(i) $\sigma(D, X)$ maps $\mathcal{S}$ into L^2 and is essentially self-adjoint on the domain $\mathcal{S}$.
(ii) For any polynomially bounded $\phi : \mathbf{R} \to \mathbf{R}$, we have $(\phi \circ \sigma)(D, X) = \phi(\sigma(D, X))$, where the operator on the right is defined by the spectral functonal calculus for self-adjoint operators.

Proof: First suppose $l_j(\xi, x) = x_j$. Then

$$\sigma(D, X)f(x) = \tau(x_1, \ldots, x_k)f(x)$$

by Proposition (2.8), so (i) follows from Lemma (2.16) and (ii) is clear since [the closure of] $\sigma(D, X)$ is already a multiplication operator. For the general case, let $\mathcal{A}$ be as in Lemma (2.17). The preceding remarks show that the result is true for $(\sigma \circ \mathcal{A})(D, X)$, and by Theorem (2.15) the latter operator is unitarily equivalent to $\sigma(D, X)$ via the map $\mu(\mathcal{A}^*)$, which preserves $\mathcal{S}$. The result therefore follows. ∎

(2.19) Corollary. *Suppose $\phi : \mathbf{R} \to \mathbf{R}$ is polynomially bounded, and $a, b \in \mathbf{R}^n$. If $\sigma(\xi, x) = \phi(a\xi + bx)$ then $\sigma(D, X) = \phi(aD + bX)$.*

The properties embodied in Theorems 2.15 and 2.18 are among the most important features that distinguish the Weyl correspondence from the standard Kohn-Nirenberg correspondence in the theory of pseudodifferential operators. (Another is Proposition (2.6).) In the Kohn-Nirenberg theory it is not true that two operators are unitarily equivalent whenever their symbols are related by a symplectic transformation, nor does one have such a nice functional calculus as is given by Theorem (2.18).

An important special case of Corollary (2.19)—which could be proved directly, without invoking the symplectic group—is that

$$(2.20) \qquad \text{if} \quad \sigma(\xi, x) = (a\xi + bx)^k \quad \text{then} \quad \sigma(D, X) = (aD + bX)^k.$$

This relation completely determines the Weyl calculus for polynomials. Indeed, if

$$\sigma_{\alpha\beta}(\xi, x) = \xi^\alpha x^\beta$$

then for all $a, b \in \mathbf{R}^n$ we have

$$(a\xi + bx)^k = \sum_{|\alpha+\beta|=k} \frac{k!}{\alpha!\beta!} a^\alpha b^\beta \sigma_{\alpha\beta}(\xi, x),$$

and therefore $\sigma_{\alpha\beta}(D, X)$, for $|\alpha+\beta| = k$, is $\alpha!\beta!/k!$ times the coefficient of $a^\alpha b^\beta$ in the expansion of $(aD + bX)^k$. In fact,

$$\sigma_{\alpha\beta}(D, X) = \frac{\alpha!\beta!}{k!} \sum_\pi Y_{\pi(1)} Y_{\pi(2)} \cdots Y_{\pi(2n)}$$

where $Y_j = D_j$ for $1 \le j \le n$, $Y_j = X_{j-n}$ for $n+1 \le j \le 2n$, and π runs over all maps from $\{1, \ldots, k\}$ into $\{1, \ldots, 2n\}$ which assume the value j exactly α_j times for $1 \le j \le n$ and β_{j-n} times for $n+1 \le j \le 2n$. This argument also shows that the Weyl correspondence is the *only* linear map from polynomials to differential operators with polynomial coefficients that satisfies (2.20).

Symbol Classes.   The Weyl correspondence, encompassing as it does all continuous linear maps from $\mathcal{S}$ to $\mathcal{S}'$, is too general to be really manageable; in order to perform effective calculations, one needs to restrict attention to smaller classes of symbols and operators. In particular, one wishes to consider classes of operators whose members can be composed with one another—for examples, operators which map $\mathcal{S}$ into itself. The following theorem describes such a class that will be sufficiently general for our purposes. (A study of more general symbol classes whose associated operators preserve $\mathcal{S}$ can be found in Antonec [5]. But beware: at the top of p. 321 in the English translation of this paper, the equation $F(\mathbf{R}^{2m}) = \bigcup_N \bigcup_M F^{N,M}(\mathbf{R}^{2n})$ should read $F(\mathbf{R}^{2n}) = \bigcap_N \bigcup_M F^{N,M}(\mathbf{R}^{2n})$.) We introduce here a bit of notation that will be used continually in this chapter:

$$\langle \xi \rangle = \left(1 + \xi^2\right)^{1/2} \quad \text{for} \quad \xi \in \mathbf{R}^n.$$

(2.21) Theorem. *Suppose $\sigma \in C^\infty(\mathbf{R}^{2n})$ and there exist $K \in \mathbf{R}$ and $\delta < 1$ such that*

(2.22) $$|D_\xi^\alpha D_x^\beta \sigma(\xi, x)| \le C_{\alpha\beta} \langle \xi \rangle^{K+\delta(|\alpha|+|\beta|)}$$

for all $\xi, x \in \mathbf{R}^n$ and all multi-indices α, β. Then:
(a) *For $f \in \mathcal{S}$, $\sigma(D, X)f$ is given by the uniformly convergent iterated integral*

$$\sigma(D, X)f(x) = \iint \sigma(\xi, \tfrac{1}{2}(x + y)) e^{2\pi i(x-y)\xi} f(y) \, dy \, d\xi.$$

(b) *$\sigma(D, X)$ maps $\mathcal{S}$ continuously into itself and extends to a continuous operator on $\mathcal{S}'$.*

Proof: Given $f \in \mathcal{S}(\mathbf{R}^n)$, set

$$g(\xi, x) = \int \sigma(\xi, \tfrac{1}{2}(x+y)) e^{2\pi i(x-y)\xi} f(y) \, dy.$$

The integral clearly converges absolutely, in view of (2.22). Moreover, substituting $u = y - x$ and denoting $-(4\pi^2)^{-1} \sum (\partial^2/\partial u_j^2)$ by L_u, for any positive integer M we have

$$
\begin{aligned}
g(\xi, x) &= \int \sigma(\xi, x + \tfrac{1}{2}u) e^{-2\pi i u\xi} f(x+u) \, du \\
&= \int \left[\frac{(1+L_u)^M e^{-2\pi i u\xi}}{\langle\xi\rangle^{2M}} \right] \sigma(\xi, x + \tfrac{1}{2}u) f(x+u) \, du \\
&= \int \frac{e^{-2\pi i u\xi}}{\langle\xi\rangle^{2M}} (1+L_u)^M \left[\sigma(\xi, x + \tfrac{1}{2}u) f(x+u) \right] du \\
&= \sum_{|\alpha+\beta| \le 2M} c_{\alpha\beta} \int \frac{e^{-2\pi i u\xi}}{\langle\xi\rangle^{2M}} D_x^\alpha \sigma(\xi, x + \tfrac{1}{2}u) D^\beta f(x+u) \, du.
\end{aligned}
$$

Therefore,

$$
\begin{aligned}
|g(\xi, x)| &\le C_M \sum_{|\alpha+\beta| \le 2M} \int \langle\xi\rangle^{-2M} \langle\xi\rangle^{K+\delta|\alpha|} |D^\beta f(x+u)| \, du \\
&\le C_M' \langle\xi\rangle^{-2M(1-\delta)+K}.
\end{aligned}
$$

Taking $M > (K+n)/2(1-\delta)$, we see that $g(\cdot, x) \in L^1(\mathbf{R}^n)$ uniformly in x. In view of the discussion leading up to (2.3), this establishes (a) and also shows that $\sigma(D, X)f$ is bounded.

Next, we recall Proposition (2.13):

$$\rho(b, -a)\sigma(D, X)\rho(-b, a) = \sigma(D + a, X + b).$$

If we differentiate this equation with respect to a_j or b_j, divide by $2\pi i$, and set $a = b = 0$, we get

$$
\begin{aligned}
-X_j \sigma(D, X) + \sigma(D, X) X_j &= (D_{\xi_j} \sigma)(D, X), \\
D_j \sigma(D, X) - \sigma(D, X) D_j &= (D_{x_j} \sigma)(D, X).
\end{aligned}
$$

Hence the preceding argument, with σ replaced by $D_{x_j}\sigma$ or $D_{\xi_j}\sigma$, or f replaced by $D_j f$ or $X_j f$, shows that $X_j \sigma(D, X)f$ and $D_j \sigma(D, X)f$ are bounded. An easy inductive argument now shows that $\sigma(D, X)f \in C^\infty$ and that $X^\alpha D^\beta \sigma(D, X)f$ is bounded for all α and β, that is, $\sigma(D, X)f \in \mathcal{S}$. The continuity of $\sigma(D, X)$ on $\mathcal{S}$ follows from examination of the above argument, or from the closed graph theorem.

Finally, we note that if σ satisfies (2.22) then so does $\bar{\sigma}$; so $\bar{\sigma}(D, X)$ is continuous on $\mathcal{S}$ and its adjoint is continuous on $\mathcal{S}'$. But $\bar{\sigma}(D, X)^* = \sigma(D, X)$ by Proposition (2.6), so we have proved (b). ∎

The proof of Theorem (2.21) also yields the following useful fact:

(2.23) Proposition. *Suppose $\{\sigma_k\}$ is a sequence of symbols that satisfy (2.22) uniformly in k and converge to σ in the C^∞ topology. Then σ satisfies (2.22), and $\sigma_k(D, X)f \to \sigma(D, X)f$ in the topology of $\mathcal{S}$ for all $f \in \mathcal{S}$.*

We now introduce the standard symbol classes with which we shall be working henceforth:

$$S^m_{\rho,\delta} = \left\{\sigma \in C^\infty(\mathbf{R}^{2n}) : |D^\alpha_\xi D^\beta_x \sigma(\xi, x)| \le C_{\alpha\beta}\langle\xi\rangle^{m-\rho|\alpha|+\delta|\beta|}\right\}.$$

Here m, ρ, and δ are real numbers, and we *always* assume that

$$0 \le \delta \le \rho \le 1 \quad \text{and} \quad \delta < 1.$$

(The conditions $\rho > 1$ and $\delta < 0$ are too restrictive to be of much interest, while the conditions $\rho < 0$, $\delta > \rho$, or $\delta \ge 1$ lead to bad behavior.) We also set

$$S^\infty_{\rho,\delta} = \bigcup_{m\in\mathbf{R}} S^m_{\rho,\delta}, \qquad S^{-\infty}_{\rho,\delta} = \bigcap_{m\in\mathbf{R}} S^m_{\rho,\delta}.$$

$S^m_{\rho,\delta}$ is a Fréchet space whose topology is defined by the obvious family of norms:

$$(2.24) \qquad \|\sigma\|_{[j]} = \sup_{|\alpha|+|\beta|\le j} \sup_{\xi,x} |D^\alpha_\xi D^\beta_x \sigma(\xi, x)| \langle\xi\rangle^{-m+\rho|\alpha|-\delta|\beta|}.$$

We denote the corresponding operator classes by $OPS^m_{\rho,\delta}$:

$$OPS^m_{\rho,\delta} = \{\sigma(D, X) : \sigma \in S^m_{\rho,\delta}\}.$$

By Theorem (2.21), the elements of $OPS^m_{\rho,\delta}$ are continuous operators on both $\mathcal{S}$ and $\mathcal{S}'$. As we shall see later, $OPS^\infty_{\rho,\delta}$ is actually an algebra of operators on $\mathcal{S}$ and $\mathcal{S}'$.

The classes $S^m_{\rho,\delta}$ are the standard pseudodifferential symbol classes of Hörmander, or rather a globalized version of them. That is, we require the estimates on $\sigma(\xi, x)$ and its derivatives to hold uniformly for $x \in \mathbf{R}^n$, whereas they are usually required to hold uniformly only for x in compact sets. However, for the most part, results about the latter symbols are easily obtained from results about the former ones by the use of cutoff functions. This is part of the common lore of pseudodifferential operators, and we shall say no more about it.

The following result, whose easy proof we leave to the reader, will be used frequently.

(2.25) Proposition. *If* $\sigma_j \in S_{\rho_j,\delta_j}^{m_j}$ *for* $j = 1,2$ *then* $\sigma_1\sigma_2 \in S_{\rho_3,\delta_3}^{m_3}$ *where* $m_3 = m_1 + m_2$, $\rho_3 = \min(\rho_1,\rho_2)$, *and* $\delta_3 = \max(\delta_1,\delta_2)$.

We recall the notion of asymptotic expansion for symbols. Suppose $\{m_j\}_0^\infty$ is a sequence of real numbers that decreases monotonically to $-\infty$, $\sigma_j \in S_{\rho,\delta}^{m_j}$, and $\sigma \in S_{\rho,\delta}^{m_0}$. Then we say that

$$\sigma \sim \sum_0^\infty \sigma_j \iff \sigma - \sum_0^{N-1} \sigma_j \in S_{\rho,\delta}^{m_N} \text{ for all } N \geq 1.$$

It is well known that every formal asymptotic series of symbols is an asymptotic expansion of some symbol; for the sake of completeness we give the proof.

(2.26) Proposition. *Suppose* $\{m_j\}_0^\infty$ *decreases to* $-\infty$ *and* $\sigma_j \in S_{\rho,\delta}^{m_j}$. *Then there exists* $\sigma \in S_{\rho,\delta}^{m_0}$ *such that* $\sigma \sim \sum_0^\infty \sigma_j$.

Proof: Pick $\phi \in C^\infty(\mathbf{R}^n)$ such that $\phi(\xi) = 1$ for $|\xi| \geq 2$ and $\phi(\xi) = 0$ for $|\xi| \leq 1$. Then for $t \geq 1$ the functions $(\xi, x) \to \phi(t^{-1}\xi)$ are easily seen to be in $S_{1,0}^0$ uniformly in t (i.e., the norms (2.24) are bounded independently of t). Thus, for each j there exists $C_j > 0$ such that

$$\left| D_\xi^\alpha D_x^\beta [\phi(t^{-1}\xi)\sigma_j(\xi,x)] \right| \leq C_j \langle \xi \rangle^{m_j - \rho|\alpha| + \delta|\beta|} \quad \text{for } |\alpha| + |\beta| \leq j \text{ and } t \geq 1.$$

Let $t_0 = 1$, and for $j \geq 1$ pick $t_j \geq 1$ large enough so that

$$C_j \langle \xi \rangle^{m_j - m_{j-1}} \leq 2^{-j} \text{ for } |\xi| \geq t_j.$$

If $|\xi| < t_j$ then $\phi(t_j^{-1}\xi) = 0$, so for $j \geq 1$ we have

$$(2.27) \quad \left| D_\xi^\alpha D_x^\beta [\phi(t_j^{-1}\xi)\sigma_j(\xi,x)] \right| \leq 2^{-j}\langle \xi \rangle^{m_j-1-\rho|\alpha|+\delta|\beta|} \quad \text{for } |\alpha| + |\beta| \leq j.$$

Now let

$$\sigma(\xi,x) = \sum_0^\infty \phi(t_j^{-1}\xi)\sigma_j(\xi,x).$$

The estimate (2.27) shows that the series on the right converges in the C^∞ topology, and moreover

$$\left(\sigma - \sum_0^{N-1} \sigma_j \right)(\xi,x)$$

$$= \sum_0^{N-1}[1 - \phi(t_j^{-1}\xi)]\sigma_j(\xi,x) + \phi(t_N^{-1}\xi)\sigma_N(\xi,x) + \sum_{N+1}^\infty \phi(t_j^{-1}\xi)\sigma_j(\xi,x).$$

The first sum vanishes for large $|\xi|$, so is in $S_{\rho,\delta}^{-\infty}$; the next term is in $S_{\rho,\delta}^{m_N}$, and (2.27) shows that the last sum is also in $S_{\rho,\delta}^{m_N}$. Thus $\sigma \in S_{\rho,\delta}^{m_0}$ and $\sigma \sim \sum \sigma_j$. ∎

Miscellaneous Remarks and Examples. The Weyl correspondence can be formulated in the following abstract context. Let $A_1, \ldots, A_k$ be bounded self-adjoint operators on a Hilbert space $\mathcal{H}$, or more generally, operators defined on a common dense domain $\mathcal{D} \subset \mathcal{H}$ of analytic vectors. Then $\xi A = \sum \xi_j A_j$ is essentially self-adjoint on $\mathcal{D}$ for all $\xi \in \mathbf{R}^k$, so we can form the unitary operators $e^{2\pi i \xi A}$ and thence define the bounded operators

$$f(A_1, \ldots, A_k) = \int \widehat{f}(\xi) e^{2\pi i \xi A} d\xi$$

for all functions f on $\mathbf{R}^k$ such that $\widehat{f} \in L^1$. The Weyl correspondence in this setting has been studied in some detail by Taylor [134] and Anderson [2], [3], [4]. We also call attention to Nelson [113], where an elegant functional calculus for non-commuting operators is developed that encompasses the Weyl correspondence for bounded operators as well as Feynman's time-ordered integrals.

A particularly interesting special case of this generalized Weyl correspondence arises by taking $\mathcal{H}$ to be $L^2(\mathbf{H}_n)$ and the operators A_j to be (a) multiplication by the coordinate functions on $\mathbf{H}_n$ and (b) the left-invariant derivatives on $\mathbf{H}_n$, to wit:

$$P_j f(p, q, t) = p_j f(p, q, t), \quad Q_j f(p, q, t) = q_j f(p, q, t), \quad T f(p, q, t) = t f(p, q, t),$$

$$D_{p_j} = \frac{1}{2\pi i} \left(\frac{\partial}{\partial p_j} - \tfrac{1}{2} q_j \frac{\partial}{\partial t} \right), \quad D_{q_j} = \frac{1}{2\pi i} \left(\frac{\partial}{\partial q_j} + \tfrac{1}{2} p_j \frac{\partial}{\partial t} \right), \quad D_t = \frac{1}{2\pi i} \frac{\partial}{\partial t}.$$

These operators, together with the identity, span a Lie algebra of Hermitian operators on $\mathcal{S}(\mathbf{H}_n)$, in which the Lie bracket is $2\pi i$ times the commutator. In fact, the nonzero Lie brackets are

$$2\pi i [T, D_{p_j}] = \tfrac{1}{2} Q_j, \quad 2\pi i [D_{q_j}, T] = \tfrac{1}{2} P_j, \quad 2\pi i [D_{p_j}, D_{q_j}] = D_t,$$
$$2\pi i [D_{p_j}, P_j] = 2\pi i [D_{q_j}, Q_j] = 2\pi i [D_t, T] = I.$$

It is easily verified that one can exponentiate to obtain a unitary representation of the corresponding Lie group G—in fact, this is just the group generated by the operators (a) multiplication by the exponentials $e_\Xi(X) = e^{2\pi i \Xi X}$ ($\Xi, X \in \mathbf{H}_n$) and (b) right translations on the group $\mathbf{H}_n$. (Thus, G might be called "the Heisenberg group of the Heisenberg group.") One can then apply the generalized Weyl correspondence to develop a theory of pseudodifferential operators on $\mathbf{H}_n$ adapted to the non-Abelian group structure on $\mathbf{H}_n$. See Dynin [40], [41] and Taylor [136].

Continuous linear maps from $\mathcal{S}(\mathbf{R}^n)$ to $\mathcal{S}'(\mathbf{R}^n)$ can be represented either in the form $\sigma(D, X)$ or in the form $\rho(F)$ for suitable $\sigma, F \in \mathcal{S}'(\mathbf{R}^{2n})$, and these representations are related by the equation $F = \widehat{\sigma}$. It is a curious fact that the Fourier transform here can be replaced by a more elementary transformation involving only composition with linear maps on $\mathbf{R}^n$ and $\mathbf{R}^{2n}$. Namely:

(2.28) Proposition. *Let* $Rf(x) = f(-x)$. *Then*

$$\sigma(D, X) = R\rho(\sigma') \quad where \quad \sigma'(p, q) = 2^{-n}\sigma(-\tfrac{1}{2}q, \tfrac{1}{2}p).$$

Equivalently,

$$\rho(F) = RF''(D, X) \quad where \quad F''(\xi, x) = 2^n F(2x, -2\xi).$$

Proof: By (1.90) and Proposition (2.5),

$$
\begin{aligned}
\langle \sigma(D, X)f, g\rangle &= \iint \sigma(\xi, x)W(f, g)(\xi, x)\, d\xi\, dx \\
&= 2^n \iint \sigma(\xi, x)V(f, Rg)(2x, -2\xi)\, d\xi\, dx \\
&= 2^{-n} \iint \sigma(-\tfrac{1}{2}q, \tfrac{1}{2}p)V(f, Rg)(p, q)\, dp\, dq \\
&= \langle \rho(\sigma')f, Rg\rangle = \langle R\rho(\sigma')f, g\rangle. \quad \blacksquare
\end{aligned}
$$

What is really going on here is the fact that twisted convolution with the constant function 1 is essentially a Fourier transform:

$$1 \natural F(p, q) = \iint F(r, s)e^{\pi i(ps - qr)}\, dr\, ds = \widehat{F}(-\tfrac{1}{2}q, \tfrac{1}{2}p).$$

In other words, with σ' as above, $1 \natural \sigma' = 2^n \widehat{\sigma}$, so that $\rho(1)\rho(\sigma') = 2^n\sigma(D, X)$. Proposition (2.28) therefore follows from the fact that $\rho(1) = 2^n R$:

$$
\begin{aligned}
\rho(1)f(x) &= \iint e^{\pi ipq + 2\pi iqx} f(x + p)\, dp\, dq \\
&= 2^n \iint e^{2\pi i(y + x)\eta} f(y)\, dy\, d\eta = 2^n f(-x),
\end{aligned}
$$

where we have set $y = p + x$ and $\eta = \tfrac{1}{2}q$.

There is a certain attraction to using the operator $F \to 2^{-n}1 \natural F$ as the basic Fourier transform on $\mathbf{R}^{2n}$, especially since the symplectic form appears in the exponent. In his writings on symbolic calculi ([74], [75], [76]), Howe does just this. He also uses the representation ρ' defined by (1.27) rather than ρ, and the upshot is that what he calls the "isotropic symbol" of the operator $\sigma(D, X)$ is $2^{-n}\sigma(\tfrac{1}{2}\xi, \tfrac{1}{2}x)$.

Warning. The transformations

$$\sigma(\xi, x) \longrightarrow 2^{-n}\sigma(-\tfrac{1}{2}x, \tfrac{1}{2}\xi) \quad \text{and} \quad \sigma(\xi, x) \longrightarrow 2^{-n}\sigma(\tfrac{1}{2}\xi, \tfrac{1}{2}x)$$

and their inverses, discussed above, are not as innocuous as they may seem. For example, they do not preserve the set of σ such that $\sigma(D, X)$ is bounded on L^2, as example (iii) below shows.

We conclude with some miscellaneous examples of the Weyl correspondence, leaving the details of the verifications as exercises for the reader.

(i) For $a, b \in \mathbf{R}^n$, let

$$\sigma_{ab}(\xi, x) = \delta(\xi - a)\delta(x - b) \qquad \text{and} \qquad T_{ab} = \sigma_{ab}(D, X).$$

Then

$$T_{ab}f(x) = 2^n e^{4\pi i(x-b)a} f(2b - x).$$

The spectrum of T_{ab} is $\{\pm 2^n\}$, and the eigenfunctions are those functions f such that $\rho(b, -a)f$ is even or odd. Moreover, for any $\sigma \in \mathcal{S}'(\mathbf{R}^{2n})$ we have

$$\sigma(\xi, x) = \iint \sigma(a, b)\delta(\xi - a)\delta(x - b)\, da\, db,$$

and hence (with the integral suitably interpreted)

$$\sigma(D, X) = \iint \sigma(a, b)T_{ab}\, da\, db.$$

(ii) Suppose $\phi, \psi \in L^2(\mathbf{R}^n)$ and $\sigma = W(\phi, \psi)$. By Propositions (1.92), (1.94d) and (2.5), we have

$$\langle \sigma(D, X)f, g \rangle = \int W(\phi, \psi)W(f, g) = \int W(f, g)\overline{W(\psi, \phi)} = \langle f, \psi \rangle \langle \phi, g \rangle,$$

and hence

(2.29) $\sigma(D, X)f = \langle f, \psi \rangle \phi \qquad \text{for} \qquad \sigma = W(\phi, \psi).$

This should be compared with Proposition (1.46) (from which it could also be deduced). In particular, if $\psi = \phi$ and $\|\phi\|_2 = 1$, $\sigma(D, X)$ is the orthogonal projection onto ϕ.

(iii) For $a \in \mathbf{R}$, let $\sigma_a(\xi, x) = e^{4\pi i a x \xi}$. Then

$$\sigma_a(D, X)f(x) = |1 - a|^{-n} f\left(\frac{1+a}{1-a}x\right) \text{ for } a \neq 1,$$

$$= \left(\int f(y)\, dy\right) \delta(x) \text{ for } a = 1.$$

In particular, $\sigma_a(D, X)$ is bounded on L^2 and on $\mathcal{S}$ if and only if $a \neq \pm 1$.

(iv) Continuing the idea of (iii), we have

$$\sigma(D, X)f(x) = \left(\int \phi(y)f(y)\, dy\right) \delta(x) \quad \text{for} \quad \sigma(\xi, x) = \phi(x)e^{4\pi i x \xi}.$$

If we take $\phi \in \mathcal{S}$ then σ satisfies (2.22) with $\delta = 1$, but $\sigma(D, X)$ does not map $\mathcal{S}$ into itself, or even into a space of honest functions!

(v) If $\sigma(\xi, x) = 2^{n/2} e^{in\pi/4} e^{-2\pi i(x^2 + \xi^2)}$ then $\sigma(D, X)$ is the Fourier transform.

(vi) If $\sigma_\alpha(\xi, x) = e^{-2\pi\alpha(x^2 + \xi^2)}$ where $\text{Re}\,\alpha \geq 0$, what is $\sigma_\alpha(D, X)$? We have just given the answer for $\alpha = i$, and similarly $\sigma_{-i}(D, X)$ is $2^{-n/2} e^{in\pi/4}$ times the inverse Fourier transform. Moreover, $2^n e^{-2\pi(x^2 + \xi^2)}$ is the Wigner distribution of $h_0(x) = 2^{n/4} e^{-\pi x^2}$, so by (2.29), $\sigma_1(D, X)$ is 2^{-n} times the orthogonal projection onto h_0. The answer in general is a bit more complicated than one might expect, but is still quite pretty. It will be revealed in Section 5.2.

2. The Kohn-Nirenberg Correspondence

In the theory of pseudodifferential operators one customarily associates to a symbol $\sigma(\xi, x)$ the operator $\sigma(D, X)_{KN}$ ("KN" for "Kohn-Nirenberg") defined by

$$(2.30) \qquad \sigma(D, X)_{KN} f(x) = \int \sigma(\xi, x) e^{2\pi i x \xi} \widehat{f}(\xi)\, d\xi.$$

(One also customarily arranges the factors of 2π differently, puts the variables (ξ, x) in the opposite order, and writes $\sigma(x, D)$ for $\sigma(D, X)_{KN}$. We apologize to the readers who find these changes of notation irritating.) Writing out the Fourier transform in (2.30), we obtain

$$(2.31) \qquad \sigma(D, X)_{KN} f(x) = \iint \sigma(\xi, x) e^{2\pi i(x-y)\xi} f(y)\, dy\, d\xi,$$

which bears an evident resemblance to (2.3). Just as with the Weyl correspondence, one can easily verify that the map $\sigma \to \sigma(D, X)_{KN}$ is an isomorphism from $\mathcal{S}'(\mathbf{R}^{2n})$ to the space of continuous linear maps from $\mathcal{S}(\mathbf{R}^n)$ to $\mathcal{S}'(\mathbf{R}^n)$. The distribution kernel of $\sigma(D, X)_{KN}$ is $K(x, y) = \mathcal{F}_1 \sigma(y - x, x)$.

The motivation for the definition (2.30) is that if σ is a polynomial in ξ, say $\sigma(\xi, x) = \sum a_\alpha(x) \xi^\alpha$, then $\sigma(D, X)_{KN} = \sum a_\alpha(x) D^\alpha$, so one obtains differential operators written in the usual way, with the differentiations on the right. (As a matter of fact, the symbol-operator correspondence used throughout most of the Kohn-Nirenberg paper [92] is not this one but the one which gives the opposite order, with the differentiations on the left. I have felt it appropriate to name $\sigma(D, X)_{KN}$ after these authors anyhow; such are the distortions of history.)

The ξ-integration in (2.31) just gives the Fourier transform of σ in its first variable, so if we set $p = y - x$ and apply the Fourier inversion formula to σ in its second variable, we obtain

$$(2.32) \qquad \sigma(D, X)_{KN} f(x) = \iint \sigma(\xi, x) e^{-2\pi i p \xi} f(x + p) \, dp \, d\xi$$

$$= \iint \widehat{\sigma}(p, q) e^{2\pi i q x} f(x + p) \, dp \, dq$$

$$= \iint \widehat{\sigma}(p, q) \left[e^{2\pi i q X} e^{2\pi i p D} f \right](x) \, dp \, dq$$

If we compare this result with (2.1) we see that for arbitrary symbols σ, $\sigma(D, X)_{KN}$ is similar to $\sigma(D, X)$ except that "the differentiations are all on the right." This also leads to the relation betwen the Weyl and Kohn-Nirenberg symbols for an operator—which, incidentally, was derived by Moyal [108] long before the invention of pseudodifferential operators.

(2.33) Proposition. If $\sigma \in S'(\mathbf{R}^{2n})$ then $\sigma(D, X)_{KN} = (T\sigma)(D, X)$ where

$$(2.34) \qquad (T\sigma)\widehat{}(p, q) = e^{-\pi i p q} \widehat{\sigma}(p, q).$$

Proof: Combining formula (2.32) with the equation

$$e^{2\pi i (pD + qX)} = e^{\pi i p q} e^{2\pi i q X} e^{2\pi i p D},$$

we obtain

$$\sigma(D, X)_{KN} = \iint \widehat{\sigma}(p, q) e^{-\pi i p q} e^{2\pi i (pD + qX)} \, dp \, dq,$$

and comparing this with (2.1), we have the desired result. This formal calculation is rigorous if, say, $\sigma \in S(\mathbf{R}^{2n})$. But multiplication by $e^{-\pi i p q}$ is a continuous isomorphism on $S'(\mathbf{R}^{2n})$, so the result follows in general by continuity. ∎

Our main goal in this section is to show that the correspondence T of this proposition is well-behaved on the symbol classes $S_{\rho, \delta}^m$. For this purpose it will be convenient to express it in a different form for $\sigma \in S$.

(2.35) Proposition. If $\sigma \in S(\mathbf{R}^{2n})$ then

$$(2.36) \qquad T\sigma(\xi, x) = 2^n \iint \sigma(\eta, y) e^{4\pi i (x - y)(\xi - \eta)} \, d\eta \, dy.$$

Proof: For $\sigma \in S$, $T\sigma$ is the convolution of σ with the inverse Fourier transform of the distribution $(p, q) \to e^{-\pi i p q}$, and this is $2^n e^{4\pi i x \xi}$ by Theorem

2 of Appendix A. Alternatively, we can invert the Fourier transform of $T\sigma$ directly:

$$
\begin{aligned}
T\sigma(\xi, x) &= \iiiint e^{2\pi i(\xi p + xq) - \pi i p q - 2\pi i(\zeta p + yq)} \sigma(\zeta, y) \, d\zeta \, dy \, dp \, dq \\
&= \iiint e^{2\pi i(x-y)q} \sigma(\zeta, y) \delta(\xi - \zeta - \tfrac{1}{2}q) \, d\zeta \, dy \, dq \\
&= \iint \sigma(\xi - \tfrac{1}{2}q, y) e^{2\pi i(x-y)q} \, dy \, dq,
\end{aligned}
$$

and this gives (2.36) on setting $\eta = \xi - \tfrac{1}{2}q$. ∎

We now come to our first main result.

(2.37) Theorem. *The operator T on $\mathcal{S}'(\mathbf{R}^{2n})$ defined by (2.34) maps all the classes $S^m_{\rho,\delta}$ ($m \in \mathbf{R}$, $0 \le \rho \le \delta \le 1$, $\delta < 1$) into themselves, and is a Fréchet space isomorphism on each of them.*

Proof: We fix m, ρ, and δ and consider the norms (2.24) on $S^m_{\rho,\delta}$. It will suffice to prove that for every j there exists $k(j)$ such that

$$
(2.38) \qquad \|T\sigma\|_{[j]} \le C_j \|\sigma\|_{[k(j)]}, \qquad \sigma \in S^m_{\rho,\delta},
$$

as the same arguments will apply to the inverse map $T^{-1}\sigma = \mathcal{F}^{-1}(e^{\pi i p q}\hat{\sigma})$. In fact, it suffices to prove (2.38) for $j = 0$, for T is translation-invariant and so commutes with D_x and D_ξ.

Step 1. We prove (2.38) for $j = 0$, assuming that $\sigma \in \mathcal{S}$. ($\mathcal{S}$ is not dense in $S^m_{\rho,\delta}$, so some work will be needed afterwards.) For $\sigma \in \mathcal{S}$ we can use the formula (2.36), and our technique will be to break up the integral in (2.36) into pieces, depending on ξ, and to extimate each piece separately. To this end, we fix $\phi \in C^\infty_c(\mathbf{R}^n)$ satisfying $\phi(\eta) = 1$ for $|\eta| \le 1$ and $\phi(\eta) = 0$ for $|\eta| \ge 2$. For each ξ, we set

$$
\begin{aligned}
\sigma_1(\eta, y) &= \sigma_1(\eta, y; \xi) = \sigma(\eta, y)\phi(3\langle\xi\rangle^{-\rho}(\eta - \xi)), \\
\sigma_2(\eta, y) &= \sigma(\eta, y) - \sigma_1(\eta, y).
\end{aligned}
$$

Thus $\sigma_1(\eta, y) = \sigma(\eta, y)$ when $|\eta - \xi| \le \tfrac{1}{3}\langle\xi\rangle^\rho$ and $\sigma_1(\eta, y) = 0$ when $|\eta - \xi| \ge \tfrac{2}{3}\langle\xi\rangle^\rho$. It is easily verified that the function

$$
(\eta, y) \longrightarrow \phi(3\langle\xi\rangle^{-\rho}(\eta - \xi))
$$

is in $S^0_{1,0}$ uniformly in ξ, and it follows that

$$
\|\sigma_1\|_{[j]} \le C_j \|\sigma\|_{[j]} \quad \text{and} \quad \|\sigma_2\|_{[j]} \le C_j \|\sigma\|_{[j]}, \quad \text{uniformly in } \xi.
$$

We proceed to the estimates. Let

$$L_\eta = \frac{-1}{16\pi^2} \sum_1^n \frac{\partial^2}{\partial \eta_j^2}, \qquad L_y = \frac{-1}{16\pi^2} \sum_1^n \frac{\partial^2}{\partial y_j^2}.$$

To estimate $T\sigma_1$, we pick a positive integer N and write

$$e^{4\pi i(x-y)(\xi-\eta)} = \left(1 + \langle\xi\rangle^{2N\rho}|x-y|^{2N}\right)^{-1}\left(1 + \langle\xi\rangle^{2N\rho}L_\eta^N\right)e^{4\pi i(x-y)(\xi-\eta)}.$$

Subsituting this into (2.36) and integrating by parts, we obtain

$$T\sigma_1(\xi,x) = 2^n \iint \frac{e^{4\pi i(x-y)(\xi-\eta)}\left(1 + \langle\xi\rangle^{2N\rho}L_\eta^N\right)\sigma_1(\eta,y)}{1 + \langle\xi\rangle^{2N\rho}|x-y|^{2N}}\,dy\,d\eta.$$

Now, $\langle\eta\rangle$ and $\langle\xi\rangle$ are comparable when $|\eta - \xi| \le \frac{2}{3}\langle\xi\rangle^\rho$, so

$$\langle\xi\rangle^{\rho|\alpha|}|D_\eta^\alpha \sigma_1(\eta,y)| \le C_\alpha \|\sigma_1\|_{[|\alpha|]}\langle\xi\rangle^m \le C_\alpha' \|\sigma\|_{[|\alpha|]}\langle\xi\rangle^m.$$

Therefore,

$$|T\sigma_1(\xi,x)| \le C\|\sigma\|_{[2N]}\langle\xi\rangle^m \iint_{|\xi-\eta|\le\frac{2}{3}\langle\xi\rangle^\rho} \left(1 + \langle\xi\rangle^{2N\rho}|x-y|^{2N}\right)^{-1}d\eta\,dy$$

$$\le C\|\sigma\|_{[2N]}\langle\xi\rangle^{m+n\rho} \int \left(1 + \langle\xi\rangle^{2N\rho}|x-y|^{2N}\right)^{-1}dy$$

$$\le C\|\sigma\|_{[2N]}\langle\xi\rangle^m \int \left(1 + |z|^{2N}\right)^{-1}dz,$$

provided that we take $N > \frac{1}{2}n$. This completes the estimate for $T\sigma_1$.
The technique for estimating $T\sigma_2$ is similar. First, we write

$$e^{4\pi i(x-y)(\xi-\eta)} = \left(1 + |x-y|^{2P}\right)^{-1}\left(1 + L_\eta^P\right)\left[|\xi-\eta|^{-2N}L_y^N e^{4\pi i(x-y)\xi-\eta)}\right]$$

and integrate by parts in (2.36) to obtain

$$T\sigma_2(\xi,x) = 2^n \iint e^{4\pi i(x-y)(\xi-\eta)}|\xi-\eta|^{-2N}L_y^N\left[\frac{(1+L_\eta^P)\sigma_2(\eta,y)}{1+|x-y|^{2P}}\right]dy\,d\eta,$$

so that

$$|T\sigma_2(\xi,x)| \le C\|\sigma_2\|_{[2N+2P]} \iint_{|\xi-\eta|\ge\frac{1}{3}\langle\xi\rangle^\rho} \frac{|\xi-\eta|^{-2N}\langle\eta\rangle^{m+2N\delta}}{1+|x-y|^{2P}}\,dy\,d\eta,$$

since derivatives of $(1 + |x - y|^{2P})^{-1}$ are bounded by $(1 + |x - y|^{2P})^{-1}$. The y-integration converges of $P > \frac{1}{2}n$, and the η-integration converges if $N > (m + m)/2(1 - \delta)$. In this case,

$$|T\sigma_2(\xi, x)| \leq C\|\sigma\|_{[2N+2P]} \int_{|\xi - \eta| \geq \frac{1}{3}\langle\xi\rangle^\rho} |\xi - \eta|^{-2N} \langle\eta\rangle^{m+2N\delta} d\eta$$

$$\leq C\|\sigma\|_{[2N+2P]}(I_1 + I_2)$$

where I_1 is the integral over the region $|\eta| \geq 2\langle\xi\rangle$ and I_2 is the integral over the region $|\eta| \leq 2\langle\xi\rangle$, $|\eta - \xi| \geq \frac{1}{3}\langle\xi\rangle^\rho$. We have

$$I_1 \leq C \int_{|\eta| \geq 2\langle\xi\rangle} |\eta|^{-2N+m+2N\delta} d\eta = C'\langle\xi\rangle^{-2N(1-\delta)+m+n},$$

which is dominated by $\langle\xi\rangle^m$ (in fact, by $\langle\xi\rangle^{-K}$ for any given K) if we take N sufficiently large. Also,

$$I_2 \leq C\langle\xi\rangle^{-2N\rho} \int_{|\eta| \leq 2\langle\xi\rangle} \langle\eta\rangle^{m+2N\delta} d\eta \leq C'\langle\xi\rangle^{-2N(\rho-\delta)+m+n},$$

which is also dominated by any power of $\langle\xi\rangle$ for N sufficiently large, provided that $\rho > \delta$. If so, we are done.

Assume then that $\delta = \rho$, and write

$$\sigma_3(\eta, y) = \sigma_2(\eta, y)\phi\left(\tfrac{1}{2}\langle\xi\rangle^{-1}\eta\right),$$
$$\sigma_4(\eta, y) = \sigma_2(\eta, y) - \sigma_3(\eta, y).$$

As before, we have $\|\sigma_j\|_{[k]} \leq C_k\|\sigma\|_{[k]}$ uniformly in ξ for $j = 3, 4$. Since σ_4 is supported in the set where $|\eta| \geq 2\langle\xi\rangle$, the above argument for I_1 yields the desired estimate for $T\sigma_4$. To estimate $T\sigma_3$, write

$$e^{4\pi i(x-y)(\xi-\eta)} = \psi(\xi, \eta, y)\left[\langle\xi\rangle^{2N\rho} L_\eta^N + \langle\xi\rangle^{-2N\rho} L_y^N\right] e^{4\pi i(x-y)(\xi-\eta)}$$

with

$$\psi(\xi, \eta, y) = \left[\langle\xi\rangle^{2N\rho}|x - y|^{2N} + \langle\xi\rangle^{-2N\rho}|\xi - \eta|^{2N}\right]^{-1},$$

and integrate by parts in (2.36) to obtain

$T\sigma_3(\xi, x)$

$$= 2^n \iint e^{4\pi i(x-y)(\xi-\eta)} \left[\langle\xi\rangle^{2N\rho} L_\eta^N + \langle\xi\rangle^{-2N\rho} L_y^N\right] \left[\sigma_3(\eta, y)\psi(\xi, \eta, y)\right] dy \, d\eta$$

$$= \sum_{|\alpha|+|\beta|=2N} C_{\alpha\beta} \iint e^{4\pi i(x-y)(\xi-\eta)} \Big[\langle\xi\rangle^{\rho|\alpha|} D_\eta^\alpha \sigma_3(\eta, y) \cdot \langle\xi\rangle^{\rho|\beta|} D_\eta^\beta \psi(\xi, \eta, y)$$

$$+ \langle\xi\rangle^{-\rho|\alpha|} D_y^\alpha \sigma_3(\eta, y) \cdot \langle\xi\rangle^{-\rho|\beta|} D_y^\beta \psi(\xi, \eta, y)\Big] dy \, d\eta.$$

Because of the fact that $\delta = \rho$, the quasi-homogeneous nature of ψ, and the relation $a^{-|\beta|}D^\beta[g(ay)] = (D^\beta g)(ay)$ (valid for any g), this shows that $|T\sigma_3(\xi, x)|$ is dominated by $\|\sigma\|_{[2N]}$ times

$$\sum_{|\alpha|+|\beta|=2N} \iint_{|\xi-\eta| \geq \frac{1}{3}\langle\xi\rangle^\rho, \, |\eta| \leq 4\langle\xi\rangle} \frac{\langle\xi\rangle^{\rho|\alpha|}\langle\eta\rangle^{m-\rho|\alpha|} + \langle\xi\rangle^{-\rho|\beta|}\langle\eta\rangle^{m+\rho|\beta|}}{\left(\langle\xi\rangle^\rho|x-y| + \langle\xi\rangle^{-\rho}|\xi-\eta|\right)^{2N+|\beta|}} dy\, d\eta$$
$$\leq (J_1 + J_2),$$

where J_1 is the integral over the region $\frac{1}{3}\langle\xi\rangle^\rho \leq |\xi - \eta| \leq \frac{1}{2}\langle\xi\rangle$ and J_2 is the integral over the region $|\xi - \eta| \geq \frac{1}{2}\langle\xi\rangle$, $|\eta| \leq 4\langle\xi\rangle$. In the region of J_1, $\langle\eta\rangle$ and $\langle\xi\rangle$ are comparable, so

$$J_1 \leq C\langle\xi\rangle^m \iint_{|\xi-\eta| \geq \frac{1}{3}\langle\xi\rangle} \left(\langle\xi\rangle^\rho|x-y| + \langle\xi\rangle^{-\rho}|\xi-\eta|\right)^{-2N} dy\, d\eta.$$

Let $\zeta = \langle\xi\rangle^{-\rho}(\xi - \eta)$ and $z = \langle\xi\rangle^\rho(x - y)$. Then $dy\, d\eta = dz\, d\zeta$, so

$$J_1 \leq C\langle\xi\rangle^m \iint_{|\zeta| \geq \frac{1}{3}} \left(|z| + |\zeta|\right)^{-2N} dz\, d\zeta \leq C'\langle\xi\rangle^m,$$

provided $N > n$. On the other hand, in the region of J_2 we have $\langle\xi\rangle^{-\rho}|\xi - \eta| \geq \frac{1}{2}\langle\xi\rangle^{1-\rho}$, so

$$J_2 \leq C \sum_{|\alpha|+|\beta|=2N} \iint_{|\eta| \leq 4\langle\xi\rangle} \frac{\langle\xi\rangle^{\rho|\alpha|}\langle\eta\rangle^{m-\rho|\alpha|} + \langle\xi\rangle^{-\rho|\beta|}\langle\eta\rangle^{m+\rho|\beta|}}{\left(\langle\xi\rangle^\rho|x-y| + \langle\xi\rangle^{1-\rho}\right)^{2n}} dy\, d\eta.$$

Since

$$\int \left(\langle\xi\rangle^\rho|x-y| + \langle\xi\rangle^{1-\rho}\right)^{-2N} dy = \langle\xi\rangle^{-n\rho} \int \left(|z| + \langle\xi\rangle^{1-\rho}\right)^{-2N} dz$$
$$\leq C\langle\xi\rangle^{-n\rho} \int_0^\infty \left(r + \langle\xi\rangle^{1-\rho}\right)^{-2N+n-1} dr = C'\langle\xi\rangle^{-n\rho+(n-2N)(1-\rho)}$$

and

$$\int_{|\eta| \leq 4\langle\xi\rangle} \langle\eta\rangle^{m+a} d\eta \leq C \int_0^{4\langle\xi\rangle} (1+r)^{m+a+n-1} dr \leq C\langle\xi\rangle^{m+a+n},$$

we obtain

$$J_2 \leq C\langle\xi\rangle^{-n\rho+(n-2N)(1-\rho)+m+n} = C\langle\xi\rangle^{m+2(n-N)(1-\rho)},$$

which is dominated by any power of $\langle\xi\rangle$ when N is sufficiently large, since $\rho = \delta < 1$. Thus Step 1 is complete.

Step 2. Suppose $\sigma \in \mathcal{S}(\mathbf{R}^{2n})$ and σ is supported in the set where $|\xi| + |x| \geq K$. We claim that for every $a \geq 0$ there exist j_a, C_a such that

$$(2.39) \qquad |D_\xi^\alpha D_x^\beta T\sigma(\xi, x)| \leq C_a \|\sigma\|_{[j_a + |\alpha| + |\beta|]} K^{-a} \quad \text{for} \quad |\xi| + |x| \leq \tfrac{1}{2}K.$$

Since T commutes with derivatives it suffices to assume $\alpha = \beta = 0$. We proceed as above: write

$$e^{4\pi i(x-y)(\xi-\eta)} = \left(1 + |x - y|^{2N} + |\xi - \eta|^{2N}\right)^{-1} \left(1 + L_\eta^N + L_y^N\right) e^{4\pi i(x-y)(\xi-\eta)}$$

and integrate by parts in (2.36) to obtain

$$T\sigma(\xi, x) = 2^n \iint e^{4\pi i(x-y)(\xi-\eta)} \left(1 + L_\eta^N + L_y^N\right) \left(\frac{\sigma(\eta, y)}{1 + |x-y|^{2N} + |\xi-\eta|^{2N}}\right) dy\, d\eta,$$

which is dominated by

$$\|\sigma\|_{[2N]} \iint_{|\eta|+|y| \geq K} \frac{\langle\eta\rangle^{m+2N\delta}}{\left(1 + |x - y| + |\xi - \eta|\right)^{2N}} dy\, d\eta.$$

If $|\xi| + |x| \leq \tfrac{1}{2}K$ and $|\eta| + |y| \geq K$ then

$$|x - y| + |\xi - \eta| \geq \tfrac{1}{2}\left(|\eta| + |y|\right),$$

so this last integral is dominated by

$$\iint_{|\eta|+|y| \geq K} \left(1 + |\eta| + |y|\right)^{|m|-2N(1-\delta)} dy\, d\eta \leq C_N K^{|m|+n-2N(1-\delta)},$$

provided $N > (|m| + n)/2(1 - \delta)$. Taking $N = (|m| + n + a)/2(1 - \delta)$, we are done.

Step 3. Finally, we prove (2.38) without the restriction that $\sigma \in \mathcal{S}$. Given $\sigma \in S_{\rho,\delta}^m$, for $0 < \epsilon < 1$ let

$$\sigma_\epsilon(\xi, x) = \phi(\epsilon\xi)\phi(\epsilon x)\sigma(\xi, x),$$

where ϕ is as in Step 1. Then $\sigma_\epsilon \to \sigma$ as $\epsilon \to 0$ in the C^∞ topology (i.e., all derivatives converge uniformly on compact sets), and $\|\sigma_\epsilon\|_{[j]} \leq c_j \|\sigma\|_{[j]}$ with c_j independent of ϵ. By Step 1 and the fact that T commutes with derivatives, $\|T\sigma_\epsilon\|_{[j]} \leq C_j \|\sigma\|_{[k(j)]}$ with C_j independent of ϵ; and by Step 2, $T\sigma_\epsilon - T\sigma_\delta \to 0$ in the C^∞ topology as $\epsilon, \delta \to 0$. Since $T\sigma_\epsilon$ obviously converges to $T\sigma$ in the sense of distributions, it follows that $T\sigma_\epsilon \to T\sigma$ in the C^∞ topology and that (2.38) holds. The proof is therefore complete. ∎

A more illuminating description of the map T relating the Kohn-Nirenberg and Weyl symbols can be obtained by expanding $e^{-\pi ipq}$ in (2.34) into a power series and *formally* taking the inverse Fourier transform term by term, which gives

$$(2.40) \qquad T\sigma(\xi, x) = \sum_0^\infty \frac{(-\pi iD_\xi D_x)^j}{j!} \sigma(\xi, x).$$

For most symbols σ this is nonsense, as the series on the right does not converge even in $\mathcal{S}'$. However, if $\sigma \in S^m_{\rho,\delta}$ with $\rho > \delta$, the kth term on the right of (2.40) is in $S^{m-j(\rho-\delta)}_{\rho,\delta}$, so the series is a perfectly good formal asymptotic sum. We shall now show that in this case, (2.40) is correct if "$=$" is replaced by "$\sim$", and moreover that (2.40) is correct as it stands if σ is a polynomial in either variable so that the series terminates.

(2.41) Theorem. *If $\sigma \in S^m_{\rho,\delta}$ with $\rho > \delta$ and $T\sigma$ is given by (2.34), then*

$$T\sigma \sim \sum_0^\infty \frac{(-\pi iD_\xi D_x)^j \sigma}{j!}.$$

If σ is a polynomial of degree k in either ξ or x, then

$$T\sigma = \sum_0^k \frac{(-\pi iD_\xi D_x)^j \sigma}{j!}.$$

Proof: Let us first dispose of the case when σ is a polynomial. If, say, $\sigma(\xi, x) = \sum_{|\alpha| \le k} a_\alpha(x) \xi^\alpha$, then

$$\mathcal{F}\left(T\sigma - \sum_0^k \frac{(-\pi iD_\xi D_x)^j}{j!} \sigma\right)(p, q) = \sum_{|\alpha| \le k} \widehat{a}_\alpha(q) D^\alpha \delta(p) r_k(-\pi ipq)$$

where $r_k(t) = e^t - \sum_0^k t^j/j!$. The terms on the right all vanish because $r_k(-\pi ipq)$ vanishes to order $k + 1$ in p at $p = 0$. Similarly if σ is a polynomial in x.

Next, suppose that $\sigma \in \mathcal{S}$. We apply Taylor's formula in the form

$$f(1) = \sum_0^k \frac{f^{(j)}(0)}{j!} + \frac{1}{k!} \int_0^1 f^{(k+1)}(t)(1 - t)^k \, dt$$

to the function $f(t) = \sigma(\eta, x + t(y - x))$ to obtain

$$\sigma(\eta, y) = \sum_{|\alpha| \le k} \frac{\partial_x^\alpha \sigma(\eta, x)}{\alpha!}(y - x)^\alpha$$

$$+ (k + 1) \sum_{|\alpha| = k+1} \int_0^1 \frac{(\partial_x^\alpha \sigma)(\eta, \, x + t(y - x))}{\alpha!}(y - x)^\alpha (1 - t)^k dt.$$

If we plug this expansion into (2.36), we get $T\sigma = P\sigma + R\sigma$ where

$$P\sigma(\xi, x) = 2^n \sum_{|\alpha| \le k} \iint \frac{\partial_x^\alpha \sigma(\eta, x)}{\alpha!} (y - x)^\alpha e^{4\pi i(x-y)(\xi-\eta)} d\eta \, dy$$

$$= 2^n \sum_{|\alpha| \le k} \iint \frac{(2\pi i D_x)^\alpha \sigma(\eta, x)}{\alpha!} \left(\frac{D_\eta}{2}\right)^\alpha e^{4\pi i(x-y)(\xi-\eta)} d\eta \, dy$$

$$= 2^n \sum_{|\alpha| \le k} \iint \frac{(-i\pi)^{|\alpha|} D_\eta^\alpha D_x^\alpha \sigma(\eta, x)}{\alpha!} e^{4\pi i(x-y)(\xi-\eta)} d\eta \, dy,$$

which by the Fourier inversion formula equals

$$\sum_{|\alpha| \le k} \frac{(-i\pi)^{|\alpha|} D_\xi^\alpha D_x^\alpha \sigma(\xi, x)}{\alpha!} = \sum_{j=0}^{k} \frac{(-i\pi D_\xi D_x)^j}{j!} \sigma(\xi, x),$$

and similarly

$$2^{-n}(k+1)^{-1} R\sigma(\xi, x)$$

$$= \sum_{|\alpha|=k+1} \iiint_0^1 \frac{(\partial_x^\alpha \sigma)(\eta, \, x+t(y-x))}{\alpha!} (y-x)^\alpha e^{4\pi i(x-y)(\xi-\eta)} (1-t)^k dt \, d\eta \, dy$$

$$= (-i\pi)^{k+1} \sum_{|\alpha|=k+1} \iiint_0^1 \frac{(D_\eta^\alpha D_x^\alpha \sigma)(\eta, \, x+t(y-x))}{\alpha!} e^{4\pi i(x-y)(\xi-\eta)} (1-t)^k dt \, d\eta \, dy.$$

It is to be noted that the ηy-integrals here are not absolutely convergent as integrals over $\mathbf{R}^{2n}$, but are prefectly well-behaved as iterated integrals in the indicated order. In short, we have

(2.42)

$$T\sigma = P\sigma + R\sigma, \quad P\sigma = \sum_{j=0}^{k} \frac{(-\pi i D_\xi D_x)^j \sigma}{j!}, \quad R\sigma(\xi, x) = T\sigma^z(\xi, x)|_{z=x},$$

where

$$\sigma^z(\eta, y) = 2^n(k+1)(-i\pi)^{k+1} \sum_{|\alpha|=k+1} \int_0^1 \frac{(D_\eta^\alpha D_x^\alpha \sigma)(\eta, \, z + t(y - z))}{\alpha!} (1 - t)^k \, dt.$$

The map $\sigma \to \sigma^z$ is clearly bounded from $S_{\rho,\delta}^m$ to $S_{\rho,\delta}^{m-(k+1)(\rho-\delta)}$ uniformly in z, for any m. Also, since T commutes with derivatives and

$$(D_{x_j} + D_{z_j})\sigma^z(\xi, x) = (D_{x_j}\sigma)^z(\xi, x), \quad D_{\xi_j}\sigma^z = (D_{\xi_j}\sigma)^z,$$

we see that

$$D_\xi^\beta D_x^\gamma [T\sigma^z(\xi, x)|_{z=x}] = T(D_\xi^\beta D_x^\gamma \sigma)^z(\xi, x)_{z=x}.$$

It therefore follows from Theorem (2.37) that the map R defined in (2.42) is bounded from $S_{\rho,\delta}^m$ to $S_{\rho,\delta}^{m-(k+1)(\rho-\delta)}$.

Now, given $\sigma \in S_{\rho,\delta}^m$, we approximate σ by $\sigma_\epsilon \in C_c^\infty$ as in Step 3 of the proof of Theorem (2.37). Then: (i) By the argument given there, $T\sigma_\epsilon \to T\sigma$ in the C^∞ topology. (ii) If P is defined as in (2.42), clearly $P\sigma_\epsilon \to P\sigma$ in the C^∞ topology. (iii) $\{\sigma_\epsilon\}_{\epsilon<1}$ is bounded in $S_{\rho,\delta}^m$, hence $\{R\sigma_\epsilon\}_{\epsilon<1}$ is bounded in $S_{\rho,\delta}^{m-(k+1)(\rho-\delta)}$ by the argument just given. It follows that $T\sigma = P\sigma + R\sigma$ with $R\sigma \in S_{\rho,\delta}^{m-(k+1)(\rho-\delta)}$, and we are done. Incidentally, this argument also gives another proof of the assertion about polynomials, for if σ is a polynomial of degree k in ξ or x then $\sigma^z = 0$ and hence $R\sigma = 0$. ∎

(2.43) Corollary. *If $\sigma \in S_{\rho,\delta}^m$ with $\rho > \delta$ then*

$$\sigma - T\sigma \in S_{\rho,\delta}^{m-(\rho-\delta)} \qquad \text{and} \qquad \sigma(D,X)_{KN} - \sigma(D,X) \in OPS_{\rho,\delta}^{m-(\rho-\delta)}.$$

This corollary shows that the Weyl and Kohn-Nirenberg correspondences on $S_{\rho,\delta}^m$ ($\rho > \delta$) give the same results modulo lower-order error terms. For this reason, they are largely interchangeable in applications to partial differential equations. However, the Weyl correspondence sometimes has advantages, one of which is the following.

Suppose $\tau \in S_{1,0}^m$ has the asymptotic expansion $\tau \sim \sum_0^\infty \tau_k$ where $\tau_k(t\xi, x) = t^{m-k}\tau_k(\xi, x)$ for $t \geq 1$ and ξ large. Then $\sigma = T\tau$ has a similar expansion, namely

$$\sigma \sim \sum_0^\infty \sigma_k \quad \text{where} \quad \sigma_k = \sum_{j=0}^k \frac{(-\pi i D_\xi D_x)^j}{j!} \tau_{k-j}.$$

Let $S = \sigma(D,X) = \tau(D,X)_{KN}$. Then $\sigma_0 = \tau_0$ is called the **principal symbol** of S, and $\sigma_1 = \tau_1 - \pi i D_\xi D_x \tau_0$ is called the **subprincipal symbol** of S. The subprincipal symbol, an object with some important applications (see Taylor [135] or Hörmander [72]), thus appears more naturally in the Weyl correspondence.

Our proofs of Theorems (2.37) and (2.41) are adaptations of some arguments in Beals–Fefferman [19]. The underlying idea is the "principle of stationary phase." In its classical form, this principle says roughly that the major contributions to an integral $\int f(x)e^{i\lambda\phi(x)}dx$ for large λ come from the critical points of ϕ, and there is an asymptotic expansion of the integral as $\lambda \to \infty$ in terms of the values of f, ϕ, and their derivatives at these critical points. In our case we have the oscillatory integral

$$T\sigma(\xi, x) = 2^n \iint \sigma(\eta, y)e^{4\pi i(x-y)(\xi-\eta)} \, d\eta \, dy.$$

The only critical point of the exponent, as a function of (η, y), is (ξ, x). The proof of Theorem (2.37) shows that for $\sigma \in S_{\rho,\delta}^m$ the major contribution to this integral for large ξ comes from the region $|\eta - \xi| \leq \frac{2}{3}\langle\xi\rangle^\rho$ (or $|\eta - \xi| \leq \frac{1}{2}\langle\xi\rangle$ in case $\rho = \delta$), the remainder being rapidly decreasing as $\xi \to \infty$. A slight elaboration of the argument, as in Beals–Fefferman [19], shows that the main contribution actually comes from the neighborhood $|\eta - \xi| \leq \frac{2}{3}\langle\xi\rangle^\rho$, $|y - x| \leq \frac{2}{3}\langle\xi\rangle^{-\delta}$ of (ξ, x). Moreover, Theorem (2.41) gives the asymptotic expansion of $T\sigma(\xi, x)$ for large ξ in terms of the values of σ and its derivatives at (ξ, x).

A detailed description of the principle of stationary phase, together with some applications to pseudodifferential and Fourier integral operators, can be found in Fedoryuk [43].

3. The Product Formula

Suppose $\sigma(D, X)$ and $\tau(D, X)$ map S into itself. Then $\sigma(D, X)\tau(D, X)$ is an operator on S, so it is $\omega(D, X)$ for some unique $\omega \in S'(\mathbf{R}^{2n})$. We call ω the **twisted product** of σ and τ and denote it by $\sigma \natural \tau$:

$$\sigma \natural \tau(D, X) = \sigma(D, X)\tau(D, X).$$

To compute $\sigma \natural \tau$, we begin by assuming that $\sigma, \tau \in S(\mathbf{R}^{2n})$. In this case we have

$$\sigma(D, X)\tau(D, X)$$
$$= \iiiint \sigma\left(\zeta, \tfrac{1}{2}(x + z)\right)\tau\left(\eta, \tfrac{1}{2}(z + y)\right)e^{2\pi i[(x-z)\zeta + (z-y)\eta]} f(y) \, dy \, d\eta \, dz \, d\zeta$$
$$= \int K(x, y)f(y) \, dy$$

where

$$K(x, y) = \iiint \sigma\left(\zeta, \tfrac{1}{2}(x + z)\right)\tau\left(\eta, \tfrac{1}{2}(z + y)\right)e^{2\pi i(x\zeta - z\zeta + z\eta - y\eta)} d\eta \, dz \, d\zeta.$$

Hence, by (2.4),

$$\sigma \natural \tau(\xi, x) = \int K(x + \tfrac{1}{2}p, \, x - \tfrac{1}{2}p)e^{-2\pi i\xi p} \, dp$$
$$= \iiiint \sigma\left(\zeta, \tfrac{1}{2}(x+z) + \tfrac{1}{4}p\right)\tau\left(\eta, \tfrac{1}{2}(x+z) - \tfrac{1}{4}p\right) \times$$
$$e^{2\pi i(x\zeta - z\zeta + z\eta - x\eta - \xi p) + \pi i p(\zeta + \eta)} \, d\eta \, dz \, d\zeta \, dp.$$

We make the change of variable

$$u = \tfrac{1}{2}(x+z) + \tfrac{1}{4}p, \qquad v = \tfrac{1}{2}(x+z) - \tfrac{1}{4}p,$$

so that

$$z = u + v - x, \qquad p = 2(u-v), \qquad dz\,dp = 4^n du\,dv,$$
$$x\zeta - z\zeta + z\eta - x\eta - \xi p + \tfrac{1}{2}p(\zeta + \eta) = 2\big[(x-u)(\xi-\eta) - (x-v)(\xi-\zeta)\big].$$

Then
(2.44a)

$$\sigma \,\natural\, \tau(\xi, x) = 4^n \iiiint \sigma(\zeta, u)\tau(\eta, v) e^{4\pi i[(x-u)(\xi-\eta)-(x-v)(\xi-\zeta)]} du\,dv\,d\eta\,d\zeta,$$

or, in more compact form,

(2.44b) $$\sigma \,\natural\, \tau(w) = 4^n \iint \sigma(w')\tau(w'') e^{4\pi i[w-w'',\,w-w']} dw'\,dw'',$$

where the bracket in the exponent is the symplectic form on $\mathbf{R}^{2n}$.

An alternative formula for $\sigma \,\natural\, \tau$ may be derived in terms of the Fourier transform. We have

$$\sigma(D, X)\tau(D, X) = \rho(\widehat{\sigma})\rho(\widehat{\tau}) = \rho(\widehat{\sigma} \,\natural\, \widehat{\tau}),$$

and therefore

$$(\sigma \,\natural\, \tau)^{\widehat{}} = \widehat{\sigma} \,\natural\, \widehat{\tau}.$$

If we write out $\mathcal{F}^{-1}(\widehat{\sigma} \,\natural\, \widehat{\tau})$ in detail, we obtain

$$\sigma \,\natural\, \tau(\xi, x) = \iiiint \widehat{\sigma}(p-r,\, q-s)\widehat{\tau}(r, s) e^{2\pi i(p\xi + qx) + \pi i(ps - qr)} dr\,ds\,dp\,dq$$

(2.45) $$= \iiiint \widehat{\sigma}(p, q)\widehat{\tau}(r, s) e^{\pi i(ps - qr)} e^{2\pi i[(p+r)\xi + (q+s)x]} dr\,ds\,dp\,dq.$$

The formulas (2.44) and (2.45) may be restated as follows. Let

$$E_1(\xi, x, \eta, y) = 4^n e^{4\pi i(x\eta - \xi y)}, \qquad E_2(\xi, x, \eta, y) = e^{\pi i(\xi y - x\eta)},$$

and for any functions ϕ and ψ on $\mathbf{R}^{2n}$ let

$$\phi \otimes \psi(\xi, x, \eta, y) = \phi(\xi, x)\psi(\eta, y).$$

Then (2.44) and (2.45) say that

(2.46)
$$\sigma \,\natural\, \tau(\xi, x) = [(\sigma \otimes \tau) * E_1](\xi, x, \xi, x)$$
$$= \mathcal{F}^{-1}[(\widehat{\sigma} \otimes \widehat{\tau})E_2](\xi, x, \xi, x).$$

The equivalence of these formulas then boils down to the fact that $E_2 = \hat{E}_1$, which follows from Theorem 2 of Appendix A.

The formula (2.44) shows that the twisted product bears a formal resemblance to the operator T of the previous section. Both are given by convolution with the exponential of an imaginary quadratic form, but in the case of the twisted product, the convolution is followed by restriction to the diagonal in $\mathbf{R}^{2n} \times \mathbf{R}^{2n}$. This extra complication is what prevents the twisted product $\sigma \,\sharp\, \tau$ from being well defined for arbitrary $\sigma, \tau \in S'$. However, on the symbol classes $S_{\rho,\delta}^m$, all is well.

(2.47) Theorem. *The map $(\sigma, \tau) \to \sigma \,\sharp\, \tau$ is continuous from $S_{\rho,\delta}^{m_1} \times S_{\rho,\delta}^{m_2}$ to $S_{\rho,\delta}^{m_1 + m_2}$ for all $m_1, m_2 \in \mathbf{R}$ when $0 \le \delta \le \rho \le 1$ and $\delta < 1$.*

Proof: The first and most difficult part is to use (2.44) to show that for $\sigma, \tau \in S$ the $S_{\rho,\delta}^{m_1 + m_2}$ norms of $\sigma \,\sharp\, \tau$ are dominated by products of the $S_{\rho,\delta}^{m_1}$ norms of σ and the $S_{\rho,\delta}^{m_2}$ seminorms of τ. The proof is almost identical to Step 1 in the proof of Theorem (2.37); we shall give only enough details to convince the reader that this is so.

We begin as before by picking $\phi \in C_c^\infty(\mathbf{R}^n)$ with $\phi(\eta) = 1$ for $|\eta| \le 1$ and $\phi(\eta) = 0$ for $|\eta| \ge 2$, and having fixed $\xi \in \mathbf{R}^{2n}$ we set

$$\omega_1(\zeta, u, \eta, v) = \sigma(\zeta, u)\tau(\eta, v)\phi\big(3\langle \xi\rangle^{-\rho}(\xi - \zeta)\big)\phi\big(3\langle \xi\rangle^{-\rho}(\xi - \eta)\big),$$

$$\omega_2(\zeta, u, \eta, v) = \sigma(\zeta, u)\tau(\eta, v) - \omega_1(\zeta, u, \eta, v).$$

Then the integral (2.44a) defining $\sigma \,\sharp\, \tau(\xi, x)$ is the sum of the two integrals obtained by substituting ω_1 and ω_2 for $\sigma\tau$. After an integration by parts, the first of these integrals is

$$\iiiint e^{4\pi i[(x-u)(\xi-\eta)-(x-v)(\xi-\zeta)]} \times$$

$$\frac{\big(1 + \langle \xi\rangle^{2N\rho}L_\eta^N\big)\big(1 + \langle \xi\rangle^{2N\rho}L_\zeta^N\big)\omega_1(\zeta, u, \eta, v)}{\big(1 + \langle \xi\rangle^{2N\rho}|x - u|^{2N}\big)\big(1 + \langle \xi\rangle^{2N\rho}|x - v|^{2N}\big)}\, du\, dv\, d\zeta\, d\eta,$$

which is dominated by

$$C\langle \xi\rangle^{m_1+m_2} \iiiint_{\max(|\xi-\eta|,|\xi-\zeta|)\le \frac{2}{3}\langle \xi\rangle^\rho} \frac{du\, dv\, d\zeta\, d\eta}{\big(1 + \langle \xi\rangle^{2N\rho}|x - u|^{2N}\big)\big(1 + \langle \xi\rangle^{2N\rho}|x - v|^{2N}\big)}.$$

If $N > \frac{1}{2}n$ this is less than $C'\langle \xi\rangle^{m_1+m_2}$, where C' incorporates suitable norms of σ and τ. The second integral, after integration by parts, is

$$\iiiint \frac{e^{4\pi i[(x-u)\xi-\eta)-(x-v)(\xi-\zeta)]}}{\big(|\xi-\zeta|^2 + |\xi-\eta|^2\big)^N} \times$$

$$(L_u + L_v)^N \left[\frac{(1+L_\zeta^N)(1+L_\eta^N)\omega_2(\zeta, u, \eta, v)}{\big(1 + |x - u|^{2N}\big)\big(1 + |x - v|^{2N}\big)}\right] du\, dv\, d\zeta\, d\eta,$$

which, for $N > \frac{1}{2}n$, is dominated by

$$C \iint_{|\xi-\zeta|^2+|\xi-\eta|^2 \geq \frac{1}{8}\langle\xi\rangle^{2\rho}} \frac{\langle\zeta\rangle^{m_1+2N\delta}\langle\eta\rangle^{m_2+2N\delta}}{\left(|\xi-\eta|^2+|\xi-\zeta|^2\right)^N} d\zeta\, d\eta.$$

Break the region of integration into two pieces: $|\eta|^2 + |\zeta|^2 \geq 4\langle\xi\rangle^2$ and $|\eta|^2 + |\zeta|^2 \leq 4\langle\xi\rangle^2$. The integral over the first region is at most

$$C \iint_{|\eta|^2+|\zeta|^2 \geq 4\langle\xi\rangle^2} \left(\langle\eta\rangle^2 + \langle\zeta\rangle^2\right)^{N(\delta-1)+(|m_1|+|m_2|)/2} d\zeta\, d\eta$$
$$\leq C'\langle\xi\rangle^{-2N(1-\delta)+|m_1|+|m_2|+2n},$$

provided N is sufficiently large. The integral over the second region is at most

$$C\langle\xi\rangle^{-2N\rho} \iint_{|\zeta|^2+|\eta|^2 \leq 4\langle\xi\rangle^2} \langle\eta\rangle^{m_1+2N\delta}\langle\zeta\rangle^{m_2+2N\delta} d\zeta\, d\eta$$
$$\leq C'\langle\xi\rangle^{-2N(\rho-\delta)+m_1+m_2+2n},$$

where again the constants C, C' involve norms of σ and τ. If $\rho > \delta$, then, we are done. We leave to the reader the extra argument for $\rho = \delta$, which still parallels the proof of Theorem (2.37).

Next, suppose $\sigma \in S^{m_1}_{\rho,\delta}$ and $\tau \in S^{m_2}_{\rho,\delta}$. With ϕ as above, let

$$\sigma_\epsilon(\xi,x) = \phi(\epsilon\xi)\phi(\epsilon x)\sigma(\xi,x), \qquad \tau_\epsilon(\xi,x) = \phi(\epsilon\xi)\phi(\epsilon x)\tau(\xi,x).$$

Then as $\epsilon \to 0$, σ_ϵ and τ_ϵ remain bounded in $S^{m_1}_{\rho,\delta}$ and $S^{m_2}_{\rho,\delta}$ and converge in the C^∞ topology to σ and τ. Steps 2 and 3 in the proof of Theorem (2.37) go through with essentially no change to show that $\sigma_\epsilon \,\natural\, \tau_\epsilon$ remains bounded in $S^{m_1+m_2}_{\rho,\delta}$ and is Cauchy in the C^∞ topology; so it converges in C^∞ to a limit $\omega \in S^{m_1+m_2}_{\rho,\delta}$. But by Proposition (2.23), $\sigma_\epsilon(D,X)$ and $\tau_\epsilon(D,X)$ converge strongly as operators on S to $\sigma(D,X)$ and $\tau(D,X)$, and hence $\sigma_\epsilon(D,X)\tau_\epsilon(D,X)$ converges strongly to $\sigma(D,X)\tau(D,X)$, while $\sigma_\epsilon \,\natural\, \tau_\epsilon(D,X)$ converges strongly to $\omega(D,X)$. It follows that $\omega = \sigma \,\natural\, \tau$, and (from the proof) that the $S^{m_1+m_2}_{\rho,\delta}$ norms of ω are bounded by the appropriate norms of σ and τ. $\blacksquare$

When $\rho > \delta$ the most useful form of the product rule is not either of the exact formulas (2.46) but the asymptotic expansion obtained by expanding the exponential E_2 in (2.46) into a power series and formally inverting the Fourier transform term by term:

$$\sigma \,\natural\, \tau(\xi,x) \sim \sum_0^\infty \frac{(\pi i)^j}{j!}(D_\xi D_y - D_\eta D_x)^j \sigma(\xi,x)\tau(\eta,y)|_{\eta=\xi,y=x}.$$

The right side can be written a bit more neatly as

$$(2.48) \qquad \sum_{0}^{\infty} \frac{(\pi i)^j}{j!} (D_{\xi,\sigma} D_{x,\tau} - D_{\xi,\tau} D_{x,\sigma})^j \sigma\tau,$$

where the subscripts σ and τ indicate that the differentiation is to be applied only to σ or τ. We observe that if $\sigma \in S_{\rho,\delta}^{m_1}$ and $\tau \in S_{\rho,\delta}^{m_2}$ then the jth term of this series is in $S_{\rho,\delta}^{m_1+m_2-j(\rho-\delta)}$.

(2.49) Theorem. *Suppose $\sigma \in S_{\rho,\delta}^{m_1}$ and $\tau \in S_{\rho,\delta}^{m_2}$ where $\rho > \delta$. The series (2.48) is an asymptotic expansion for $\sigma \,\natural\, \tau$. Moreover, if σ and τ are both polynomials in ξ, the series (2.48) terminates with the $(m_1 + m_2)$th term and is exactly equal to $\sigma \,\natural\, \tau$.*

Proof: Again, the proof is almost identical to the proof of Theorem (2.41), so we shall only give a sketch. Assuming first that $\sigma, \tau \in \mathcal{S}$, we write

$$\sigma(\zeta, u)\tau(\eta, v) = \sum_{|\alpha|+|\beta|\leq k} \frac{\partial_x^\alpha \sigma(\zeta, x)\partial_x^\beta \tau(\eta, x)}{\alpha!\beta!}(u-x)^\alpha(v-x)^\beta + \text{ remainder},$$

plug this expression into (2.44a), observe that

$$(u-x)^\alpha(v-x)^\beta e^{4\pi i[(x-u)(\xi-\eta)-(x-v)(\xi-\zeta)]}$$
$$= \frac{(-1)^{|\beta|}}{2^{|\alpha|+|\beta|}} D_\eta^\alpha D_\zeta^\beta e^{4\pi i[(x-u)(\xi-\eta)-(x-v)(\xi-\zeta)]},$$

integrate by parts, and use the Fourier inversion formula to obtain

$$\sigma \,\natural\, \tau(\xi, x) = \sum_{|\alpha|+|\beta|\leq k} \frac{(\pi i)^{|\alpha|+|\beta|}(-1)^{|\alpha|}}{\alpha!\beta!} D_\xi^\beta D_x^\alpha \sigma(\xi, x) D_\xi^\alpha D_x^\beta \tau(\xi, x) + R_k(\xi, x)$$
$$= \sum_{j=0}^{k} \frac{(\pi i)^j}{j!}(D_{\xi,\sigma} D_{x,\tau} - D_{\xi,\tau} D_{x,\sigma})^j \sigma(\xi, x)\tau(\xi, x) + R_k(\xi, x).$$

As in the proof of Theorem (2.41), an application of Theorem (2.48) shows that the map $(\sigma, \tau) \to R_k$ is continuous from $S_{\rho,\delta}^{m_1} \times S_{\rho,\delta}^{m_2}$ to $S_{\rho,\delta}^{m_1+m_2-(k+1)(\rho-\delta)}$, and the proof of the validity of the asymptotic expansion is completed by approximating $\sigma, \tau \in S_{\rho,\delta}^m$ by $\sigma_\epsilon, \tau_\epsilon \in C_c^\infty$. If σ and τ are polynomials in ξ, then so are $\sigma \,\natural\, \tau$ (since $\sigma(D, X)\tau(D, X)$ is a differential operator) and the series (2.48). They agree to infinite order in ξ and hence are equal. ∎

With this establishment of a workable formula for products, the map $\sigma \to$ $\sigma(D, X)$ graduates from being a mere "correspondence" to being a "calculus" on $S_{\rho,\delta}^m$ for $\rho > \delta$.

It is also true that (2.48) is a finite sum and equals $\sigma \sharp \tau$ when σ and τ are both polynomials in x. (Of course, in this case they do not belong to our symbol classes. A precise result, whose formulation we leave to the reader, may be obtained from Theorem (2.49) by conjugating with the Fourier transform.) The other situation in which the series (2.48) terminates is when either σ or τ is a polynomial in both variables. If, say, σ is a polynomial in ξ and x, then $\sigma(D, X)$ is a differential operator with polynomial coefficients and so can be composed in either order with $\tau(D, X)$ for any τ, and we can define $\sigma \sharp \tau$ and $\tau \sharp \sigma$ to be the symbols of these compositions.

(2.50) Theorem. *If either σ or τ is a polynomial in ξ and x of degree m, then the series (2.48) terminates with the mth term and equals $\sigma \sharp \tau$.*

Proof: Suppose, to be definite, that σ is a polynomial of degree m. Then $\hat{\sigma}$ is a linear combination of derivatives of δ of order $\leq m$, so the twisted convolution $\hat{\sigma} \natural \hat{\tau}$ can be defined in an appropriate distributional sense. (It suffices to assume that $\tau \in S$ and then to pass to limits.) Writing $e^t = \sum_0^m t^k/k! + R_m(t)$, we have

$$\hat{\sigma} \natural \hat{\tau}(p, q) = \iint \hat{\sigma}(r, s)\hat{\tau}(p - r, q - s)e^{\pi i(rq - sp)} dr\, ds$$

$$= \iint \hat{\sigma}(r, s)\hat{\tau}(p - r, q - s) \left[\sum_{k=0}^m \frac{(\pi i)^k}{k!}(rq - sp)^k + R_m(\pi i(rq - sp)) \right] dr\, ds.$$

The integral involving R_m is zero because R_m vanishes to order $m + 1$ at $r = s = 0$. Moreover, we have

$$(rq - sp)^k = \left[r(q - s) - s(p - r) \right]^k = \sum_{|\alpha+\beta|=k} \frac{(-1)^{|\beta|}k!}{\alpha!\beta!} r^\alpha(q - s)^\alpha s^\beta(p - r)^\beta.$$

It follows that $\hat{\sigma} \natural \hat{\tau}$ is a sum of untwisted convolutions, namely

$$\hat{\sigma} \natural \hat{\tau} = \sum_{k=0}^m \sum_{|\alpha+\beta|=k} \frac{(\pi i)^k(-1)^{|\beta|}}{\alpha!\beta!}(p^\alpha q^\beta \hat{\sigma}) * (p^\beta q^\alpha \hat{\tau}).$$

On applying the inverse Fourier transform to both sides, we obtain the desired result. ∎

An important observation is that the even-order terms of the expansion (2.48) are invariant under the interchange of σ and τ, while the odd-order terms change sign. Therefore:

$$\tfrac{1}{2}(\sigma \,\natural\, \tau + \tau \,\natural\, \sigma) \sim \sum_{k=0}^{\infty} \frac{(-1)^k \pi^{2k}}{(2k)!} (D_{\xi,\sigma} D_{x,\tau} - D_{\xi,\tau} D_{x,\sigma})^{2k} \sigma\tau,$$

$$\sigma \,\natural\, \tau - \tau \,\natural\, \sigma \sim 2i \sum_{k=0}^{\infty} \frac{(-1)^k \pi^{2k+1}}{(2k+1)!} (D_{\xi,\sigma} D_{x,\tau} - D_{\xi,\tau} D_{x,\sigma})^{2k+1} \sigma\tau.$$

The things of greatest interest here are the leading terms of these expansions. By examining them, we deduce the following corollary of Theorems (2.49) and (2.50).

(2.51) Corollary. Let $S = \sigma(D,X)$ and $T = \tau(D,X)$. If $\sigma \in S_{\rho,\delta}^{m_1}$, $\tau \in S_{\rho,\delta}^{m_2}$, then

(i) $\qquad \tfrac{1}{2}(ST + TS) = (\sigma\tau)(D,X) \mod OPS_{\rho,\delta}^{m_1+m_2-2(\rho-\delta)}$,

(ii) $\qquad ST - TS = (2\pi i)^{-1}\{\sigma,\tau\}(D,X) \mod OPS_{\rho,\delta}^{m_1+m_2-3(\rho-\delta)}$.

If σ or τ is a polynomial of degree ≤ 1, then

$$\tfrac{1}{2}(ST + TS) = (\sigma\tau)(D,X).$$

If σ or τ is a polynomial of degree ≤ 2, then

$$ST - TS = (2\pi i)^{-1}\{\sigma,\tau\}(D,X).$$

Thus, the Weyl calculus for symbols in $S_{\rho,\delta}^m$ has the multiplicative properties of an ideal quantization procedure—the pointwise product becomes the Jordan product and the Poisson bracket becomes $2\pi i$ times the commutator—modulo errors of order $2(\rho - \delta)$ less. This is an improvement on the Kohn-Nirenberg calculus, in which the errors are only of order $\rho - \delta$ less (as we shall shortly see). We can reformulate these results in a more quantum-mechanical spirit by putting in Planck's constant. We recall that the quantum momentum operator is $P = hD$, so the quantum observable corresponding to the classical observable $\sigma(\xi, x)$ should be $\sigma(P, X) = \sigma(hD, X)$, which is given by

$$\sigma(hD,X)f(x) = h^{-n} \int \sigma\left(\xi, \tfrac{1}{2}(x+y)\right) e^{2\pi i(x-y)\xi/h} f(y)\, dy\, d\xi.$$

As the reader may verify, the calculations at the beginning of this section, with factors of h inserted appropriately, lead to the following analogue of the product formulas (2.46). Define the h-twisted product $\natural_h$ by

$$(\sigma \,\natural_h\, \tau)(hD,X) = \sigma(hD,X)\tau(hD,X).$$

Then

$$(\sigma \,\natural_h\, \tau)(\xi, x) = [(\sigma \otimes \tau) * E_1^h](\xi, x, \xi, x),$$
$$= \mathcal{F}^{-1}[(\hat{\sigma} \otimes \hat{\tau})E_2^h](\xi, x, \xi, x),$$

where

$$E_1^h(\xi, x, \eta, y) = (2/h)^{2n} e^{4\pi i(x\eta - \xi y)/h},$$
$$E_2^h(\xi, x, \eta, y) = e^{\pi i h(\xi y - x\eta)}.$$

Expanding E_2^h in its Taylor series and formally inverting the Fourier transform termwise, we obtain

$$\sigma \,\natural_h\, \tau \sim \sum_0^\infty \frac{(\pi i h)^j}{j!}(D_{\xi,\sigma}D_{x,\tau} - D_{\xi,\tau}D_{x,\sigma})^j \sigma\tau.$$

Here the right hand side is to be interpreted as an asymptotic series in powers of h which will approximate $\sigma \,\natural_h\, \tau$ (for reasonable σ and τ) when h is near 0. In this setting, the product and Poisson bracket of symbols appear naturally as the classical limits (i.e., limits as $h \to 0$) of the Jordan product and commutator of quantum observables:

$$\sigma\tau = \tfrac{1}{2}(\sigma \,\natural_h\, \tau + \tau \,\natural_h\, \sigma) + O(h^2),$$
$$\{\sigma, \tau\} = \frac{2\pi i}{h}(\sigma \,\natural_h\, \tau - \tau \,\natural_h\, \sigma) + O(h^2).$$

A detailed analysis of the asymptotics of the Weyl correspondence as $h \to 0$ can be found in Voros [148], [149].

Incidentally, $\frac{2\pi i}{h}(\sigma \,\natural_h\, \tau - \tau \,\natural_h\, \sigma)$ is sometimes called the **Moyal bracket** of σ and τ, after Moyal [108].

With the results we have in hand, we can easily derive the asymptotic expansions for products and adjoints in the Kohn-Nirenberg calculus.

(2.52) Theorem. *If* $\sigma, \tau \in S_{\rho,\delta}^\infty$ *with* $\rho > \delta$, *then*

$$\sigma(D, X)_{KN}\, \tau(D, X)_{KN} = \omega(D, X)_{KN} \quad \text{and} \quad \big(\sigma(D, X)_{KN}\big)^* = \chi(D, X)_{KN}$$

where

$$\omega \sim \sum_{|\alpha| \geq 0} \frac{(2\pi i)^{|\alpha|}}{\alpha!}(D_\xi^\alpha \sigma)(D_x^\alpha \tau) \quad \text{and} \quad \chi \sim \sum_{|\alpha| \geq 0} \frac{(2\pi i)^{|\alpha|}}{\alpha!} D_\xi^\alpha D_x^\alpha \overline{\sigma}.$$

Proof: We have

$$\sigma(D, X)_{KN} = (T\sigma)(D, X) \quad \text{with} \quad T\sigma \sim e^{-\pi i D_\xi D_x}\sigma$$

and

$$\sigma \,\natural\, \tau \sim e^{\pi i (D_\xi D_y - D_\eta D_x)} \sigma(\xi, x) \tau(\eta, y)|_{\eta = \xi, y = x}.$$

Here the exponentials are to be interpreted as shorthand for the asymptotic series in Theorems (2.41) and (2.49). These exponentials can be multiplied with one another via the formal (Cauchy) product of their series expansions, and since D_ξ, D_x, etc., all commute, the product of the exponentials is the exponential of the sum. It follows that

$$\sigma(D, X)_{KN} \, \tau(D, X)_{KN} = \omega(D, X)_{KN}$$

where

$$\omega = T^{-1}(T\sigma \,\natural\, T\tau)$$
$$\sim e^{\pi i (D_\xi + D_\eta)(D_x + D_y)} e^{\pi i (D_\xi D_y - D_\eta D_x)} e^{-\pi i D_\xi D_x} e^{-\pi i D_\eta D_y} \sigma(\xi, x) \tau(\eta, y)|_{\eta = \xi, y = x}$$
$$= e^{2\pi i D_\xi D_y} \sigma(\xi, x) \tau(\eta, y)|_{\eta = \xi, y = x} = \sum_{|\alpha| \ge 0} \frac{(2\pi i)^{|\alpha|}}{\alpha!} (D_\xi^\alpha \sigma)(D_x^\alpha \tau).$$

Moreover, by Proposition (2.6),

$$\left(\sigma(D, X)_{KN} \right)^* = \left((T\sigma)(D, X) \right)^* = \overline{T\sigma}(D, X) = \chi(D, X)_{KN}$$

where

$$\chi = T^{-1}(\overline{T\sigma}) \sim e^{\pi i D_\xi D_x} e^{\pi i D_\xi D_x} \overline{\sigma} = e^{2\pi i D_\xi D_x} \overline{\sigma}$$
$$= \sum_{|\alpha| \ge 0} \frac{(2\pi i)^{|\alpha|}}{\alpha!} D_\xi^\alpha D_x^\alpha \overline{\sigma}. \quad \blacksquare$$

Unfortunately, we have no asymptotic expansion for products in the case $\rho = \delta$. Operators with $\rho = \delta$ (especially $\rho = \delta = \frac{1}{2}$) arise in a number of problems coming from hypoelliptic differential equations and noncommutative harmonic analysis, however, and in these contexts, workable but more complicated symbolic calculi have been developed to handle such operators. See Nagel–Stein [111], Dynin [40], [41], Taylor [136], Beals–Greiner–Vauthier [20], and Gaveau–Greiner–Vauthier [54].

4. Basic Pseudodifferential Theory

In this section we develop some basic results about pseudodifferential operators in the context of the Weyl calculus. These results are all very well known for the Kohn-Nirenberg calculus, and their adaptation to the Weyl calculus is quite routine; we are including it for the sake of completeness. In fact, once this material is developed for either the Weyl or the Kohn-Nirenberg calculus, most of it can be readily transferred to the other one by applying the results of the previous two sections.

We begin by examining the structure of the distribution kernels of operators on $OPS_{\rho,\delta}^m$ with $\rho > 0$.

(2.53) Theorem. *Suppose $\sigma \in S_{\rho,\delta}^m$ with $\rho > 0$, and let $K(x,y)$ be the kernel of $\sigma(D,X)$. Then $(x-y)^\alpha K(x,y)$ is of class C^k, and its derivatives of order $\leq k$ are bounded, provided $|\alpha| > (m+n+k)/\rho$. In particular, $K(x,y)$ is C^∞ off the diagonal and is rapidly decreasing as $x - y \to \infty$.*

Proof: We recall that K is related to σ by

$$K(u + \tfrac{1}{2}v,\, u - \tfrac{1}{2}v) = \int \sigma(\xi, u) e^{2\pi i v \xi} d\xi$$

(where the integral is interpreted in the distributional sense). Let

$$K_\alpha(u,v) = v^\alpha K(u + \tfrac{1}{2}v,\, u - \tfrac{1}{2}v) = (-1)^{|\alpha|} \int D_\xi^\alpha \sigma(\xi, u) e^{2\pi i v \xi} d\xi.$$

Since $\sigma \in S_{\rho,\delta}^m$ we have $D_\xi^\alpha \sigma(\cdot, u) \in L^1$ uniformly in u if $|\alpha| > (m+n)/\rho$, so K_α is bounded and continuous for $|\alpha| > (m+n)/\rho$. Similarly,

$$D_u^\beta D_v^\gamma K_\alpha(u,v) = (-1)^{|\alpha|} \int D_u^\beta D_\xi^\alpha \sigma(\xi, u) \xi^\gamma e^{2\pi i v \xi} d\xi$$

is bounded and continuous provided that $|\alpha| > (m + n + |\gamma| + \delta|\beta|)/\rho$. Setting $u = \tfrac{1}{2}(x+y)$ and $v = x - y$, we obtain the desired result. ∎

(2.54) Corollary. *Suppose $\sigma \in S_{\rho,\delta}^{-\infty}$. Then $K(x,\cdot) \in S$ for each x, and $x \to K(x,\cdot)$ is a smooth map from $\mathbf{R}^n$ to S. Hence, $\sigma(D,X)$ extends to a continuous map from S' to C^∞.*

(2.55) Corollary. *If $\sigma \in S_{\rho,\delta}^m$ with $m < -n(1-\rho)-\rho$, then $|K(x,y)| \leq \phi(x-y)$ where $\phi \in L^1(\mathbf{R}^n)$.*

Proof: The hypothesis is that $n - 1 > (m+n)/\rho$, so for C sufficiently large we can take $\phi(x) = C|x|^{-n+1}$ for $|x| < 1$, $\phi(x) = C|x|^{-n-1}$ for $|x| > 1$. ∎

(2.56) Corollary. *If $\sigma \in S_{\rho,\delta}^m$ with $m < -n(1-\rho)-\rho$ then $\sigma(D,X)$ is bounded on L^2.*

Proof: This follows from Corollary (2.55) and Young's inequality (see [50, Theorem 6.18]). ∎

Remark. The hypothesis $m < -n(1-\rho)-\rho$ in Corollaries (2.55) and (2.56) is not sharp. It can be replaced by $m < -n(1-\rho)$ in Corollary (2.55) and by $m \leq 0$ in Corollary (2.56). We shall prove the latter result shortly.

We recall that the **singular support** of a distribution f is the smallest closed set outside of which f is C^∞; it is denoted sing supp f. An operator T on (tempered) distributions is called **pseudolocal** if it decreases singular supports:

$$\text{sing supp} \, Tf \subset \text{sing supp} \, f.$$

(2.57) Theorem. *If* $\sigma \in S_{\rho,\delta}^m$ *with* $\rho > 0$ *then* $\sigma(D, X)$ *is pseudolocal.*

Proof: We first show that

$$(2.58) \qquad \text{sing supp} \, \sigma(D,X)g \subset \text{supp} \, g \quad \text{for} \quad g \in \mathcal{S}'.$$

Suppose $x_0 \notin \text{supp} \, g$, say $\text{dist}(x_0, \text{supp} \, g) = 4\epsilon > 0$. Pick $\phi \in C_c^\infty$ with $\phi(z) = 1$ for $|z| < \epsilon$ and $\phi(z) = 0$ for $|z| > 2\epsilon$, let K be the kernel of $\sigma(D,X)$, and let $K_1(x,y) = \phi(x-y)K(x,y)$ and $K_2 = K - K_1$. Then we have $\sigma(D,X) = T_1 + T_2$ where T_1 and T_2 are the operators with kernels K_1 and K_2. Clearly

$$\text{supp} \, T_1 g \subset \{\, x : \text{dist}(x, \text{supp} \, g) \leq 2\epsilon \,\},$$

and in particular $T_1 g$ vanishes near x_0. On the other hand, by Theorem (2.53), $K_2 \in C^\infty(\mathbf{R}^{2n})$ and $K_2(x, \cdot) \in \mathcal{S}(\mathbf{R}^n)$ for each x, so $T_2 g \in C^\infty$. Thus $\sigma(D,X)g$ is C^∞ near x_0, which proves (2.58).

Now suppose $f \in \mathcal{S}'$ and $x_0 \notin \text{sing supp} \, f$. Choose $\psi \in C_c^\infty$ with $\psi = 1$ near x_0 and $\psi = 0$ on sing supp f. Then

$$\sigma(D,X)f = \sigma(D,X)(\psi f) + \sigma(D,X)((1 - \psi)f).$$

The first term is in $\mathcal{S}$ because $\psi f \in \mathcal{S}$, and the second term is C^∞ near x_0 by (2.58). ∎

Remark. The hypothesis $\rho > 0$ is necessary in Theorems (2.53) and (2.57). For example, the operator $Tf(x) = f(x + a)$ $(a \neq 0 \in \mathbf{R}^n)$ is in $S_{0,0}^0$, as its symbol is $\sigma(\xi, x) = e^{2\pi i a \xi}$; but its kernel is $\delta(x + a - y)$, whose singularities are not on the diagonal, and it is not pseudolocal.

We now turn to questions of boundedness and positivity on L^2. Our results in this section will be based on the following theorem, which essentially asserts the existence of "twisted square roots" under appropriate hypotheses. In the next two sections we shall obtain sharper results by deeper arguments.

(2.59) Theorem. *Suppose* $\sigma \in S_{\rho,\delta}^0$ *with* $\rho > \delta$ *and the range of* σ *is in* $[\epsilon, \infty)$ *for some* $\epsilon > 0$. *Then there exists a real* $\tau \in S_{\rho,\delta}^0$ *such that* $\tau \natural \tau - \sigma \in S_{\rho,\delta}^{-\infty}$.

Proof: Since the functions $t \to \sqrt{t}$ and $t \to 1/\sqrt{t}$ are C^∞ on the interval $[\epsilon, \sup |\sigma|]$, it is easily verified that $\sqrt{\sigma}$ and $1/\sqrt{\sigma}$ are in $S_{\rho,\delta}^0$. Let $\tau_0 = \sqrt{\sigma}$; then by Theorem (2.50), $\rho_1 \equiv \tau_0 \,\natural\, \tau_0 - \sigma \in S_{\rho,\delta}^{-2(\rho-\delta)}$. By induction we can choose $\tau_j \in S_{\rho,\delta}^{-2j(\rho-\delta)}$ for $j = 1, 2, \ldots$ such that

$$\rho_{k+1} \equiv \left(\sum_0^k \tau_j\right) \,\natural\, \left(\sum_0^k \tau_j\right) - \sigma \in S_{\rho,\delta}^{-2(k+1)(\rho-\delta)}.$$

Indeed, having chosen $\tau_1, \ldots, \tau_{k-1}$, it suffices to take $\tau_k = -\rho_k/2\sqrt{\sigma}$, for

$$\left(\sum_0^k \tau_j\right) \,\natural\, \left(\sum_0^k \tau_j\right) = \left(\sum_0^{k-1} \tau_j\right) \,\natural\, \left(\sum_0^{k-1} \tau_j\right) + \tau_0 \,\natural\, \tau_k + \tau_k \,\natural\, \tau_0 \bmod S_{\rho,\delta}^{-2(k+1)(\rho-\delta)}$$

$$= \sigma + \rho_k + 2\tau_k\sqrt{\sigma} \bmod S_{\rho,\delta}^{-2(k+1)(\rho-\delta)}.$$

Finally, we invoke Proposition (2.26) to find $\tau \in S_{\rho,\delta}^0$ such that $\tau \sim \sum_0^\infty \tau_j$; then for all k,

$$\tau \,\natural\, \tau - \sigma = \left(\sum_0^k \tau_j\right) \,\natural\, \left(\sum_0^k \tau_j\right) - \sigma \bmod S_{\rho,\delta}^{-2(k+1)(\rho-\delta)},$$

so that $\tau \,\natural\, \tau - \sigma \in S_{\rho,\delta}^{-\infty}$. ∎

(2.60) Theorem. *If $\sigma \in S_{\rho,\delta}^0$ with $\rho > \delta$, then $\sigma(D, X)$ is bounded on $L^2(\mathbf{R}^n)$.*

Proof: Let $S = \sigma(D, X)$ and $\tau = \overline{\sigma} \,\natural\, \sigma$. Then τ is bounded (being in $S_{\rho,\delta}^0$) and real (being the symbol of the Hermitian operator S^*S). Pick $C > \sup \tau$. By Theorem (2.59), there exist real $\omega_0 \in S_{\rho,\delta}^0$ and $\omega_\infty \in S_{\rho,\delta}^{-\infty}$ such that $C - \tau = \omega_0 \,\natural\, \omega_0 + \omega_\infty$. Then

$$\langle \omega_0(D, X)^2 f, f \rangle = \|\omega_0(D, X)f\|_2^2 \geq 0$$

for all $f \in \mathcal{S}$ since $\omega_0(D, X)$ is Hermitian. Moreover, $\omega_\infty(D, X)$ is bounded on L^2 by Corollary (2.56), so

$$\langle (C - S^*S)f, f \rangle = \langle \omega_0(D, X)^2 f, f \rangle + \langle \omega_\infty(D, X)f, f \rangle \geq -C'\|f\|_2^2.$$

Thus, for all $f \in \mathcal{S}$,

$$\|Sf\|_2^2 = \langle S^*Sf, f \rangle \leq (C + C')\|f\|_2^2,$$

so S extends boundedly to L^2. ∎

We can generalize this result by throwing in derivatives. Recall that the Sobolev space H_s is defined for $s \in \mathbf{R}$ by

$$H_s = \left\{ f \in \mathcal{S}' : \widehat{f} \text{ is a function and } \|f\|_{(s)}^2 = \int \langle \xi \rangle^{2s} |\widehat{f}(\xi)|^2 d\xi < \infty \right\}.$$

We also set

$$H_\infty = \bigcap_{s \in \mathbf{R}} H_s, \qquad H_{-\infty} = \bigcup_{s \in \mathbf{R}} H_s.$$

(See [50] for a fuller explanation of H_s.) It is obvious from the definition that the operator $\langle D \rangle^t$ (i.e., $\sigma(D, X)$ where $\sigma(\xi, x) = \langle \xi \rangle^t$) is an isomorphism from H_s to H_{s-t} for all $s, t \in \mathbf{R}$. Thus, an operator T is bounded from H_s to H_{s-m} if and only if $\langle D \rangle^{s-m} T \langle D \rangle^{-s}$ is bounded on $H_0 = L^2$. Moreover, $\langle D \rangle^t \in OPS_{1,0}^t$, so

$$\langle D \rangle^{s-m} \sigma(D, X) \langle D \rangle^{-s} \in S_{\rho,\delta}^0 \text{ whenever } \sigma \in S_{\rho,\delta}^m.$$

From these observations we have immediately:

(2.61) Corollary. *If $\sigma \in S_{\rho,\delta}^m$ with $\rho > \delta$ then $\sigma(D, X)$ is bounded from H_s to H_{s-m} for all $s \in \mathbf{R}$.*

This result can also be localized. If $f \in \mathcal{S}'$, we say that "$f \in H_s$ at x_0" if f agrees with an element of H_s on a neighborhood of x_0.

(2.62) Corollary. *If $\sigma \in S_{\rho,\delta}^m$ and $f \in H_s$ at x_0 then $\sigma(D, X)f \in H_{s-m}$ at x_0.*

Proof: If $f = g + h$ where $g \in H_s$ and $h = 0$ on a neighborhood U of x_0, choose $\phi \in C_c^\infty(U)$ with $\phi = 1$ near x_0; then

$$\sigma(D, X)f = \sigma(D, X)g + \phi\sigma(D, X)h + (1 - \phi)\sigma(D, X)h.$$

We have $\sigma(D, X)g \in H_{s-m}$ by Corollary (2.61); $\phi\sigma(D, X)h \in C_c^\infty \subset H_{s-m}$ by Theorem (2.57), and $(1 - \phi)\sigma(D, X)h = 0$ near x_0. ∎

The other important consequence of Theorem (2.59) is the following positivity result.

(2.63) Gårding's Inequality. *Suppose $\sigma \in S_{\rho,\delta}^m$ with $\rho > \delta$, and for some $A, B > 0$,*

$$\operatorname{Re} \sigma(\xi, x) \geq A\langle \xi \rangle^m \quad \text{for} \quad \langle \xi \rangle \geq B.$$

Then for any $\epsilon > 0$ and $a > 0$ there is a constant $C_{a\epsilon}$ such that

$$\operatorname{Re}\langle \sigma(D, X)f, f \rangle \geq (A - \epsilon)\|f\|_{(m/2)}^2 + C_{a\epsilon}\|f\|_{((m-a)/2)}^2 \quad \text{for} \quad f \in \mathcal{S}.$$

Proof: First we deal with the case $m = 0$. Since $\sigma(D, X)^* = \overline{\sigma}(D, X)$ we have

$$\operatorname{Re}\langle \sigma(D, X)f, f \rangle = \langle (\operatorname{Re} \sigma)(D, X)f, f \rangle,$$

so we may as well assume that σ is real. We then have $\sigma(\xi, x) \geq A$ for $\langle \xi \rangle \geq B$, so we can find a C^∞ function $\phi(\xi)$ that vanishes for $\langle \xi \rangle \geq 2B$ such that $\sigma(\xi, x) + \phi(\xi) \geq A$ everywhere. We apply Theorem (2.59) to the symbol

$$\sigma(\xi, x) + \phi(\xi) - (A - \epsilon)$$

to find a real $\tau_\epsilon \in S^0_{\rho, \delta}$ and $\omega_\epsilon \in S^{-\infty}_{\rho, \delta}$ such that

$$\sigma = (A - \epsilon) + \tau_\epsilon \,\sharp\, \tau_\epsilon + \omega_\epsilon.$$

(Here ϕ is incorporated into ω_ϵ.) It follows that

$$
\begin{aligned}
\langle \sigma(D, X)f,\, f \rangle &= (A - \epsilon)\|f\|_2^2 + \|\tau_\epsilon(D, X)f\|_2^2 + \langle \omega_\epsilon(D, X)f,\, f \rangle \\
&\geq (A - \epsilon)\|f\|_{(0)}^2 + 0 - \|\omega_\epsilon(D, X)f\|_{(a/2)}\|f\|_{(-a/2)} \\
&\geq (A - \epsilon)\|f\|_{(0)}^2 - C_{a\epsilon}\|f\|_{(-a/2)}^2.
\end{aligned}
$$

Here we have used the Schwarz inequality $|\langle g, f \rangle| \leq \|g\|_{(s)}\|f\|_{(-s)}$ and the fact that $\omega_\epsilon(D, X)$ is bounded from H_s to H_t for all s, t (Corollary (2.61)).

The proof is thus complete when $m = 0$. For the general case, let

$$\sigma_1 = \langle \xi \rangle^{-m/2} \,\sharp\, \sigma \,\sharp\, \langle \xi \rangle^{-m/2}, \quad \text{i.e.,} \quad \sigma_1(D, X) = \langle D \rangle^{-m/2}\sigma(D, X)\langle D \rangle^{-m/2}.$$

Then $\sigma_1 \in S^0_{\rho, \delta}$ and $\sigma_1 = \langle \xi \rangle^{-m}\sigma \bmod S^{-(\rho - \delta)}_{\rho, \delta}$, from which it follows that $\operatorname{Re}\sigma_1(\xi, x) \geq A$ for $\langle \xi \rangle \geq B'$ for sufficiently large B'. So we can apply the result for $m = 0$ to σ_1, obtaining

$$
\begin{aligned}
\operatorname{Re}\langle \sigma(D, X)f,\, f \rangle &= \operatorname{Re}\langle \sigma_1(D, X)\langle D \rangle^{m/2}f,\, \langle D \rangle^{m/2}f \rangle \\
&\geq (A - \epsilon)\|\langle D \rangle^{m/2}f\|_{(0)}^2 - C_{a\epsilon}\|\langle D \rangle^{m/2}f\|_{(-a/2)}^2 \\
&= (A - \epsilon)\|f\|_{(m/2)}^2 - C_{a\epsilon}\|f\|_{((m-a)/2)}^2. \quad \blacksquare
\end{aligned}
$$

Remark. As far as one can tell from this proof, the constant $C_{a\epsilon}$ might blow up as $\epsilon \to 0$. However, it does not do so when $a \leq \rho - \delta$, as we shall see in Section 2.6.

We now consider invertibility in the Weyl calculus. We say that $\sigma \in S^m_{\rho, \delta}$ is **elliptic** (of order m) if there exist $A, B > 0$ such that

$$|\sigma(\xi, x)| \geq A\langle \xi \rangle^m \quad \text{for} \quad \langle \xi \rangle \geq B.$$

(This condition is weaker than the hypothesis of Gårding's inequality, which is called "strong ellipticity.")

(2.64) Theorem. *Suppose $\sigma \in S^m_{\rho,\delta}$ with $\rho > \delta$. Then σ is elliptic if and only if there exists $\tau \in S^{-m}_{\rho,\delta}$ such that $1 - \sigma \,\natural\, \tau$ and $1 - \tau \,\natural\, \sigma$ are both in $S^{-\infty}_{\rho,\delta}$.*

Proof: If such a τ exists, we have $1 - \sigma\tau \in S^{-(\rho-\delta)}_{\rho,\delta}$, and therefore

$$\tfrac{1}{2} \leq |\sigma(\xi,x)\tau(\xi,x)| \leq C\langle\xi\rangle^{-m}|\sigma(\xi,x)| \text{ for } \langle\xi\rangle \text{ large.}$$

Thus σ is elliptic. Conversely, suppose $|\sigma(\xi,x)| \geq A\langle\xi\rangle^m$ for $\langle\xi\rangle \geq B$. Choose $\phi \in C^\infty(\mathbf{R}^n)$ with $\phi(\xi) = 0$ for $\langle\xi\rangle \leq B$ and $\phi(\xi) \geq 1$ for $\langle\xi\rangle \geq 2B$, and set

$$\tau_0(\xi,x) = \frac{\phi(\xi)}{\sigma(\xi,x)} \quad \text{and} \quad \omega_1 = 1 - \sigma \,\natural\, \tau_0.$$

Of course $\tau_0(\xi,x)$ is taken to be zero when $\langle\xi\rangle \leq B$, and a routine calculation using the ellipticity of σ shows that $\tau_0 \in S^{-m}_{\rho,\delta}$. Moreover, since $1-\phi = 1-\sigma\tau_0 = \omega_1 \bmod S^{-(\rho-\delta)}_{\rho,\delta}$ and $1 - \phi \in C^\infty_c$, we have $\omega_1 \in S^{-(\rho-\delta)}_{\rho,\delta}$. Let ω_j be the j-fold twisted product of ω_1 with itself, and set $\tau_j = \tau_0 \,\natural\, \omega_j$. Then $\tau_j \in S^{-m-j(\rho-\delta)}_{\rho,\delta}$, and

$$\sigma \,\natural\, \left(\sum_0^k \tau_j\right) = (\sigma \,\natural\, \tau_0) \,\natural\, \sum_0^k \omega_j = (1-\omega_1) \,\natural\, \sum_0^k \omega_j = 1 - \omega_{k+1}$$

since $\omega_1 \,\natural\, \omega_j = \omega_{j+1}$. It follows that if we pick $\tau \sim \sum_0^\infty \tau_j$ according to Proposition (2.26) and set $\omega_\infty = 1 - \sigma \,\natural\, \tau$, for all k we have

$$\omega_\infty = \omega_{k+1} - \sigma \,\natural\, \left(\tau - \sum_0^k \tau_j\right) \in S^{-m-(k+1)(\rho-\delta)}_{\rho,\delta},$$

so $\omega_\infty \in S^{-\infty}_{\rho,\delta}$. Similarly, if we set $\tau'_j = \omega_j \,\natural\, \tau_0$, choose $\tau' \sim \sum_0^\infty \tau'_j$, and set $\omega'_\infty = 1 - \tau' \,\natural\, \sigma$, we have $\omega'_\infty \in S^{-\infty}_{\rho,\delta}$. But

$$\tau' \,\natural\, (1 - \omega_\infty) = \tau' \,\natural\, \sigma \,\natural\, \tau = (1 - \omega'_\infty) \,\natural\, \tau,$$

so

$$\tau' - \tau = \tau' \,\natural\, \omega_\infty - \omega'_\infty \,\natural\, \tau \in S^{-\infty}_{\rho,\delta},$$

and hence $1 - \tau \,\natural\, \sigma \in S^{-\infty}_{\rho,\delta}$. $\blacksquare$

(2.65) Corollary. *Suppose $\sigma \in S^m_{\rho,\delta}$ is elliptic, $S = \sigma(D,X)$, and $f \in \mathcal{S}'$. Then:*

(i) $\text{sing supp } Sf = \text{sing supp } f.$

(ii)   *If $f \in H_{-\infty}$ and $Sf \in H_s$ then $f \in H_{s+m}$.*

(iii)   *If $Sf \in H_s$ at x_0 then $f \in H_{s+m}$ at x_0.*

Proof: (i) Let $T = \tau(D, X)$ be as in Theorem (2.64). We have $(I - TS)f \in C^\infty$ by Corollary (2.54), so applying Theorem (2.57) twice we get

$$\text{sing supp } f = \text{sing supp } TSf \subset \text{sing supp } Sf \subset \text{sing supp } f.$$

(ii) We have $f = T(Sf) + (I - TS)f$, and by Corollary (2.61), T maps H_s to H_{s+m} and $I - TS$ maps $H_{-\infty}$ to H_{s+m}.

(iii) This follows from (ii) as Corollary (2.62) does from Corollary (2.61). ∎

Wave Front Sets. We now briefly review the definitions and basic properties of wave front sets, for reference in Chapter 3. We work in the context of tempered distributions; this is merely a concession to the fact that our pseudo-differential operators are defined on S', and the definitions and results can easily be extended to arbitrary distributions. Also, we shall restrict attention to the symbol classes $S^m_{1,0}$ and leave the extension to $S^m_{\rho,\delta}$ to the reader.

Throughout this discussion we shall employ the convention that for any symbols σ and τ,

$$S = \sigma(D, X), \qquad T = \tau(D, X).$$

Also, we shall be thinking of points (ξ, x) in phase space as specifying a direction ξ at the point x in position space, so we shall usually want $\xi \neq 0$. Accordingly, we set

$$\mathbf{R}^{2n}_* = \{\, (\xi, x) \in \mathbf{R}^{2n} : \xi \neq 0 \,\},$$

and the letter ξ will implicitly denote a nonzero element of $\mathbf{R}^n$.

First, some definitions concerning pseudodifferential operators. A set $U \subset \mathbf{R}^{2n}_*$ is called **conic** if $(t\xi, x) \in U$ for all $t > 0$ whenever $(\xi, x) \in U$. If $\sigma \in S^m_{1,0}$, the **characteristic variety** of the operator $S = \sigma(D, X)$, denoted char S, is the smallest closed conic set outside of which σ is elliptic (i.e., of true order m), and the **microsupport** of S, denoted $\mu \operatorname{supp} S$, is the smallest closed conic set outside of which σ and its derivatives are rapidly decreasing. In other words, char S and $\mu \operatorname{supp} S$ are the complements of the following sets:

$$\mathbf{R}^{2n}_* \backslash (\operatorname{char} S) = \{\, (\xi_0, x_0) : |\sigma(\xi, x)| \geq C|\xi|^m$$
$$\text{for } \xi \text{ large and } (\xi, x) \text{ in a conic neighborhood of } (\xi_0, x_0) \,\},$$

$$\mathbf{R}^{2n}_* \backslash (\mu \operatorname{supp} S) = \{\, (\xi_0, x_0) : |D^\alpha_\xi D^\beta_x \sigma(\xi, x)| \leq C_{\alpha\beta M} |\xi|^{-M}$$
$$\text{for } M > 0 \text{ and } (\xi, x) \text{ in a conic neighborhood of } (\xi_0, x_0) \,\}.$$

The union of char S and $\mu \operatorname{supp} S$ is alsways $\mathbf{R}^{2n}_*$, but in general they have a large overlap. Incidentally, we follow Treves [139] in using the term "microsupport"; $\mu \operatorname{supp} S$ is called the "essential support" of S by Taylor [135] and the "wave front set" of S by Hörmander [72].

Before proceeding, we set down a couple of useful lemmas.

(2.66) Lemma. *For any $\sigma \in S_{1,0}^m$ we have*

$$\text{char}\,\sigma(D,X) = \text{char}\,\sigma(D,X)_{KN}, \qquad \mu\,\text{supp}\,\sigma(D,X) = \mu\,\text{supp}\,\sigma(D,X)_{KN}.$$

Proof: This follows easily from Theorem (2.41). ∎

(2.67) Lemma. *For any $\sigma,\tau \in S_{1,0}^m$ we have*

$$\mu\,\text{supp}\,ST \subset \mu\,\text{supp}\,S \cap \mu\,\text{supp}\,T.$$

Proof: If σ and τ are rapidly decreasing on a conic neighborhood of (ξ_0, x_0) then so is $\sigma \,\sharp\, \tau$, by Theorem (2.50). ∎

(2.68) Lemma. *If U is a conic neighborhood of (ξ_0, x_0), there exists $S \in OPS_{1,0}^0$ such that $(\xi_0, x_0) \notin \text{char}\,S$ and $\mu\,\text{supp}\,S \subset U$.*

Proof: Choose a C^∞ function $\phi(\xi)$ supported in $\{\xi : (\xi, x_0) \in U, |\xi| \geq \frac{1}{2}|\xi_0|\}$ such that $\phi(\xi) = 1$ for ξ near ξ_0 and $\phi(t\xi) = \phi(\xi)$ for $t \geq 1$ and $|\xi| \geq |\xi_0|$. Since ϕ is determined by its values on the compact set $\{\xi : \frac{1}{2}|\xi_0| \leq |\xi| \leq |\xi_0|\}$, there is a neighborhood V of x_0 such that $\text{supp}\,\phi \subset \{\xi : (\xi, x) \in U\}$ for all $x \in V$. Choose $\psi \in C_c^\infty(V)$ with $\psi = 1$ near x_0; then $S = \sigma(D,X)$ will do the job if we take $\sigma(\xi, x) = \phi(\xi)\psi(x)$. ∎

If $f \in \mathcal{S}'$, the **wave front set** of f, denoted $WF(f)$, is roughly the set of all $(\xi, x) \in \mathbf{R}_*^{2n}$ such that f fails to be smooth at x in the direction ξ. There are two common definitions of $WF(f)$, one of which is the following and the other is the characterization given in Theorem (2.72) below.

$$WF(f) = \bigcap \{\,\text{char}\,S : S \in OPS_{1,0}^0 \text{ and } Sf \in C^\infty\,\}.$$

If this definition is to reflect the intuitive idea described above, it should be the case that f is smooth at x precisely when $(\xi, x) \notin WF(f)$ for all ξ. We now show that this is so.

(2.69) Theorem. *Let $\pi : \mathbf{R}_*^{2n} \to \mathbf{R}^n$ be defined by $\pi(\xi, x) = x$. Then*

$$\pi\big(WF(f)\big) = \text{sing supp}\,f.$$

Proof: If $x_0 \notin \text{sing supp}\,f$, choose $\phi \in C_c^\infty(\mathbf{R}^n)$ with $\phi = 1$ near x_0 and $\phi f \in C^\infty$ and set $\sigma(\xi, x) = \phi(x)$. Then $Sf = \phi f \in C^\infty$, $S \in OPS_{1,0}^0$, and $\text{char}\,S = \{((\xi, x) : \phi(x) = 0\}$. It follows that $(\xi, x_0) \notin WF(f)$ for all ξ.

On the other hand, if $x_0 \notin \pi\big(WF(f)\big)$, for each ξ with $|\xi| = 1$ there exists $S_\xi \in OPS_{1,0}^0$ with $S_\xi f \in C^\infty$ and $(\xi, x_0) \notin \text{char}\,S_\xi$. Since characteristic varieties are conic and the unit sphere is compact, there exist $S_1, \ldots, S_N$ such that $S_j f \in C^\infty$ and $(\xi, x_0) \notin \bigcap \text{char}\,S_j$ for all $\xi \neq 0$. Then $S = \sum_1^N S_j^* S_j$ satisfies $Sf \in C^\infty$; also, its symbol σ satisfies $\sigma = \sum |\sigma_j|^2 \mod S_{1,0}^{-1}$, hence

$\sigma(\xi, x) \geq C > 0$ for all (ξ, x) with, say, $|x - x_0| < 4\epsilon$. Thus there exists an elliptic symbol τ such that $\tau(\xi, x) = \sigma(\xi, x)$ for $|x - x_0| < 2\epsilon$. I claim that $Tf = \tau(D, X)f$ is C^∞ near x_0, whence it follows from Corollary (2.65) that $x_0 \notin \text{sing supp } f$.

It suffices to prove that $(T - S)f$ is C^∞ near x_0. Choose $\phi \in C_c^\infty$ with $\phi = 1$ near x_0 and $\phi(x) = 0$ for $|x - x_0| \geq \epsilon$. Then

$$(T - S)f = (T - S)(\phi f) + (T - S)([1 - \phi]f).$$

But $(T - S)([1 - \phi]f)$ is C^∞ near $x - 0$ by Theorem (2.57), and $(T - S)(\phi f)$ is C^∞ by Corollary (2.54) because the symbols of $T - S$ and $\phi(X)$ have disjoint supports and hence $(T - S)\phi(X) \in OPS_{1,0}^{-\infty}$ by Lemma (2.67). ∎

In view of this result, the next theorem and its corollary are refinements of Theorem (2.57) and Corollary (2.65).

(2.70) Theorem. *If $S \in OPS_{1,0}^m$ and $f \in \mathcal{S}'$ then*

$$WF(Sf) \subset WF(f) \cap \mu \text{ supp } S.$$

Proof: If $(\xi_0, x_0) \notin \mu \text{ supp } S$, by Lemma (2.68) there exists $T \in OPS_{1,0}^0$ such that

$$(\xi_0, x_0) \notin \text{char } T, \qquad \mu \text{ supp } T \cap \mu \text{ supp } S = \emptyset.$$

Then $TS \in OPS_{1,0}^{-\infty}$ by Lemma (2.67), so $TSf \in C^\infty$ and hence $(\xi_0, x_0) \notin WF(Sf)$.

If $(\xi_0, x_0) \notin WF(f)$, there exists $T \in OPS_{1,0}^0$ with $Tf \in C^\infty$ and $(\xi_0, x_0) \notin \text{char } T$. I claim there exist $P, Q \in OPS_{1,0}^0$ with $(\xi_0, x_0) \notin \text{char } P$ and $PS = QT$ mod $OPS_{1,0}^{-\infty}$. Granted this, we have $PSf - QTf \in C^\infty$, hence $PSf \in C^\infty$ since Q is pseudolocal, and so $(\xi_0, x_0) \notin WF(Sf)$.

To prove the claim, choose an elliptic symbol τ_0 that agrees with τ, the symbol of T, on a conic neighborhood U of (ξ_0, x_0). By Lemma (2.68), there exists $P \in OPS_{1,0}^0$ with $(\xi_0, x_0) \notin \text{char } P$ and $\mu \text{ supp } P \subset U$, and by Theorem (2.64) there exists $Q_0 \in OPS_{1,0}^0$ with $I - Q_0 T_0 \in OPS_{1,0}^{-\infty}$. Let $Q = PSQ_0$. Then

$$QT = PSQ_0 T_0 + PSQ_0(T - T_0).$$

The first term on the right equals PS mod $OPS_{1,0}^{-\infty}$, and the second term is in $OPS_{1,0}^{-\infty}$ by Lemma (2.67) since P and $T - T_0$ have disjoint microsupports. ∎

(2.71) Corollary. *Suppose $S \in OPS_{1,0}^m$ is elliptic. Then $WF(Sf) = WF(f)$ for all $f \in \mathcal{S}'$. Moreover, let us say that $f \in H_s$ at (ξ_0, x_0) if $f = g + h$ where $g \in H_s$ and $(\xi_0, x_0) \notin WF(h)$. Then $f \in H_{s+m}$ at (ξ_0, x_0) whenever $Sf \in H_s$ at (ξ_0, x_0).*

Proof: Let T be an inverse of S mod $OPS_{1,0}^{-\infty}$ as in Theorem (2.64). Then by Theorem (2.70),

$$WF(f) = WF(TSf) \subset WF(Sf) \subset WF(f).$$

Moreover, if $Sf = g + h$ with $g \in H_s$ and $(\xi_0, x_0) \notin WF(h)$, then

$$f = TSf + (I - TS)f = Tg + Th + (I - TSf).$$

We have $Tg \in H_{s+m}$ by Corollary (2.65), $(\xi_0, x_0) \notin WF(Th)$ by Theorem (2.70), and $(I - TS)f \in C^\infty$ by Corollary (2.54). ∎

Finally, we come to the characterization of wave front sets in terms of Fourier transforms.

(2.72) Theorem. $(\xi_0, x_0) \notin WF(f)$ if and only if there exists $\phi \in C_c^\infty$ such that $\phi(x_0) = 1$ and $\widehat{\phi f}$ agrees with an element of S on some conic neighborhood U of (ξ_0, x_0).

Proof: If such ϕ, U exist, choose $\sigma \in C^\infty(\mathbf{R}^n)$ such that $\sigma(\xi_0) = 1$, $\sigma(t\xi) = \sigma(\xi)$ for $t \geq 1$ and $|\xi| \geq |\xi_0|$, and $\mathrm{supp}\,\sigma \subset U$. Then $\sigma\widehat{\phi f} \in S$, so $\sigma(D)\phi(X)f \in S$. But $S = \sigma(D)\phi(X) \in OPS^0_{1,0}$ and $(\xi_0, x_0) \notin \mathrm{char}\,S$, so $(\xi_0, x_0) \notin WF(f)$.

Conversely, suppose $(\xi_0, x_0) \notin WF(f)$. Then there is a neighborhood V of x_0 with $(\xi_0, x) \notin WF(f)$ for all $x \in V$. Choose $\phi \in C_c^\infty(V)$ with $\phi(x_0) = 1$; then by Theorem (2.70) applied to $\phi(X)$, $(\xi_0, x) \notin WF(\phi f)$ for all x. Now, the projection of $WF(\phi f) \cap \{(\xi, x) : |\xi| = 1\}$ on ξ-space is compact and does not contain $\xi_0/|\xi_0|$, so there exists $\sigma = \sigma(\xi) \in S^0_{1,0}$ with $\sigma = 1$ on a conic neighborhood U of ξ_0 (except near the origin) and $\sigma = 0$ on a conic neighborhood W of the set of ξ such that $(\xi, x) \in WF(\phi f)$ for some x. Then $\mu\,\mathrm{supp}\,\sigma(D)$ does not intersect $WF(\phi f)$, so $WF(\sigma(D)\phi f) = \emptyset$ by Theorem (2.70) and hence $\sigma(D)\phi f \in C^\infty$ by Theorem (2.69). Moreover, $\sigma(D)\phi f = (\mathcal{F}^{-1}\sigma) * \phi f$, ϕf is compactly supported, and $\mathcal{F}^{-1}\sigma$ agrees with an element of S away from the origin by Theorem (2.53). It follows that $\sigma(D)\phi f \in S$, so $\sigma\widehat{\phi f} \in S$ and hence $\widehat{\phi f}$ agrees with an element of S on U. ∎

5. The Calderón-Vaillancourt Theorems

In this section we study the extent to which the Weyl correspondence preserves boundedness. The boundedness of σ on $\mathbf{R}^{2n}$ is not necessary for the boundedness of $\sigma(D, X)$ on $L^2(\mathbf{R}^n)$, for $\sigma(D, X)$ is bounded (actually, Hilbert-Schmidt) whenever $\sigma \in L^2(\mathbf{R}^{2n})$. Nor is it sufficient, as is shown by the example $\sigma(\xi, x) = e^{4\pi i x\xi}$, for which $\sigma(D, X)f = (\int f)\delta$. However, if we assume some boundedness conditions not only on σ but also on some of its derivatives, we get a positive result. Indeed, in the previous section we proved that operators in $S^0_{\rho,\delta}$ are bounded on L^2 when $\rho > \delta$, and a slightly more careful formulation of the proof would show that one only needs the estimates

$$|D^\alpha_\xi D^\beta_x \sigma(\xi, x)| \leq C_{\alpha\beta}\langle\xi\rangle^{-\rho|\alpha| + \delta|\beta|}$$

to hold for finitely many α and β. We now show that this result remains true for the borderline case $\rho = \delta$, as was originally established for the Kohn-Nirenberg calculus by Calderón and Vaillancourt [27], [28]. We first consider the case $\rho = 0$, following the argument in Howe [74].

(2.73) Theorem. *Suppose σ is of class C^{2n+1} on $\mathbf{R}^{2n}$ and*

$$(2.74) \qquad \|\sigma\|_* \equiv \sum_{|\alpha|+|\beta|\leq 2n+1} \|D_\xi^\alpha D_x^\beta \sigma\|_\infty < \infty.$$

Then $\sigma(D,X)$ is bounded on $L^2(\mathbf{R}^n)$.

The heart of the proof is the following lemma.

(2.75) Lemma. *Let*

$$Q = [-1,1]^{2n} \quad \text{and} \quad \mathcal{A} = \{\, \sigma \in \mathcal{S}(\mathbf{R}^{2n}) : \widehat{\sigma} \in C_c^\infty(Q) \,\}.$$

There exists $C > 0$ such that

$$(2.76) \qquad \|\sigma(D,X)\| \leq C\|\sigma\|_\infty \text{ for all } \sigma \in \mathcal{A},$$

where $\|\sigma(D,X)\|$ denotes the operator norm on L^2.

We shall give two proofs of this lemma. The first one, due to Howe [74], is self-contained modulo the results of Sections 1.3 and 1.4 but is rather magical. The second one, which is new but related to the ideas in Howe [76], is perhaps more transparent but relies on Cotlar's lemma.

First proof: If $\phi, f \in L^2(\mathbf{R}^n)$, define the operator $T_{\phi,f}$ by $T_{\phi,f}g = \langle g, \phi \rangle f$. We recall from Proposition (1.46) that $T_{\phi,f} = \rho(\overline{V(\phi,f)})$. Hence

$$\rho\big(\overline{V(\rho(p,q)\phi,\ \rho(p,q)f)}\big)\phi = T_{\rho(p,q)\phi,\rho(p,q)f}\phi$$
$$= \langle \phi,\ \rho(p,q)\phi \rangle \rho(p,q)f = \overline{V(\phi,\phi)(p,q)}\rho(p,q)f.$$

But from Proposition (1.44) we also have

$$V\big(\rho(p,q)\phi,\ \rho(p,q)f\big)(r,s) = e^{2\pi i(rq-sp)}V(\phi,f)(r,s).$$

Let us now take

$$\phi(x) = 2^{n/4}e^{-\pi x^2}, \qquad V(\phi,\phi)(p,q) = e^{-(\pi/2)(p^2+q^2)},$$

and set

$$F_{pq}(r,s) = e^{(\pi/2)(p^2+q^2)}e^{2\pi i(rq-sp)}V(\phi,f)(r,s).$$

Then the above equations yield

$$\rho(p,q)f = \rho(\overline{F}_{pq})\phi.$$

Now suppose $\hat{\sigma} \in C_c^\infty(Q)$. Multiply both sides of this last equation by $\hat{\sigma}$ and integrate:

$$\sigma(D,X)f = \rho(\hat{\sigma})f = \rho\left(\iint \hat{\sigma}(p,q)\overline{F}_{pq}\,dp\,dq\right)\phi.$$

Pick $G \in C_c^\infty(\mathbf{R}^{2n})$ such that $\hat{G}$ is real and $\hat{G}(p,q) = e^{(\pi/2)(p^2+q^2)}$ on Q. Then

$$\iint \hat{\sigma}(p,q)\overline{F}_{pq}(r,s)\,dp\,dq = V(\phi,f)(r,s)\iint \hat{\sigma}(p,q)\hat{G}(p,q)e^{2\pi i(ps-qr)}\,dp\,dq$$

$$= V(\phi,f)(r,s)\mathcal{F}^{-1}(\widehat{\sigma G})(s,-r) = V(\phi,f)(r,s)(\sigma * G)(s,-r).$$

Since $\|V(\phi,f)\|_2 = \|\phi\|_2\|f\|_2$ (Proposition (1.42)), $\|\rho(H)\| \le \|H\|_2$ (Theorem (1.30)), and $\|\phi\|_2 = 1$, the last two equations yield

$$\|\sigma(D,X)f\|_2 \le \|\iint\hat{\sigma}(p,q)\overline{F}_{pq}\,dp\,dq\|_2 \le \|\sigma * G\|_\infty\|V(\phi,f)\|_2$$

$$= \|\sigma * G\|_\infty\|f\|_2 \le \|G\|_1\|\sigma\|_\infty\|f\|_2.$$

So (2.76) is true with $C = \|G\|_1$. ∎

Second proof: We use the same circle of ideas as above, but arranged differently. Again let

$$\phi(x) = 2^{n/4}e^{-\pi x^2}, \qquad \Phi(p,q) = V(\phi,\phi)(p,q) = e^{-(\pi/2)(p^2+q^2)},$$

and, given $F \in \mathcal{S}(\mathbf{R}^{2n})$, define the operator T_{pq} on $L^2(\mathbf{R}^n)$ by

$$T_{pq} = F(p,q)\rho(-p,-q)\rho(\Phi)\rho(p,q).$$

Since $\rho(\Phi)$ is the orthogonal projection onto ϕ by Proposition (1.46), and $\rho(p,q)$ is unitary, we have $\|T_{pq}\| \le \|F\|_\infty$. Also, T_{pq}^* is the same as T_{pq} except that F is replaced by $\overline{F}$, so by Proposition (1.48c),

$$T_{pq}T_{rs}^* = F(p,q)\overline{F}(r,s)e^{\pi i(qr-ps)}\rho(-p,-q)\rho(\Phi)\rho(p-r,q-s)\rho(\Phi)\rho(r,s)$$

$$= F(p,q)\overline{F}(r,s)e^{\pi i(qr-ps)}e^{-(\pi/2)[(p-r)^2+(q-s)^2]}\rho(-p,-q)\rho(\Phi)\rho(r,s),$$

and hence

$$\|T_{pq}T_{rs}^*\| \le \|F\|_\infty^2 e^{-(\pi/2)[(p-r)^2+(q-s)^2]}.$$

Therefore,

$$\iint \|T_{pq}T_{rs}^*\|^{1/2}\,dr\,ds \le \|F\|_\infty \iint e^{-(\pi/4)(r^2+s^2)}\,dr\,ds = 4^n\|F\|_\infty,$$

and likewise for $T_{pq}^* T_{rs}$. It now follows from Cotlar's lemma (cf. Appendix B) that the operator

$$T = \iint T_{pq}\,dp\,dq = \iint F(p,q)\rho(-p,-q)\rho(\Phi)\rho(p,q)\,dp\,dq$$

is bounded on L^2 with norm $\leq 4^n\|F\|_\infty$. But by Proposition (1.31),

$$\rho(-p,-q)\rho(\Phi)\rho(p,q) = \rho(\Psi) \quad \text{where} \quad \Psi(p',q') = e^{2\pi i(p'q-q'p)}\Phi(p',q'),$$

so that

$$T = \rho(\widetilde{F}\Phi) \quad \text{where} \quad \widetilde{F}(p',q') = \widehat{F}(q',-p') = \iint e^{2\pi i(p'q-q'p)}F(p,q)\,dp\,dq.$$

Now, given $\sigma \in \mathcal{S}(\mathbf{R}^{2n})$ with $\widehat{\sigma} \in C_c^\infty(Q)$, choose $G \in \mathcal{S}$ with $\widehat{G}(p,q) = e^{(\pi/2)(p^2+q^2)}$ on Q and define F by $F(\xi,x) = (\sigma * G)(-x,\xi)$. Then

$$T = \rho(\widetilde{F}\Phi) = \rho(\widehat{\sigma}\widehat{G}\Phi) = \rho(\widehat{\sigma}) = \sigma(D,X),$$

so

$$\|\sigma(D,X)\| = \|T\| \leq 4^n\|\sigma * G\|_\infty \leq 4^n\|G\|_1\|\sigma\|_\infty. \quad \blacksquare$$

(2.77) Lemma. The estimate (2.76) holds for all $\sigma \in \mathcal{S}$ such that $\widehat{\sigma}$ is supported in some translate of Q.

    Proof: Suppose $\operatorname{supp}\widehat{\sigma} \subset Q+(p_0,q_0)$. Let

$$\tau(\xi,x) = e^{-2\pi i(p_0\xi+q_0 x)}\sigma(\xi,x),$$

so that $\widehat{\tau}(p,q) = \widehat{\sigma}(p+p_0, q+q_0)$. Then $\operatorname{supp}\widehat{\tau} \subset Q$, and

$$\sigma(D,X) = \rho(\widehat{\sigma}) = \iint \tau(p,q)\rho(p+p_0, q+q_0)\,dp\,dq$$

$$= \iint \widehat{\tau}(p,q)\rho(\tfrac{1}{2}p_0, \tfrac{1}{2}q_0)\rho(p,q)\rho(\tfrac{1}{2}p_0, \tfrac{1}{2}q_0)\,dp\,dq$$

$$= \rho(\tfrac{1}{2}p_0, \tfrac{1}{2}q_0)\tau(D,X)\rho(\tfrac{1}{2}p_0, \tfrac{1}{2}q_0).$$

Hence, by Lemma (2.75),

$$\|\sigma(D,X)\| = \|\tau(D,X)\| \leq C\|\tau\|_\infty = C\|\sigma\|_\infty. \quad \blacksquare$$

(2.78) Lemma. There exists $C > 0$ such that for all $\sigma \in \mathcal{S}$,

$$\|\sigma(D,X)\| \leq C\|\sigma\|_*,$$

where $\|\sigma\|_*$ is defined in (2.74).

Proof: For any function F on $\mathbf{R}^{2n}$, let

$$F_{ab}(p,q) = F(p-a, q-b).$$

Let χ be the characteristic function of the cube $Q' = \left[-\frac{1}{2}, \frac{1}{2}\right]^{2n}$, pick a nonnegative $\psi \in C_c^\infty(Q')$ with $\int \psi = 1$, and set $\Phi = \chi * \psi$. Then $\Phi \in C_c^\infty(Q)$, and since $\sum_{(a,b)\in\mathbf{Z}^{2n}} \chi_{ab} \equiv 1$, $\{\Phi_{ab} : (a,b) \in \mathbf{Z}^{2n}\}$ is a partition of unity. Hence for any $\sigma \in \mathcal{S}$ we have

$$\sigma = \sum_{(a,b)\in\mathbf{Z}^{2n}} \sigma_{(a,b)} \quad \text{where} \quad \sigma_{(\alpha,b)} = \mathcal{F}^{-1}(\hat{\sigma}\Phi_{ab}) = \sigma * \mathcal{F}^{-1}\Phi_{ab}.$$

Now,

$$\sigma_{(a,b)}(\xi, x) = \iint \sigma(\xi - \eta,\, x - y)\check{\Phi}(\eta, y)e^{2\pi i(a\eta + by)}\, d\eta\, dy$$

where $\check{\Phi} = \mathcal{F}^{-1}\Phi$, so by integration by parts,

$$a^\alpha b^\beta \sigma_{(a,b)}(\xi, x) = (-1)^{|\alpha|+|\beta|} \iint D_\eta^\alpha D_y^\beta[\sigma(\xi - \eta,\, x - y)\check{\Phi}(\eta, y)]e^{2\pi i(a\eta + by)}\, d\eta\, dy$$

and hence

$$|a^\alpha b^\beta|\, \|\sigma_{(a,b)}\|_\infty \leq C_{\alpha\beta} \sum_{|\gamma|\leq|\alpha|,\, |\delta|\leq|\beta|} \|D_\xi^\gamma D_x^\delta \sigma\|_\infty$$

where $C_{\alpha\beta}$ is related to the L^1 norms of $\check{\Phi}$ and its derivatives. Therefore,

$$\|\sigma_{(a,b)}\|_\infty \leq C\big(1 + |a| + |b|\big)^{-(2n+1)} \sum_{|\gamma|+|\delta|\leq 2n+1} \|D_\xi^\gamma D_x^\delta \sigma\|_\infty$$

$$\leq C\big(1 + |a| + |b|\big)^{-(2n+1)} \|\sigma\|_*,$$

so by Lemma (2.77),

$$\|\sigma(D, X)\| \leq \sum_{\mathbf{Z}^{2n}} \|\sigma_{(a,b)}(D, X)\| \leq C \sum_{\mathbf{Z}^{2n}} \|\sigma_{(a,b)}\|_\infty$$

$$\leq C \sum_{\mathbf{Z}^{2n}} (1 + |a| + |b|)^{-(2n+1)} \|\sigma\|_*.$$

(The number $2n + 1$ is of course the smallest integer that makes this sum converge.) ∎

Proof of Theorem (2.73): Clearly the conclusion of Lemma (2.78) remains valid for all σ in the closure of S under $\| \ \|_*$, which includes all compactly supported σ such that $\|\sigma\|_* < \infty$. Pick $\Psi \in C_c^\infty(\mathbf{R}^{2n})$ with $\Psi = 1$ near the origin, and given a symbol σ with $\|\sigma\|_* < \infty$, set $\sigma_\epsilon(\xi,x) = \sigma(\xi,x)\Psi(\epsilon\xi,\epsilon x)$. Then $\|\sigma_\epsilon\|_*$ is easily seen to be bounded independently of ϵ, and hence so is $\|\sigma_\epsilon(D,X)\|$ by Lemma (2.78). If $f \in L^2(\mathbf{R}^n)$, then, $\|\sigma_\epsilon(D,X)f\|_2$ remains bounded as $\epsilon \to 0$. On the other hand, $\sigma_\epsilon \to \sigma$ in $S'(\mathbf{R}^{2n})$, so $\sigma_\epsilon(D,X)f \to \sigma(D,X)f$ in $S'(\mathbf{R}^n)$. It follows that $\sigma_\epsilon(D,X)f \to \sigma(D,X)f$ weakly in L^2, so by the uniform boundedness principle,

$$\|\sigma(D,X)\| \le \limsup \|\sigma_\epsilon(D,X)\| \le C\|\sigma\|_*. \quad \blacksquare$$

We now turn to the case $\rho > 0$, for which we shall follow the argument of Beals [16]. We shall need a particular partition of unity on $[0,\infty)$ that will also be used later.

(2.79) Lemma. *There is a sequence $\{\phi_j\}_{-1}^\infty \subset C_c^\infty([0,\infty))$ such that $\phi_j \ge 0$ for all j, $\sum_{-1}^\infty \phi_j = 1$ on $[0,\infty)$, ϕ_{-1} is supported in $[0,2]$, ϕ_j is supported in $[2^{j-1}, 2^{j+2}]$ for $j \ge 0$, and $\phi_j(t) = \phi_0(2^{-j}t)$ for $j \ge 0$.*

Proof: Pick a nonnegative $\psi \in C_c^\infty$ supported in $[\frac{1}{2}, 2]$ and satisfying $\int \psi = 1$. For $j \ge 0$, let χ_j be the characteristic function of $[2^j, 2^{j+1})$ and let ϕ_j be the convolution of χ_j with ψ on the multiplicative group $(0,\infty)$:

$$\phi_j(t) = \int \chi_j(ts^{-1})\psi(s)s^{-1}\,ds.$$

Then $\phi_j(t) = \phi_0(2^{-j}t)$ since the same is true of χ_j; $\operatorname{supp}\phi_j \subset [2^{j-1}, 2^{j+2}]$, and $\sum_0^\infty \phi_j = 1$ on $[2,\infty)$ since $\sum_0^\infty \chi_j = 1$ on $[1,\infty)$. Setting $\phi_{-1} = 1 - \sum_0^\infty \phi_j$, we are done. $\blacksquare$

(2.80) Theorem. *Suppose $0 < \rho < 1$, and let*

$$I = \left\{ (\alpha,\beta) : |\alpha|+|\beta| \le 2n+1, \text{ or } |\alpha| \le 2\left(\left[\frac{n}{2}\right]+1\right) \text{ and } |\beta| \le 2\left(\left[\frac{n}{1-\rho}\right]+1\right) \right\},$$

where $[t]$ denotes the greatest integer in t. If σ is a function on $\mathbf{R}^{2n}$ such that $D_\xi^\alpha D_x^\beta \sigma$ is continuous for $(\alpha,\beta) \in I$ and satisfies

$$(2.81) \qquad |D_\xi^\alpha D_x^\beta \sigma(\xi,x)| \le A_{\alpha\beta}\langle\xi\rangle^{\rho(|\beta|-|\alpha|)}, \qquad (\alpha,\beta) \in I,$$

then $\sigma(D,X)$ is bounded on $L^2(\mathbf{R}^n)$, and

$$\|\sigma(D,X)\| \le C \sum_{(\alpha,\beta)\in I} A_{\alpha\beta}.$$

Proof: As in the proof of Theorem (2.73), it suffices to establish the estimate on $\|\sigma(D, X)\|$ for $\sigma \in S$. Given $\sigma \in S$, let $\{\phi_j\}_1^\infty$ be as in Lemma (2.79), and set

$$\sigma_j(\xi, x) = \sigma(\xi, x)\phi_j(\langle\xi\rangle^\rho).$$

It is easily verified that $|D^k\phi_j(t)| \leq C_k 2^{-jk}$ and hence that σ_j satisfies estimates of the type (2.81) uniformly in j. Since σ_j is supported in the set where $2^{(j-1)/\rho} \leq \langle\xi\rangle \leq 2^{(j+2)/\rho}$, these estimates imply that

$$(2.82) \qquad |D_\xi^\alpha D_x^\beta \sigma_j(\xi, x)| \leq C A_{\alpha\beta} 2^{j(|\alpha|-|\beta|)}, \qquad (\alpha, \beta) \in I,$$

which in turn implies that, if we set

$$\sigma_j'(\xi, x) = \sigma_j(2^j\xi, 2^{-j}x),$$

we have

$$|D_\xi^\alpha D_x^\beta \sigma_j'(\xi, x)| \leq C A_{\alpha,\beta}, \qquad (\alpha, \beta) \in I.$$

Thus, by Theorem (2.73), $\sigma_j'(D, X)$ is bounded on L^2 uniformly in j, with bound depending only on the constants $A_{\alpha\beta}$ with $|\alpha| + |\beta| \leq 2n + 1$. But a simple calculation shows that $\sigma_j'(D, X) = V_j^{-1}\sigma_j(D, X)V_j$ where V_j is the unitary operator on L^2 defined by $V_j f(x) = 2^{nj/2} f(2^j x)$. Conclusion: the operators

$$S_j = \sigma_j(D, X)$$

are bounded on L^2 uniformly in j, and $\|S_j\| \leq C \sum_I A_{\alpha\beta}$.

We now wish to apply Cotlar's lemma (cf. Appendix B) to conclude that $\sigma(D, X) = \sum S_j$ is bounded on L^2. To do so, we must estimate $\|S_j^* S_k\|$ and $\|S_j S_k^*\|$. It suffices to consider $S_j^* S_k$ (since replacing S_j by S_j^* merely means replacing σ by $\overline{\sigma}$), to assume that $k \geq j$ (since $S_k^* S_j = (S_j^* S_k)^*$), and in fact to assume that $k \geq j + 4$ (since we already have the estimates $\|S_j^* S_k\| \leq (C \sum_I A_{\alpha\beta})^2$). In this case, we have

$$S_j^* S_k f(x) = \int K(x, y) f(y)\, dy,$$

where

$$K(x, y) = \iiint \overline{\sigma}_j\left(\xi, \tfrac{1}{2}(x + z)\right)\sigma_k\left(\eta, \tfrac{1}{2}(z + y)\right) e^{2\pi i(x\xi - z\xi + z\eta - y\eta)}\, dz\, d\eta\, d\xi.$$

The reader who has followed the arguments of Sections 2.2 and 2.3 will hardly need to be told the next step: we set $L = \sum_1^n D_j^2$ and integrate by parts to obtain

$$K(x, y) = \iiint \frac{e^{2\pi i(x\xi - z\xi + z\eta - y\eta)}}{(x - z)^{2M}(y - z)^{2M}} \times$$

$$(1 + L_\xi)^M(1 + L_\eta)^M \left\{ \frac{L_z^N\left[\overline{\sigma}_j\left(\xi, \tfrac{1}{2}(x + z)\right)\sigma_k\left(\eta, \tfrac{1}{2}(z + y)\right)\right]}{|\xi - \eta|^{2N}} \right\} dz\, d\eta\, d\xi.$$

By construction of σ_j, the integrand vanishes unless

$$2^{(j-1)/\rho} \le \langle \xi \rangle \le 2^{(j+2)/\rho} \quad \text{and} \quad 2^{(k-1)/\rho} \le \langle \eta \rangle \le 2^{(k+2)/\rho},$$

and since $k \ge j + 4$ this cannot happen unless

$$(2.83) \qquad |\xi - \eta| \ge 2^{(k-1)/\rho} - 2^{(j+2)/\rho} \ge 2^{(k-1)/\rho} - 2^{(k-2)/\rho} = C2^{k/\rho}.$$

In particular, $|\xi - \eta| \ge C$, and where this condition is satisfied the derivatives of $|\xi - \eta|^{-2N}$ are bounded by $|\xi - \eta|^{-2N}$ itself. Therefore, from the estimates (2.82) we obtain

$$|K(x,y)| \le \iiint_R C_{MN} 2^{kN} \langle x - z \rangle^{-2M} \langle y - z \rangle^{-2M} |\xi - \eta|^{-2N} dz\, d\eta\, d\xi,$$

where

$$R = \left\{ (z, \eta, \xi) : z \in \mathbf{R}^n,\ 2^{j-1} \le \langle \xi \rangle^\rho \le 2^{j+2},\ 2^{k-1} \le \langle \eta \rangle^\rho \le 2^{k+2} \right\}$$

and C_{MN} depends only on the constants $A_{\alpha\beta}$ in (2.81) for $|\alpha| \le 2M$ and $|\beta| \le 2N$. To estimate the z-integral we note that for all x, y, z we have either $|x - z| \ge \frac{1}{2}|x - y|$ or $|y - z| \ge \frac{1}{2}|x - y|$, so

$$\int \frac{dz}{\langle x-z \rangle^{2M} \langle y-z \rangle^{2M}} \le C \left[\int \frac{dz}{\langle x-y \rangle^{2M} \langle y-z \rangle^{2M}} + \int \frac{dz}{\langle x-z \rangle^{2M} \langle x-y \rangle^{2M}} \right]$$

$$\le C \langle x - y \rangle^{-2M}$$

provided $M > \frac{1}{2}n$. To estimate the $\xi\eta$-integral we use (2.83):

$$\int_{2^{j-1} \le \langle \xi \rangle^\rho \le 2^{j+2}} \int_{2^{k-1} \le \langle \eta \rangle^\rho \le 2^{k+2}} \frac{d\xi\, d\eta}{|\xi - \eta|^{2N}} \le C \frac{2^{jn/\rho} 2^{kn/\rho}}{2^{2Nk/\rho}} \le C 2^{2k(n-N)/\rho}.$$

Therefore,

$$|K(x,y)| \le C 2^{2k[N(\rho-1)+n]/\rho} \langle x - y \rangle^{-2M},$$

where C depends only on the constants $A_{\alpha\beta}$ with $|\alpha| \le 2M$ and $|\beta| \le 2N$. Assuming again that $M > \frac{1}{2}n$, Young's inequality ([50], Theorem 6.18) shows that

$$\|S_j^* S_k\| \le C \left[\int \langle x \rangle^{-2M} dx \right] 2^{2k[N(\rho-1)+n]/\rho},$$

and the exponent on the right will be negative provided that $N > n/(1 - \rho)$. The desired result therefore follows from Cotlar's lemma. ∎

This argument can be generalized to yield L^2 boundedness of $\sigma(D, X)$ for much wider classes of symbols σ; see Beals [16]. Other general theorems of this nature can be found in Hörmander [70], [72] and Unterberger [140].

Our estimate on the number of derivatives of σ which must be controlled in order to ensure boundedness of $\sigma(D, X)$ can probably be improved a bit. Cordes [32] has given a different proof of Theorem (2.73) that requires only the boundedness of $D_\xi^\alpha D_x^\beta \sigma$ for $|\alpha|, |\beta| \le 2([n/2]+1)$. This condition on α and β is almost weaker than ours but not quite comparable. (If it is satisfied, $|\alpha|+|\beta|$ can be as big as $2n+4$ when n is even.) The sharpest results for the Kohn-Nirenberg calculus are due to Coifman and Meyer [30]. They give several criteria for the boundedness of $\sigma(D, X)_{KN}$ involving minimal assumptions on the derivatives of σ, of which the one most directly related to Theorems (2.73) and (2.80) is the following: $\sigma(D, X)_{KN}$ is bounded on L^2 provided that $D_\xi^\alpha D_x^\beta \sigma$ is continuous and dominated by $\langle \xi \rangle^{\rho(|\beta|-|\alpha|)}$ for some $\rho \in [0, 1)$ and all α, β of order at most $[n/2] + 1$. This theorem does not immediately imply the corresponding result for $\sigma(D, X)$, however, because in passing from the Kohn-Nirenberg calculus to the Weyl calculus one needs to estimate some additional derivatives—roughly n of them, according to the proof of Theorem (2.37).

What about L^p boundedness for $p \ne 2$? It is classical that the operators in $OPS_{1,0}^0$ are bounded on L^p for $1 < p < \infty$, and this remains true for $OPS_{1,\delta}^0$ for $\delta < 1$; proofs can be found in Taylor [135]. On the other hand, operators in $OPS_{\rho,\delta}^0$ are generally not bounded on L^p for $p \ne 2$ when $\rho < 1$; see Beals [17]. Other conditions on symbols that guarantee L^p boundedness for $1 < p < \infty$ can be found in Nagel–Stein [111] and Beals [18]. (Even the simplest pseudodifferential operators tend to be unbounded on L^1 and L^∞.)

6. The Sharp Gårding Inequality

In this section we address the question: if σ is a nonnegative function on $\mathbf{R}^{2n}$, is $\sigma(D, X)$ a positive operator? (We say that an operator $T : S \to S'$ is **positive** if $\langle Tf, f \rangle \ge 0$ for all $f \in S$. If $T \in OPS_{\rho,\delta}^m$ then T is bounded from $H_{m/2}$ to $H_{-m/2} \cong H_{m/2}^*$; hence $\langle Tf, f \rangle$ makes sense for all $f \in H_{m/2}$, and if $\langle Tf, f \rangle \ge 0$ for $f \in S$ then the same is true for all $f \in H_{m/2}$.)

The literal answer to this question is no. As with boundedness, positivity of σ is neither necessary nor sufficient for positivity of $\sigma(D, X)$, as the following two examples show.

Example 1. Let $\sigma(\xi, x) = 2\pi(\xi^2 + x^2) - n$. By (1.83b), the Hermite functions h_α form an orthonormal eigenbasis for $\sigma(D, X)$, with eigenvalues $2|\alpha|$. Thus $\sigma(D, X)$ is positive although σ is not.

Example 2. Let $n = 1$, and let

$$\sigma(\xi, x) = \xi x, \qquad \tau(\xi, x) = \xi^2 x^2, \qquad S = \sigma(D, X), \qquad T = \tau(D, X).$$

It is an easy exercise to show that

$$S = \frac{1}{2\pi i}\left(x\frac{d}{dx} + \frac{1}{2}\right), \qquad T = S^2 - \frac{1}{16\pi^2}.$$

I claim that for any $\epsilon > 0$ there exists $f \in \mathcal{S}$ with $\|Sf\|_2 \leq \epsilon\|f\|_2$; if we take $\epsilon < (4\pi)^{-1}$, we then have

$$\langle Tf, f\rangle = \|Sf\|_2^2 - (16\pi^2)^{-1}\|f\|_2^2 < 0,$$

despite the fact that $\tau \geq 0$. The idea behind the claim is that if $f(x) = |x|^{-1/2}$ then $Sf = 0$, and we have merely to modify f a bit to put it into $\mathcal{S}$. Indeed, pick $\phi \in C^\infty(\mathbf{R})$ with $\phi(x) = 1$ for $x \in [\frac{1}{2}, 2]$ and $\phi(x) = 0$ for $x \notin [\frac{1}{4}, 4]$, and let $f_\epsilon(x) = x^{-1/2}\phi(x^\epsilon)$. Then

$$\|f_\epsilon\|_2^2 \geq \int_{2^{-1/\epsilon}}^{2^{1/\epsilon}} \frac{dx}{x} = \frac{2}{\epsilon}\log 2.$$

On the other hand,

$$Sf_\epsilon = \frac{\epsilon}{2\pi i}x^{\epsilon-(1/2)}\phi'(x^\epsilon),$$

and $\phi'(x^\epsilon)$ vanishes except for $\frac{1}{4} \leq x^\epsilon \leq \frac{1}{2}$ and $2 \leq x^\epsilon \leq 4$. Hence

$$\|Sf_\epsilon\|_2^2 \leq \frac{C\epsilon^2}{4\pi^2}\left(\int_{4^{-1/\epsilon}}^{2^{-1/\epsilon}} + \int_{2^{1/\epsilon}}^{4^{1/\epsilon}}\right)x^{2\epsilon-1}dx = C'\epsilon,$$

so that $\|Sf\|_2 \leq C''\epsilon\|f\|_2$.

Thus the Weyl correspondence does not completely preserve positivity. A more reasonable expectation is that if σ is in a nice symbol class and $\sigma \geq 0$ then $\sigma(D, X) \geq 0$ modulo a lower-order error. (This is the case in Example 2 above: we have $T \in OPS_{1,0}^2$ and $T + (16\pi^2)^{-1} = S^2 \geq 0$.) An assertion of this sort can be very easily proved if we strengthen the hypothesis somewhat, as follows. Suppose $\sigma \in S_{\rho,\delta}^m$ with $\rho > \delta$ and $\sigma(\xi, x) \geq \epsilon\langle\xi\rangle^m$ for some $\epsilon > 0$. It is then easy to check that $\sqrt{\sigma} \in S_{\rho,\delta}^{m/2}$, and $\sqrt{\sigma}(D, X)$ is Hermitian since $\sqrt{\sigma}$ is real. Thus $\sigma(D, X)$ equals the positive operator $[\sqrt{\sigma}(D, X)]^2$ modulo an error in $OPS_{\rho,\delta}^{m-(\rho-\delta)}$.

This argument breaks down in the limiting case $\epsilon > 0$ because the square root function is not differentiable at the origin. Nonetheless, as Hörmander [69] showed, the conclusion is still true; and that is the main result of this section. Our proof is a distillation of an argument in Taylor [136]; it is also related to the proofs in Córdoba–Fefferman [33] and Hörmander [72], as we shall explain in Section 3.1.

(2.84) Theorem. *Suppose $\sigma \in S^m_{\rho,\delta}$ with $\rho > \delta$, and $\sigma \geq 0$. Then $\sigma = \tau + \omega$ where $\tau(D, X) \geq 0$ and $\omega \in S^{m-(\rho-\delta)}_{\rho,\delta}$.*

The proof will require several steps. We shall assume for simplicity that $\rho = 1$ and $\delta = 0$ and indicate the necessary modifications for the general case at the end. For $a > 0$, consider the Gaussian functions

$$\phi_a(x) = (2a)^{n/4} e^{-\pi a x^2}, \qquad \Phi_a(\xi, x) = W\phi_a(\xi, x) = 2^n e^{-2\pi(ax^2 + a^{-1}\xi^2)}.$$

The key to the proof is the following lemma.

(2.85) Lemma. *If $\sigma \geq 0$ then $(\sigma * \Phi_a)(D, X) \geq 0$ for all $a > 0$.*

Proof: Since Φ_a is real and even, convolution with it is a Hermitian operator. Hence, by Proposition (2.5), for all $f \in S$ we have

$$\langle (\sigma * \Phi_a)(D, X)f, f \rangle = \int (\sigma * \Phi_a)(Wf) = \int \sigma(Wf * \Phi_a) = \int \sigma(Wf * W\phi_a).$$

But this is ≥ 0 since $\sigma \geq 0$ and $Wf * W\phi_a \geq 0$ by Proposition (1.99). Alternatively, we could observe that $\Phi_a(\xi - \eta, \, x - y)$, as a function of η and y, is the symbol of the orthogonal projection onto $e^{2\pi i(\xi X - x D)}\phi_a$ by (1.95b) and (2.29). Thus $(\sigma * \Phi_a)(D, X)$ is a linear combination of orthogonal projections with nonnegative coefficients, so it is a positive operator. ∎

At this point an optimist might hope that $\sigma - \sigma * \Phi_a \in S^{m-1}_{1,0}$ when $\sigma \in S^m_{1,0}$. But life is not quite that simple; what is true is the following.

(2.86) Lemma. *There exists $C > 0$ such that for all $\sigma \in S^m_{1,0}$ and all multi-indices α, β,*

$$(2.87) \qquad \begin{aligned} &|D^\alpha_\xi D^\beta_x(\sigma - \sigma * \Phi_a)(\xi, x)| \\ &\leq C\|\sigma\|_{[|\alpha|+|\beta|+2]} \left(a\langle\xi\rangle^{m-2-|\alpha|} + \langle\xi\rangle^{m-1-|\alpha|} + a^{-1}\langle\xi\rangle^{m-|\alpha|} \right) \end{aligned}$$

whenever $|\xi| \geq a/4$. (Here $\|\sigma\|_{[j]}$ is defined by (2.24).)

Proof: Since differentiation commutes with convolution, it suffices to assume $\alpha = \beta = 0$. We begin by using the fact that $\int \Phi_a = 1$ to write

$$\begin{aligned} (\sigma - \sigma * \Phi_a)(\xi, x) &= \iint [\sigma(\xi, x) - \sigma(\xi - \eta, \, x - y)]\Phi_a(\eta, y) \, d\eta \, dy \\ &= I_1 + I_2, \end{aligned}$$

where I_1 and I_2 are the integrals over the regions $|\eta| \leq \frac{1}{2}|\xi|$ and $|\eta| > \frac{1}{2}|\xi|$ respectively. To estimate I_1, we apply Taylor's formula to σ:

$$\begin{aligned} \sigma(\xi, x) - \sigma(\xi - \eta, \, x - y) = &-\sum \frac{\partial \sigma}{\partial \xi_j}(\xi, x)\eta_j - \sum \frac{\partial \sigma}{\partial x_j}(\xi, x)y_j \\ &- \frac{1}{2}\sum \left[\frac{\partial^2 \sigma}{\partial \xi_j \partial \xi_k}(\zeta, z)\eta_j\eta_k + 2\frac{\partial^2 \sigma}{\partial \xi_j \partial x_k}(\zeta, z)\eta_j y_k + \frac{\partial^2 \sigma}{\partial x_j \partial x_k}(\zeta, z)y_j y_k \right], \end{aligned}$$

where (ζ, z) is some point on the line segment from (ξ, x) to $(\xi - \eta, \ x - y)$. If we plug this into I_1, the first-order terms integrate to zero because $\eta_j \Phi_a(\eta, y)$ and $y_j \Phi_a(\eta, y)$ are odd functions and the region $|\eta| \le \frac{1}{2}|\xi|$ is symmetric about the origin. Moreover, when $|\eta| \le \frac{1}{2}|\xi|$, $\langle \zeta \rangle$ and $\langle \xi \rangle$ are comparable, so in estimating the second-order terms we can replace $\langle \zeta \rangle$ by $\langle \xi \rangle$, obtaining

$$|I_1| \le C\|\sigma\|_{[2]} \iint \left(\langle \xi \rangle^{m-2} |\eta|^2 + \langle \xi \rangle^{m-1} |\eta||y| + \langle \xi \rangle^m |y|^2 \right) e^{-2\pi(a^{-1}\eta^2 + ay^2)} d\eta \, dy.$$

On replacing (η, y) by $(a^{1/2}\eta, \ a^{-1/2}y)$, the integral becomes

$$\iint \left(a\langle \xi \rangle^{m-2} |\eta|^2 + \langle \xi \rangle^{m-1} |\eta||y| + a^{-1} \langle \xi \rangle^m |y|^2 \right) e^{-2\pi(\eta^2 + y^2)} d\eta \, dy$$
$$\le C\left(a\langle \xi \rangle^{m-2} + \langle \xi \rangle^{m-1} + a^{-1} \langle \xi \rangle^m \right),$$

which is what we want. To estimate I_2, we use the simple bound

$$|\sigma(\xi, x) - \sigma(\xi - \eta, \ x - y)| \le \|\sigma\|_{[0]} (\langle \xi \rangle^m + \langle \xi - \eta \rangle^m) \le C\|\sigma\|_{[0]} \langle \xi \rangle^m \langle \eta \rangle^{|m|}$$

and replace (η, y) by $(a^{1/2}\eta, \ a^{-1/2}y)$ as above to obtain

$$|I_2| \le C\|\sigma\|_{[0]} \langle \xi \rangle^m \max(a^{|m|/2}, 1) \iint_{|\eta| \ge |\xi|/2\sqrt{a}} \langle \eta \rangle^{|m|} e^{-2\pi(\eta^2 + y^2)} d\eta \, dy$$
$$\le C'\|\sigma\|_{[0]} \langle \xi \rangle^m \max(a^{|m|/2}, 1) e^{-\pi \xi^2/4a}.$$

When $|\xi| \ge a/4$ this is dominated by $\|\sigma\|_{[0]} \langle \xi \rangle^{m-1}$, so we are done. ∎

Now, the parameter a is at our disposal, and for a given ξ the right side of (2.87) is minimized by taking $a = \langle \xi \rangle$, when it reduces to

$$C\|\sigma\|_{[|\alpha|+|\beta|+2]} \langle \xi \rangle^{m-1-|\alpha|}.$$

This is just what we want, but of course making a depend on ξ ruins the proof of Lemma (2.85). This line of argument can actually be salvaged, as we shall show in Section 3.1, but there is an easier way out. Namely, if σ happens to be supported in a set where $\langle \xi \rangle$ is essentially constant, we can take a to be that constant; and any σ can be broken up into pieces which each have this property. The estimates we need are the following:

(2.88) Lemma. *Suppose* $\operatorname{supp} \sigma \subset \{(\xi, x) : \frac{1}{2}a \le |\xi| \le 4a\}$ *where* $a \ge 1$. *Then*

$$|D_\xi^\alpha D_x^\beta (\sigma - \sigma * \Phi_a)(\xi, x)| \le C \|\sigma\|_{[|\alpha|+|\beta|+2]} \psi(\xi) \langle \xi \rangle^{m-1-|\alpha|},$$

where C *is independent of* σ *and* a *and*

$$\psi(\xi) = \begin{cases} 1 & \text{for } \frac{1}{4}a \le |\xi| \le 8a, \\ e^{-|\xi|} & \text{for } |\xi| > 8a, \\ e^{-a/4} & \text{for } |\xi| < \frac{1}{4}a. \end{cases}$$

Proof: The estimate for $\frac{1}{4}a \le |\xi| \le 8a$ is valid by Lemma (2.86), since (for $a \ge 1$) $|\xi|$, $\langle \xi \rangle$, and a are all comparable in this range. In the other regions we have $\sigma = 0$, so we need to estimate $|D_\xi^\alpha D_x^\beta (\sigma * \Phi_a)|$; and replacing σ by $D_\xi^\alpha D_x^\beta \sigma$, we can assume $\alpha = \beta = 0$. We have

$$|\sigma * \Phi_a(\xi, x)| = \left| \iint \sigma(\eta, y) e^{-2\pi[a^{-1}(\xi-\eta)^2 + a(x-y)^2]} \, d\eta \, dy \right|$$

$$\le \|\sigma\|_{[0]} \int e^{-2\pi a y^2} dy \int_{a/2 \le |\eta| \le 4a} \langle \eta \rangle^m e^{-2\pi a^{-1}(\xi-\eta)^2} \, d\eta$$

$$\le C\|\sigma\|_{[0]} a^{m+(n/2)} \sup_{a/2 \le |\eta| \le 4a} e^{-2\pi a^{-1}(\xi-\eta)^2}.$$

Suppose $\frac{1}{2}a \le |\eta| \le 4a$. If $|\xi| > 8a$ we have $|\xi - \eta| \ge \frac{1}{2}|\xi|$ and hence $e^{-2\pi a^{-1}(\xi-\eta)^2} \le e^{-8\pi|\xi|}$, while if $|\xi| \le \frac{1}{4}a$ we have $|\xi - \eta| \ge \frac{1}{4}a$ and hence $e^{-2\pi a^{-1}(\xi-\eta)^2} \le e^{-\pi a/8}$. The desired result is now immediate. ∎

Proof of Theorem (2.84): Given $\sigma \in S_{1,0}^m$, by using the partition of unity $\{\phi_j\}$ in Lemma (2.79) we write $\sigma = \sum_{-1}^\infty \sigma_j$, where $\sigma_j(\xi, x) = \sigma(\xi, x)\phi_j(|\xi|)$. Then σ_j is supported in the set where $2^{j-1} \le |\xi| \le 2^{j+2}$ for $j \ge 0$, and $\|\sigma_j\|_{[k]}$ is bounded uniformly in j for each k. Moreover, $\sum_0^\infty (\sigma_j - \sigma_j * \Phi_{2^j})$ converges in the C^∞ topology to an element of $S_{1,0}^{m-1}$. Indeed, if $2^J \le |\xi| \le 2^{J+1}$ where $J \ge 0$, by Lemma (2.88) we have

$$\sum_0^\infty |(\sigma_j - \sigma_j * \Phi_{2^j})(\xi, x)| = \left(\sum_0^{J-2} + \sum_{J-1}^{J+2} + \sum_{J+3}^\infty \right) |(\sigma_j - \sigma_j * \Phi_{2^j})(\xi, x)|$$

$$\le C \left(J e^{-|\xi|} + 1 + \sum_{J+3}^\infty e^{-2^{j-2}} \right) \langle \xi \rangle^{m-1},$$

which is dominated by $\langle\xi\rangle^{m-1}$ since $J \approx \log_2 |\xi|$; likewise for derivatives. Hence, if we set

$$\omega = \sigma_{-1} + \sum_0^\infty (\sigma_j - \sigma_j * \Phi_{2^j}) \quad \text{and} \quad \tau = \sigma - \omega = \sum_0^\infty \sigma_j * \Phi_{2^j},$$

we have $\omega \in S_{1,0}^{m-1}$, $\tau \in S_{1,0}^m$, and by Lemma (2.85),

$$\langle \tau(D,X)f,\; f\rangle = \sum_0^\infty \langle (\sigma_j * \Phi_{2^j})(D,X)f,\; f\rangle \geq 0.$$

Thus the proof for $\rho = 1$, $\delta = 0$ is complete.

The argument in the general case is essentially identical. The estimate (2.87) becomes

$$|(\sigma - \sigma * \Phi_a)(\xi, x)| \leq C\|\sigma\|_{[2]} \left(a\langle\xi\rangle^{m-2\rho} + \langle\xi\rangle^{m-\rho+\delta} + a^{-1}\langle\xi\rangle^{m+2\delta} \right),$$

and similarly for derivatives; this is valid for $|\xi|^{2-\epsilon} \geq \frac{1}{4}a$ for any $\epsilon > 0$, and in particular for $|\xi|^{\rho+\delta} \geq \frac{1}{4}a$. The right side is minimized by taking $a = \langle\xi\rangle^{\rho+\delta}$. Thus, in Lemma (2.88) we set $b = a^{1/(\rho+\delta)}$ and assume that σ is supported in $\frac{1}{2}b \leq |\xi| \leq 4b$; we estimate $\sigma - \sigma * \Phi_{b^{\rho+\delta}}$ in the regions $|\xi| < \frac{1}{4}b$, $\frac{1}{4}b \leq |\xi| \leq 8b$, and $|\xi| > 8b$ by $\langle\xi\rangle^{m-(\rho-\delta)}$ times $\exp(-\epsilon|\xi|^{2-\rho-\delta})$, 1, and $\exp(-\epsilon b^{2-\rho-\delta})$ respectively (where ϵ is a suitably small positive number). These are good estimates since $\rho + \delta < 2$. Lastly, we define σ_j as above and find that

$$\sum_0^\infty (\sigma_j - \sigma_j * \Phi_{2^{j(\rho+\delta)}}) \in S_{\rho,\delta}^{m-(\rho-\delta)},$$

and the proof is complete. ∎

(2.89) Remark. In this proof, instead of using $\Phi_a(\xi, x) = 2^n e^{-2\pi(a^{-1}\xi^2 + ax^2)}$, we could equally well have used $\Psi_a = W\psi_a$ where $\psi_a(x) = a^{n/4}\psi(a^{1/2}x)$ and ψ is any even function in $\mathcal{S}(\mathbf{R}^n)$ with $\|\psi\|_2 = 1$. These conditions on ψ imply that Ψ_a is even, belongs to $\mathcal{S}(\mathbf{R}^{2n})$, and satisfies $\int \Psi_a = 1$; moreover, $\Psi_a(\xi, x) = \Psi_1(a^{-1/2}\xi,\, a^{1/2}x)$. If Φ_a is replaced by Ψ_a, then, the proofs of Lemma (2.85) and of the estimate for I_1 in Lemma (2.86) are unchanged; the estimates for I_2 and the "tail ends" in Lemma (2.88) are a bit less transparent and involve large negative powers instead of negative exponentials, but the final result is the same. We leave the details to the reader, with the admonition that this point will be of significance in Section 3.1.

Theorem (2.84) is often called the "sharp Gårding inequality"; the actual inequality in question is the following.

(2.90) Corollary. *Suppose $\sigma \in S^m_{\rho,\delta}$ with $\rho > \delta$.*
(a) *If $\sigma \geq 0$, then for some $C \geq 0$ we have*

$$\langle \sigma(D,X)f, \, f \rangle \geq -C\|f\|^2_{((m-\rho+\delta)/2)} \text{ for } f \in \mathcal{S}.$$

(b) *If $\operatorname{Re}\sigma(\xi,x) \geq A\langle\xi\rangle^m$ for large ξ, then for some $C \geq 0$ we have*

$$\operatorname{Re}\langle \sigma(D,X)f, \, f \rangle \geq A\|f\|^2_{(m/2)} - C\|f\|^2_{((m-\rho+\delta)/2)} \text{ for } f \in \mathcal{S}.$$

Proof: For (a), set $s = (m - \rho + \delta)/2$ and write $\sigma = \tau + \omega$ as in Theorem (2.84). Then by Corollary (2.61),

$$\langle \sigma(D,X)f, \, f \rangle \geq \langle \omega(D,X)f, \, f \rangle \geq -\|\omega(D,X)f\|_{(-s)}\|f\|_{(s)} \geq -C\|f\|^2_{(s)}.$$

For (b), apply (a) to the symbol

$$\sigma'(\xi, x) = \operatorname{Re}\sigma(\xi, x) + \phi(\xi) - A\langle\xi\rangle^m$$

where $\phi \in C^\infty_c$ is large enough to make $\sigma' \geq 0$. The result follows since $\phi(D) \in OPS^{-\infty}_{1,0}$ and $\langle \langle D \rangle^m f, \, f \rangle = \|f\|_{(m/2)}$. ∎

We have implicitly been considering operators on complex-valued functions on $\mathbf{R}^n$, but much of what we have done extends to the vector-valued case. Indeed, let $\mathcal{H}$ be a separable Hilbert space, and let $\mathcal{L}(\mathcal{H})$ be the algebra of bounded operators on $\mathcal{H}$; we consider functions on $\mathbf{R}^n$ with values in $\mathcal{H}$ and symbols with values in $\mathcal{L}(\mathcal{H})$. The spaces $\mathcal{S}_\mathcal{H}$, $L^2_\mathcal{H}$, $(S^m_{\rho,\delta})_{\mathcal{L}(\mathcal{H})}$, etc., are defined in the same way as in the scalar case, with norms replacing absolute values in appropriate spots, and every $\sigma \in (S^m_{\rho,\delta})_{\mathcal{L}(\mathcal{H})}$ defines an operator $\sigma(D,X)$ on $\mathcal{S}_\mathcal{H}$ just as before. Most of the theory of this chapter goes through with only minor changes, the main pitfall being that pointwise multiplication of symbols is no longer commutative. (Thus, for example, if $\sigma_j \in (S^{m_j}_{\rho,\delta})_{\mathcal{L}(\mathcal{H})}$ for $j = 1,2$ then $[\sigma_1(D,X), \sigma_2(D,X)]$ is generally of order $m_1 + m_2$ rather than $m_1 + m_2 - 1$, as the leading term of its symbol is $\sigma_1\sigma_2 - \sigma_2\sigma_1$.) The extension of Theorem (2.84) to the vector case, due to Lax and Nirenberg [95], is the following.

(2.91) Theorem. *Suppose $\sigma \in (S^m_{\rho,\delta})_{\mathcal{L}(\mathcal{H})}$ with $\rho > \delta$, and $\sigma(\xi, x)$ is a positive operator on $\mathcal{H}$ for all ξ, x. Then $\sigma = \tau + \omega$ where $\omega \in (S^{m-(\rho-\delta)}_{\rho,\delta})_{\mathcal{L}(\mathcal{H})}$ and $\tau(D,X)$ is a positive operator on $L^2_\mathcal{H}$.*

Proof: The only point that requires comment is Lemma (2.85), as the remainder of the proof goes through without change. Choose an orthonormal basis $\{e_i\}$ for $\mathcal{H}$, and let

$$f_i(x) = \langle f(x), \, e_i \rangle_\mathcal{H}, \qquad \sigma_{ij}(\xi, x) = \langle \sigma(\xi, x)e_j, \, e_i \rangle_\mathcal{H}.$$

Then, as in the proof of Lemma (2.85),

$$\langle (\sigma * \Phi_a)(D, X)f, f \rangle_{L^2_{\mathcal{H}}} = \sum_{ij} \langle (\sigma_{ij} * \Phi_a)(D, X)f_j, f_i \rangle_{L^2}$$

$$= \sum_{ij} \int (\sigma_{ij} * \Phi_a) W(f_j, f_i) = \sum_{ij} \int \sigma_{ij}[W(f_j, f_i) * \Phi_a].$$

The argument that proves Proposition (1.99) shows that

$$W(f_j, f_i) * \Phi_a(\xi, x) = \psi_j(\xi, x)\overline{\psi_i(\xi, x)},$$

where

$$\psi_j(\xi, x) = \langle f_j, \rho(-x, \xi)\phi_a \rangle_{L^2}.$$

Since $\sigma(\xi, x)$ is a positive operator, $(\sigma_{ij}(\xi, x))$ is a positive matrix, so

$$\sum_{ij} \int \sigma_{ij} \psi_j \overline{\psi_i} \geq 0. \quad \blacksquare$$

This argument can be rephrased in coordinate-free terms as follows. If $f \in L^2_{\mathcal{H}}$, the Wigner distribution Wf may be defined as the $\mathcal{L}(\mathcal{H})$-valued function defined by

$$\langle Wf(\xi, x)u, v \rangle_{\mathcal{H}} = W(\langle f, v \rangle_{\mathcal{H}}, \langle f, u \rangle_{\mathcal{H}})(\xi, x), \qquad u, v \in \mathcal{H}.$$

(Here the right side is the ordinary Wigner transform of the scalar functions $f_u(x) = \langle f(x), u \rangle_{\mathcal{H}}$ and $f_v(x) = \langle f(x), v \rangle_{\mathcal{H}}$.) $Wf(\xi, x)$ is trace-class for each (ξ, x), for if $\{e_i\}$ is an orthonormal basis for $\mathcal{H}$ and $f = \sum f_i e_i$,

$$\sum |\langle Wf(\xi, x)e_i, e_i \rangle_{\mathcal{H}}| = \sum |Wf_i(\xi, x)| \leq \sum \|f_i\|_2^2 = \|f\|_{L^2_{\mathcal{H}}}^2.$$

For any $\mathcal{L}(\mathcal{H})$-valued symbol σ, then, Proposition (2.5) implies that

$$\langle \sigma(D, X)f, f \rangle_{L^2_{\mathcal{H}}} = \iint \operatorname{tr}[\sigma(\xi, x)Wf(\xi, x)] \, d\xi \, dx.$$

Moreover, it is easily verified that for $u \in \mathcal{H}$,

$$(Wf * \Phi_a)(\xi, x)u = \langle u, v(\xi, x) \rangle_{\mathcal{H}} v(\xi, x)$$

where

$$v(\xi, x) = \langle f, \rho(-x, \xi)\phi_a \rangle_{L^2} \in \mathcal{H}.$$

Thus, for any orthonormal basis $\{e_i\}$ of $\mathcal{H}$,

$$\langle(\sigma * \Phi_a)(D, X)f, f\rangle_{L^2_{\mathcal{H}}} = \int \text{tr}[(\sigma * \Phi_a)(Wf)] = \int \text{tr}[\sigma(Wf * \Phi_a)]$$

$$= \int \sum \langle \sigma(Wf * \Phi_a)e_i, e_i\rangle_{\mathcal{H}} = \int \sum \langle \sigma v, e_i\rangle_{\mathcal{H}}\langle e_i, v\rangle_{\mathcal{H}} = \int \langle \sigma v, v\rangle_{\mathcal{H}} \geq 0.$$

In the scalar case, Theorem (2.84) can be improved: it has been shown by Fefferman and Phong [45] (see also Hörmander [72]) that the error term ω can be taken to be in $S_{\rho,\delta}^{m-2(\rho-\delta)}$. However, this is not known to be true in the vector case; in fact, it is stated without proof in Hörmander [72, p. 176] that it is false. Thus the fact that our proof works in the vector case is significant.

Theorem (2.84) can be sharpened in other ways, for example by weakening the hypothesis that $\sigma \geq 0$. It turns out that $\sigma(D, X)$ will still be positive provided that the set where $\sigma < 0$ is, in a certain geometric sense related to the uncertainty principle, "not too big." The exploitation of this idea has led to some very deep results; see Fefferman–Phong [46] and Fefferman [44].

7. The Wick and Anti-Wick Correspondences

In quantum field theory one frequently has to manipulate products of annihilation and creation operators. We recall that in the Fock model these are the operators A_j and A_j^* on $\mathcal{F}_n$ defined by

$$A_j F(z) = \frac{1}{\sqrt{\pi}} \frac{\partial F}{\partial z_j}, \qquad A_j^* F(z) = \sqrt{\pi} z_j F(z),$$

and in the Schrödinger model they are the operators

$$B^{-1} A_j B = \sqrt{\pi}(X_j + iD_j) = \sqrt{\pi} Z_j, \qquad B^{-1} A_j^* B = \sqrt{\pi}(X_j - iD_j) = \sqrt{\pi} Z_j^*$$

on $L^2(\mathbf{R}^n)$, where B denotes the Bargmann transform.

We work in the Schrödinger model for the moment. A product of Z_j's and Z_j^*'s is said to be **Wick ordered**, or in **Wick normal form**, if it is of the form $Z^{*\alpha} Z^\beta$, that is, if all the Z's occur to the right of all the Z^*'s. This specifies the ordering completely, since the Z's and the Z^*'s commute amongst themselves. Moreover, any product of Z's and Z^*'s can be written uniquely as a linear combination of Wick ordered products, by virtue of the commutation relations $[Z_j, Z_k^*] = \pi^{-1}\delta_{jk}$.

This suggests the following quantization procedure for polynomials. We define the complex coordinates

$$z = x + i\xi, \qquad \overline{z} = x - i\xi$$

on $\mathbf{R}^{2n}$; and given any polynomial

$$p(z,\bar{z}) = \sum a_{\alpha\beta} z^{\alpha} \bar{z}^{\beta}$$

on $\mathbf{R}^{2n}$, we define the operator

$$p(Z,Z^*)_W = \sum a_{\alpha\beta} Z^{*\beta} Z^{\alpha}.$$

We call $p(Z,Z^*)_W$ the **Wick ordered operator** corresponding to p, and we call $p(\bar{z},z)$ [sic] the **Wick symbol** of $p(Z,Z^*)_W$. (The replacement of z by $\bar{z}$ is a matter of convention, intended to make this definition concordant with formula (2.92) below.) Similarly, we can consider the opposite ordering in which all the Z's are placed to the left:

$$p(Z,Z^*)_{AW} = \sum a_{\alpha\beta} Z^{\alpha} Z^{*\beta}.$$

We call $p(Z,Z^*)_{AW}$ the **anti-Wick ordered operator** corresponding to p, and we call $p(\bar{z},z)$ the **anti-Wick symbol** of $p(Z,Z^*)_{AW}$. It is easily verified that the Wick and anti-Wick correspondences both map the set of polynomials on $\mathbf{R}^{2n}$ bijectively onto the set of differential operators with polynomial coefficients.

In order to extend these correspondences to more general symbols, it is convenient to move over to Fock space. Thus, we define the operators T_p^W and T_p^{AW} on $\mathcal{F}_n$ by

$$T_p^W = Bp(Z,Z^*)_W B^{-1}, \qquad T_p^{AW} = Bp(Z,Z^*)_{AW} B^{-1}.$$

Thus, if $p(z,\bar{z}) = \sum a_{\alpha\beta} z^{\alpha} \bar{z}^{\beta}$ we have

$$T_p^W F(z) = \sum \frac{a_{\alpha\beta}}{\pi^{|\alpha+\beta|/2}} A^{*\beta} A^{\alpha} F(z) = \sum \frac{a_{\alpha\beta}}{\pi^{|\alpha|}} z^{\beta} \partial_z^{\alpha} F(z),$$

$$T_p^{AW} F(z) = \sum \frac{a_{\alpha\beta}}{\pi^{|\alpha+\beta|/2}} A^{\alpha} A^{*\beta} F(z) = \sum \frac{a_{\alpha\beta}}{\pi^{|\alpha|}} \partial_z^{\alpha} \big(z^{\beta} F(z) \big).$$

We can rewrite this by using the reproducing formula

$$F(z) = \int e^{\pi z \bar{w}} F(w) e^{-\pi |w|^2} \, dw,$$

as follows:

$$T_p^W F(z) = \sum \frac{a_{\alpha\beta}}{\pi^{|\alpha|}} \int z^{\beta} \partial_z^{\alpha} e^{\pi z \bar{w}} F(w) e^{-\pi |w|^2} \, dw$$

$$= \sum a_{\alpha\beta} \int z^{\beta} \bar{w}^{\alpha} e^{\pi z \bar{w}} F(w) e^{-\pi |w|^2} \, dw,$$

and

$$T_p^{AW} F(z) = \sum \frac{a_{\alpha\beta}}{\pi^{|\alpha|}} \partial_z^\alpha \int e^{\pi z \overline{w}} w^\beta F(w) e^{-\pi |w|^2} \, dw$$
$$= \sum a_{\alpha\beta} \int \overline{w}^\alpha w^\beta e^{\pi z \overline{w}} F(w) e^{-\pi |w|^2} \, dw.$$

In other words,

(2.92) $$T_p^W F(z) = \int p(\overline{w}, z) e^{\pi z \overline{w}} F(w) e^{-\pi |w|^2} \, dw,$$

(2.93) $$T_p^{AW} F(z) = \int p(\overline{w}, w) e^{\pi z \overline{w}} F(w) e^{-\pi |w|^2} \, dw.$$

These formulas now make sense for non-polynomial p. In (2.92), p should be an entire function on $\mathbf{C}^n \times \mathbf{C}^n$, while in (2.93), p can be a more or less arbitrary (non-holomorphic) function on $\mathbf{C}^n$. In both cases, p should satisfy suitable growth restrictions at infinity so that the integrals will converge.

In this way we can pass from symbols to operators. In the case of the Wick correspondence it is also easy to reverse the process. Namely, suppose T is an operator on $\mathcal{F}_n$, possibly unbounded, such that the domains of T and T^* contain all the coherent states $E_w(z) = e^{\pi z \overline{w}}$. (This will be the case, for example, if the domains of the corresponding operators $B^{-1}TB$ and $B^{-1}T^*B$ on $L^2(\mathbf{R}^n)$ contain $\mathcal{S}$, since $B^{-1}E_w \in \mathcal{S}$.) We define the **Wick symbol** of T to be

$$\sigma_T^W(\overline{z}, z) = e^{-\pi |z|^2} \langle T E_z, E_z \rangle_{\mathcal{F}} = e^{-\pi |z|^2} T E_z(z).$$

This is a smooth function on $\mathbf{C}^n$ which is the restriction to the set $\{w = z\}$ of the entire function

$$\sigma_T^W(\overline{w}, z) = e^{-\pi z \overline{w}} \langle T E_w, E_z \rangle_{\mathcal{F}} = e^{-\pi z \overline{w}} T E_w(z) = e^{-\pi z \overline{w}} \, \overline{T^* E_z(w)}$$

of $(\overline{w}, z) \in \mathbf{C}^{2n}$. (As we noted in Proposition (1.68), $\sigma_T^W(\overline{z}, z)$ determines $\sigma_T^W(\overline{w}, z)$ uniquely.) If $F \in \mathrm{Dom}(T)$, we have

$$TF(z) = \langle TF, E_z \rangle_{\mathcal{F}} = \langle F, T^* E_z \rangle_{\mathcal{F}} = \int \sigma_T^W(\overline{w}, z) e^{\pi z \overline{w}} F(w) e^{-\pi |w|^2} \, dw,$$

which is formula (2.92) with $p = \sigma_T^W$. Thus, the maps $p \to T_p^W$ and $T \to \sigma_T^W$ are mutually inverse, and σ_T^W is just $e^{-\pi z \overline{w}}$ times the integral kernel of T.

The anti-Wick representation (2.93) is of quite a different nature; the operator T_p^{AW} thus defined is what is usually called a Toeplitz operator. To be precise, let p be a measurable function on $\mathbf{C}^n$ which satisfies

(2.94) $$|p(\overline{z}, z)| \le C e^{\delta |z|^2} \text{ for some } \delta < \tfrac{1}{2}\pi.$$

(We write $p(\bar{z}, z)$ to emphasize the fact that p need not be holomorphic.) If $F \in \mathcal{F}_n$, we can form the function pF on $\mathbf{C}^n$; if this function is in $L^2(\mathbf{C}^n, e^{-\pi|z|^2} dz)$, we can project it orthogonally onto $\mathcal{F}_n$ by integrating against the kernel $e^{\pi z \bar{w}}$, and the result is precisely $T_p^{AW} F$. Thus T_p^{AW} is defined on

$$\mathrm{Dom}(T_p^{AW}) = \left\{ F \in \mathcal{F}_n : \int |p(\bar{z}, z) F(z)|^2 e^{-\pi|z|^2} dz < \infty \right\},$$

which includes all polynomials and all coherent states when p satisfies (2.94). Moreover, we have:

(2.95) Lemma. *If p_1 and p_2 satisfy (2.94) and $T_{p_1}^{AW} = T_{p_2}^{AW}$ then $p_1 = p_2$.*

Proof: Let $p = p_1 - p_2$, so $T_p^{AW} = T_{p_1}^{AW} - T_{p_2}^{AW}$. From the preceding description of T_p^{AW} it is clear that $T_p^{AW} F = 0$ precisely when $pF \perp G$ in $L^2(\mathbf{C}^n, e^{-\pi|z|^2} dz)$ for all $G \in \mathcal{F}_n$. Taking F and G to be the monomials z^α and z^β, we see that if $T_p^{AW} = 0$ then

$$\int p(\bar{z}, z) z^\alpha \bar{z}^\beta e^{-\pi|z|^2} dz = 0 \text{ for all } \alpha, \beta.$$

But the linear span of $\{z^\alpha \bar{z}^\beta\}_{\alpha,\beta}$—namely, the set of all polynomials in Re z and Im z— is dense in $L^2(\mathbf{C}^n, e^{-\pi|z|^2} dz)$, as we know from the theory of Hermite functions. Therefore $p = 0$. ∎

This lemma shows that the operator $T = T_p^{AW}$ uniquely determines the function p, which we therefore call the **anti-Wick symbol** of T and denote by σ_T^{AW}. However, not all reasonable operators on $\mathcal{F}_n$, not even all bounded operators, have anti-Wick symbols (i.e., are representable in the form (2.93)). For example, if T is orthogonal projection onto the constant functions, $TF(z) = F(0)$, then the only possible way to write T in the form (2.93) is to take p to be the δ-function at the origin. But even allowing p to be a distribution doesn't solve the problem in general, as we shall see shortly.

We now calculate the relationships between the Wick, anti-Wick, and Weyl symbols of an operator.

(2.96) Proposition. *Suppose the operator T on $\mathcal{F}_n$ has anti-Wick symbol σ_T^{AW}. Then the Wick symbol of T is*

$$\sigma_T^W(\bar{z}, z) = \int e^{-\pi|z-w|^2} \sigma_T^{AW}(\bar{w}, w) \, dw.$$

Proof: We have

$$\sigma_T^W(\bar{z}, z) = e^{-\pi|z|^2} T E_z(z) = e^{-\pi|z|^2} \int \sigma_T^{AW}(\bar{w}, w) e^{\pi z \bar{w}} e^{\pi w \bar{z}} e^{-\pi|w|^2} dw$$

$$= \int e^{-\pi|z-w|^2} \sigma_T^{AW}(\bar{w}, w) \, dw. \quad ∎$$

(2.97) Proposition. *Suppose $\sigma \in S'(\mathbf{R}^{2n})$ and $\sigma(D, X)$ maps S into itself. Then the Wick symbol of $T = B\sigma(D, X)B^{-1}$ is*

$$\sigma_T^W(\bar{z}, z) = 2^n \int e^{-2\pi[(s-\xi)^2+(r-x)^2]}\sigma(\xi, x)\, d\xi\, dx, \quad \text{where } z = r - is.$$

Proof: We recall from (1.72) that

$$B^{-1}E_z = e^{(\pi/2)|z|^2}\rho(-r, s)\phi_0, \quad \text{where } \phi_0(x) = 2^{n/4}e^{-\pi x^2} \text{ and } z = r - is.$$

Hence, by (1.95) and Proposition (2.5),

$$\sigma_T^W(\bar{z}, z) = e^{-\pi|z|^2}\langle B\sigma(D, X)B^{-1}E_z, E_z\rangle = e^{-\pi|z|^2}\langle \sigma(D, X)B^{-1}E_z, B^{-1}E_z\rangle$$

$$= e^{-\pi|z|^2}\iint \sigma(\xi, x)W(B^{-1}E_z)(\xi, x)\, d\xi\, dx$$

$$= \iint \sigma(\xi, x)W\phi_0(\xi-s, x-r)\, d\xi\, dx$$

$$= 2^n \iint e^{-2\pi[(\xi-s)^2+(x-r)^2]}\sigma(\xi, x)\, d\xi\, dx. \quad \blacksquare$$

The operators $\sigma_T^{AW} \to \sigma_T^W$ and $\sigma \to \sigma_T^W$ in these propositions belong to the heat-diffusion semigroup $\{H_t\}_{t>0}$ on $\mathbf{R}^{2n}$ defined by

$$H_t f = f * \gamma_t, \qquad \gamma_t(\xi, x) = (4\pi t)^{-n}e^{-(\xi^2+x^2)/4t},$$

or

$$(H_t f)\widehat{}(p, q) = e^{-t(p^2+q^2)}\widehat{f}(p, q).$$

Indeed, if we identify $\mathbf{R}^{2n}$ with $\mathbf{C}^n$ via $z = x - i\xi$, we have

$$\sigma_T^W = H_\pi \sigma_T^{AW} \quad \text{and} \quad \sigma_T^W = H_{\pi/2}\sigma$$

when T is as in Propositions (2.96) and (2.97) respectively. From this it also follows that under these conditions,

$$\sigma = H_{\pi/2}\sigma_T^{AW}.$$

Thus, the Weyl symbol of $B^{-1}TB$ is obtained by smoothing out the anti-Wick symbol of T by $H_{\pi/2}$, and the Wick symbol is obtained in the same way from the Weyl symbol. In particular, a necessary condition for $T = B\sigma(D, X)B^{-1}$ to have an anti-Wick symbol (even a distributional one) is that σ should be the restriction to $\mathbf{R}^{2n}$ of an entire function on $\mathbf{C}^{2n}$.

If p is a harmonic polynomial on $\mathbf{R}^{2n}$, then p is a steady-state solution of the heat equation and hence is invariant under the semigroup $\{H_t\}$. Propositions (2.96) and (2.97) therefore yield the following result, due to Geller [57].

(2.98) Corollary. *Let $\mathbf{R}^{2n}$ be identified with $\mathbf{C}^n$ by $(\xi, x) \leftrightarrow x - i\xi$. If p is a harmonic polynomial on $\mathbf{R}^{2n}$ then*

$$T_p^W = T_p^{AW} = Bp(D, X)B^{-1}.$$

The algebra of Wick-ordered products and its application to quantum field theory are due to Wick [155], but the general theory of Wick and anti-Wick symbols was developed by Berezin [21], who showed that a number of properties of operators on $\mathcal{F}_n$ can be profitably studied in terms of these symbols. The following propositions, which we quote from [21] without proof, will give the flavor of Berezin's results. In them, T will denote an operator on $\mathcal{F}_n$ such that the domains of T and T^* contain all coherent states E_z; σ_T^W will denote its Wick symbol; and σ_T^{AW} will denote its anti-Wick symbol if it exists.

(i) For T to be bounded it is necessary that σ_T^W be bounded and sufficient that σ_T^{AW} exist and be bounded; moreover,

$$\|\sigma_T^W\|_\infty \leq \|T\| \leq \|\sigma_T^{AW}\|_\infty.$$

(ii) For T to be trace class it is necessary that σ_T^W be in $L^1(\mathbf{C}^n, dz)$ and sufficient that σ_T^{AW} exist and be in $L^1(\mathbf{C}^n, dz)$; moreover,

$$\mathrm{tr}(T) = \int \sigma_T^W(\bar{z}, z)\, dz = \int \sigma_T^{AW}(\bar{z}, z)\, dz.$$

(iii) If T is self-adjoint and σ_T^{AW} exists, then

$$\int \exp[-t\sigma_T^W(\bar{z}, z)]\, dz \leq \mathrm{tr}(e^{-tT}) \leq \int \exp[-t\sigma_T^{AW}(\bar{z}, z)]\, dz.$$

The Wick and anti-Wick correspondences can be formulated in the setting of an arbitrary Hilbert space $\mathcal{H}$ of functions with a reproducing kernel. That is, suppose that $\mathcal{H} \subset L^2(X, \mu)$ for some measure space (X, μ), and that for each $x \in X$ the evaluation functional $f \to f(x)$ is bounded on $\mathcal{H}$ so that there exists $E_x \in \mathcal{H}$ with $f(x) = \langle f, E_x \rangle_{\mathcal{H}}$ for $f \in \mathcal{H}$. If T is an operator on $\mathcal{H}$, its Wick symbol is the function

$$\sigma(x) = \|E_x\|_{\mathcal{H}}^{-2} \langle TE_x, E_x \rangle_{\mathcal{H}},$$

and if $T = PM_\phi$ where M_ϕ is multiplication by the function ϕ and P is orthogonal projection onto $\mathcal{H}$, ϕ is the anti-Wick symbol of T. For the development of the theory in this context and some applications, see Berezin [22]. (In this paper, Wick and anti-Wick symbols are called covariant and contravariant symbols.) See also Berezin [23] and Unterberger [142] for discussion of more general classes of quantization procedures.

CHAPTER 3.
WAVE PACKETS AND WAVE FRONTS

By the term "wave packet" we mean a function ϕ on $\mathbf{R}^n$ that is "well localized in phase space," that is, such that ϕ and $\hat{\phi}$ are both concentrated in reasonably small sets. The archetypical examples of wave packets are the Gaussians $\phi(x) = e^{2\pi i p x}e^{-\pi a(x-q)^2}$, which are the extremals for the uncertainty inequalities (1.39). However, for some purposes it is appropriate to consider any Schwartz class function as a wave packet.

The main themes of this chapter are (i) expressing an "arbitrary" function as a superposition of wave packets, and (ii) deriving information about a distribution f by studying the integrals $\int f\phi$ for suitable wave packets ϕ. More precisely, suppose $\phi \neq 0 \in \mathcal{S}(\mathbf{R}^n)$. We obtain a family of wave packets by subjecting ϕ to dilations,

$$\phi^t(x) = t^{n/4}\phi(t^{1/2}x), \qquad t > 0,$$

and then to translations in phase space,

$$e^{2\pi i(pX-qD)}\phi^t(x) = t^{n/4}e^{-\pi ipq+2\pi ipx}\phi\big(t^{1/2}(x-q)\big).$$

(The use of $t^{1/2}$ rather than t in the dilations will turn out to be convenient below; the factor of $t^{n/4}$ makes $\|\phi^t\|_2$ independent of t.) For any $f \in \mathcal{S}'$, then, we can consider the function

$$P_\phi^f(p,q,t) = \langle f, e^{2\pi i(pX-qD)}\phi^t\rangle.$$

(The notation here is only provisional.) Various things can be done with this construction:

(i) If $t = 1$, P_ϕ^f is essentially the Fourier-Wigner transform $V(f,\phi)$.

(ii) If $t = \langle p\rangle^\alpha$ for some $\alpha > 0$, the resulting function of p and q is a generalization of the wave packet transform of Córdoba and Fefferman [33]. Both this transform and the Fourier-Wigner transform can be used to expand f as a superposition of wave packets. We shall examine these constructions, and their relation to pseudodifferential operators, in Section 3.1.

(iii) If (p, q) is restricted to lie in a conic neighborhood of (ξ_0, x_0), the asymptotic behavior of $P_\phi^f(p, q, t)$ as $p, t \to \infty$ reveals information about the microlocal behavior of f at (ξ_0, x_0). This is the subject of Sections 3.2 and 3.3.

(iv) One of the first attempts to expand general functions in terms of wave packets was made by Gabor [53], who, working in dimension $n = 1$, took $\phi(x) = 2^{1/4} e^{-\pi x^2}$ and suggested using $\{e^{2\pi i(pX - qD)}\phi : p, q \in \mathbf{Z}\}$ as a basis for L^2. This idea turns out to be less successful than one might hope, but it will provide us with a few pages of entertainment in Section 3.4.

1. Wave Packet Expansions

Suppose $\phi \in \mathcal{S}(\mathbf{R}^n)$ and $\|\phi\|_2 = 1$. We say that ϕ is **centered** at $(a, b) \in \mathbf{R}^{2n}$ if a and b are the expected values of momentum and position in the state ϕ:

$$\int \xi_j |\widehat{\phi}(\xi)|^2 d\xi = a_j, \qquad \int x_j |\phi(x)|^2 dx = b_j.$$

We define the phase-space translates ϕ_{pq} of ϕ by

$$(3.1) \qquad \phi_{pq}(x) = e^{2\pi i(pX - qD)}\phi(x) = e^{-\pi i pq + 2\pi i px}\phi(x - q).$$

The point of using $e^{2\pi i(pX - qD)}$ rather than $e^{2\pi i(pD + qX)}$ is that if ϕ is centered at (a, b) then ϕ_{pq} is centered at $(a + p, b + q)$.

The simplest sort of wave packet transform is the map $P_\phi : L^2(\mathbf{R}^n) \to L^2(\mathbf{R}^{2n})$ defined by

$$(3.2) \qquad P_\phi f(p, q) = \langle f, \phi_{pq} \rangle = \langle f, \rho(-q, p)\phi \rangle = V(f, \phi)(q, -p).$$

Since we assumed that $\|\phi\|_2 = 1$, it follows from Proposition (1.42) that P_ϕ is an isometry from $L^2(\mathbf{R}^n)$ into $L^2(\mathbf{R}^{2n})$. Moreover, if $G \in L^2(\mathbf{R}^{2n})$,

$$\langle P_\phi f, G \rangle = \iiint f(x)\overline{\phi_{pq}(x)G(p, q)}\, dx\, dp\, dq,$$

so that

$$(3.3) \qquad \begin{aligned} P_\phi^* G(x) &= \iint \phi_{pq}(x)G(p, q)\, dp\, dq \\ &= \iint e^{-\pi i pq + 2\pi i px}\phi(x - q)G(p, q)\, dp\, dq. \end{aligned}$$

This needs to be taken with a grain of salt, since ϕ_{pq} is not in L^2 as a function of (p, q). It is, however, bounded, so (3.3) makes sense for $G \in L^1 \cap L^2$ and can be extended to arbitrary $G \in L^2$ by taking limits. At any rate, since P_ϕ is an isometry, we have $P_\phi^* P_\phi = I$, which gives an expansion of f as a superposition of the ϕ_{pq}'s:

$$(3.4) \qquad f = P_\phi^* P_\phi f = \iint P_\phi f(p, q) \phi_{pq} \, dp \, dq.$$

Moreover, if $f \in \mathcal{S}(\mathbf{R}^n)$ then $P_\phi f \in \mathcal{S}(\mathbf{R}^{2n})$ and the integral in (3.4) converges in $\mathcal{S}$.

Remark. If ϕ is the standard Gaussian, $\phi(x) = 2^{n/4} e^{-\pi x^2}$, this construction is just the Bargmann transform in disguise:

$$P_\phi f(p, q) = e^{-(\pi/2)(p^2 + q^2)} Bf(q - ip).$$

We wish to apply wave packet expansions to pseudodifferential operators. Intuitively, one would hope that if ϕ is a wave packet centered at $(0, 0)$, so that ϕ_{pq} is centered at (p, q), then $\sigma(D, X)\phi_{pq}$ will be approximately $\sigma(p, q)\phi_{pq}$, and hence the expansion (3.4) will approximately diagonalize $\sigma(D, X)$. To see how well this works, we make a very crude calculation of $\sigma(D, X)\phi_{pq}$ where $\sigma \in S^0_{\rho, \delta}$ and ϕ is the best sort of wave packet, namely $\phi(x) = e^{-\pi a x^2}$ for some $a > 0$:

$$\sigma(D, X)\phi_{pq}(x) = \iint \sigma(\xi, \tfrac{1}{2}(x + y)) e^{2\pi i(x-y)\xi} e^{-\pi i pq + 2\pi i py} e^{-\pi a(y-q)^2} dy \, d\xi$$

$$\sim \iint \sigma(\xi, \tfrac{1}{2}(x + q)) e^{2\pi i(x-y)\xi} e^{-\pi i pq + 2\pi i py} e^{-\pi a(y-q)^2} dy \, d\xi$$

$$= a^{-n/2} \int \sigma(\xi, \tfrac{1}{2}(x + q)) e^{2\pi i(x-q)\xi} e^{\pi i pq} e^{-\pi a^{-1}(\xi - p)^2} d\xi$$

$$\sim a^{-n/2} \int \sigma(p, \tfrac{1}{2}(x + q)) e^{2\pi i(x-q)\xi} e^{\pi i pq} e^{-\pi a^{-1}(\xi - p)^2} d\xi$$

$$= \sigma(p, \tfrac{1}{2}(x + q)) e^{-\pi a(x-q)^2} \sim \sigma(p, q)\phi_{pq}(x).$$

Here we have used the fact that $e^{-\pi a(y-q)^2}$ is negligibly small for $|y - q| \gg a^{-1/2}$ to replace y by q in σ, and likewise ξ by p and x by q later on. This will be a good approximation provided that $\sigma(\xi, x)$ is essentially constant in x on the ball $|x - q| \leq a^{-1/2}$ and essentially constant in ξ on the ball $|\xi - p| \leq a^{1/2}$, that is, provided that $|D_x \sigma| \ll a^{1/2}$ and $|D_\xi \sigma| \ll a^{-1/2}$ for (ξ, x) near (p, q). Since $\sigma \in S^0_{\rho, \delta}$, we therefore want

$$\langle p \rangle^\delta \ll a^{1/2}, \qquad \langle p \rangle^{-\rho} \ll a^{-1/2}.$$

To make this work, a will have to depend on p, and in fact (assuming $\rho > \delta$), it can best be achieved for large p by taking $a \approx \langle p \rangle^{\rho+\delta}$. This suggests that we will want to modify the transform $P_\phi f(p, q)$ by dilating ϕ by a factor depending on p.

Let us start over and approach the problem from a different angle. We hope that the expansion (3.4) will diagonalize $\sigma(D, X)$ approximately, i.e., that $P_\phi^* M_\sigma P_\phi$ will be almost equal to $\sigma(D, X)$, where

$$M_\sigma = \text{multiplication by } \sigma.$$

But in fact, we have:

(3.5) Proposition. *If $\phi \in S$ and $\sigma \in S_{\rho,\delta}^\infty$ then*

$$P_\phi^* M_\sigma P_\phi = (\sigma * W\phi)(D, X).$$

Proof: For any $f, g \in S$, by (1.93), (1.95), Proposition (2.5), and the fact that $W\phi$ is real, we have

$$\langle P_\phi^* M_\sigma P_\phi f, g \rangle = \langle M_\sigma P_\phi f, P_\phi g \rangle = \iint \sigma(p,q) \langle f, \phi_{pq} \rangle \overline{\langle g, \phi_{pq} \rangle} \, dp \, dq$$

$$= \iint \sigma(p,q) \langle W(f,g), W\phi_{pq} \rangle \, dp \, dq$$

$$= \iiiint \sigma(p,q) W(f,g)(\xi, x) W\phi(\xi-p, \, x-q) d\xi \, dx \, dp \, dq$$

$$= \iint (\sigma * W\phi)(\xi, x) W(f,g)(\xi, x) \, d\xi \, dx = \langle (\sigma * W\phi)(D, X)f, g \rangle. \quad \blacksquare$$

So the question is: how good an approximation to σ is $\sigma * W\phi$? But the attentive reader will realize that we have already answered this question in the course of proving the sharp Gårding inequality. Indeed, suppose $\phi \in S$ is even and $\|\phi\|_2 = 1$, and set $\phi^a(x) = a^{n/4}\phi(a^{1/2}x)$. Then we have shown (cf. Remark (2.89)) that $\sigma - \sigma * W\phi^a \in S_{\rho,\delta}^{m-(\rho-\delta)}$ whenever $\sigma \in S_{\rho,\delta}^m$ is supported in a set where $\langle \xi \rangle^{\rho+\delta} \approx a$. The restriction $\langle \xi \rangle^{\rho+\delta} \approx a$ is essentially the same as the relation $\langle p \rangle^{\rho+\delta} \approx a$ obtained in the preceding argument, and we can remove it by cutting σ up into pieces that are supported on sets where $\langle \xi \rangle$ is essentially constant. In short, combining our proof of Theorem (2.84) with Remark (2.89) and Proposition (3.5), we obtain the following theorem.

(3.6) Theorem. *Suppose $\phi \in S$ is even with $\|\phi\|_2 = 1$, and $\sigma \in S^m_{\rho,\delta}$ with $\rho > \delta$. Let $\{\zeta_j\}^\infty_{-1}$ be the partition of unity in Lemma (2.79), and for $j \geq 0$ set*

$$\sigma_j(\xi, x) = \zeta_j(|\xi|)\sigma(\xi, x), \qquad \phi_j(x) = 2^{nj(\rho+\delta)/4}\phi(2^{j(\rho+\delta)/2}x).$$

Then

$$\sigma - \sum_0^\infty \sigma_j * W\phi_j \in S^{m-(\rho-\delta)}_{\rho,\delta}$$

and hence

$$\sigma(D, X) - \sum_0^\infty P^*_{\phi_j} M_{\sigma_j} P_{\phi_j} \in OPS^{m-(\rho-\delta)}_{\rho,\delta}.$$

Thus we have achieved an approximate diagonalization of $\sigma(D, X)$, at least in a piecemeal fashion. We next observe that if we replace the cutoff functions ζ_j by the characteristic functions of which they are the smoothed-out versions—namely, set $\sigma'_j(\xi, x) = \chi_j(|\xi|)\sigma(\xi, x)$ where χ_j is the characteristic function of $[2^j, 2^{j+1})$ for $j \geq 0$ and of $[0, 1)$ for $j = -1$—then $\sigma = \sum_{-1}^\infty \sigma'_j$ and

$$\sum P^*_{\phi_j} M_{\sigma'_j} P_{\phi_j} f(x) = \iint \langle f, \widetilde{\phi}_{pq}\rangle \sigma(p, q)\widetilde{\phi}_{pq}(x)\, dp\, dq$$

where ϕ_j is as in Theorem (3.6) and

$$\widetilde{\phi}_{pq}(x) = (\phi_j)_{pq}(x) \text{ for } 2^j \leq |p| < 2^{j+1}.$$

Ths sum $\sum P^*_{\phi_j} M_{\sigma_j} P_{\phi_j}$ in Theorem (3.6) is a regularized version of this operator. This suggests that we could achieve much the same effect if, instead of performing a dyadic decomposition of momentum space and using a different dilation factor on ϕ in each annulus, we allowed the dilation factor on ϕ_{pq} to depend continuously on p in the following fashion.

For any $t > 0$, we set

$$\phi^t(x) = t^{n/4}\phi(t^{1/2}x).$$

Given $\alpha \geq 0$, we consider the functions

$$(\phi^{\langle p\rangle^\alpha})_{pq}(x) = e^{-\pi ipq + 2\pi ipx}\langle p\rangle^{\alpha n/4}\phi(\langle p\rangle^{\alpha/2}(x - q))$$

and define the transform

(3.7) $$Q^\alpha_\phi f(p, q) = \langle f, (\phi^{\langle p\rangle^\alpha})_{pq}\rangle.$$

When ϕ is a Gaussian and $\alpha = 1$, Q_ϕ^α is essentially the wave packet transform of Córdoba and Fefferman [33]. As before, we have

$$\langle Q_\phi^\alpha f, G \rangle = \iiint f(x)\overline{(\phi^{\langle p \rangle^\alpha})_{pq}(x)G(p,q)}\, dx\, dp\, dq,$$

so that

$$(Q_\phi^\alpha)^* G(x) = \iint (\phi^{\langle p \rangle^\alpha})_{pq}(x)G(p,q)\, dp\, dq,$$

where the integral converges absolutely when G decays sufficiently rapidly at infinity. Q_ϕ^α is no longer an isometry when $\alpha \neq 0$, so $(Q_\phi^\alpha)^* Q_\phi^\alpha \neq I$, but it is reasonable to hope that $(Q_\phi^\alpha)^* Q_\phi^\alpha$ will be a small perturbation of I, and more generally that $(Q_\phi^\alpha)^* M_\sigma Q_\phi^\alpha$ will be a small perturbation of $\sigma(D, X)$. We now show that this is indeed the case, by a modification of the argument used in Hörmander [72, Section 18.1] to prove the sharp Gårding inequality. We begin with the analogue of Proposition (3.5).

(3.8) Lemma. *If $\sigma \in S_{\rho,\delta}^\infty$, $\phi \in S$, and Q_ϕ^α is defined by (3.7), then*

$$(Q_\phi^\alpha)^* M_\sigma Q_\phi^\alpha = (T\sigma)(D, X),$$

where

$$(3.9) \qquad T\sigma(\xi, x) = \iint \sigma(p,q) W\phi\big(\langle p \rangle^{-\alpha/2}(\xi - p),\, \langle p \rangle^{\alpha/2}(x - q)\big)\, dp\, dq.$$

   Proof: For any $a > 0$ and $(p, q) \in \mathbf{R}^{2n}$, by Proposition (1.94) we have

$$W\big((\phi^a)_{pq}\big)(\xi, x) = W\phi\big(a^{-1/2}(\xi - p),\, a^{1/2}(x - q)\big).$$

Hence, as in the proof of Proposition (3.5),

$$\langle (Q_\phi^\alpha)^* M_\sigma Q_\phi^\alpha f,\, g \rangle = \langle M_\sigma Q_\phi^\alpha f,\, Q_\phi^\alpha g \rangle$$

$$= \iint \sigma(p,q)\langle f, (\phi^{\langle p \rangle^\alpha})_{pq}\rangle \overline{\langle g, (\phi^{\langle p \rangle^\alpha})_{pq}\rangle}\, dp\, dq$$

$$= \iint \sigma(p,q)\langle W(f,g), W(\phi^{\langle p \rangle^\alpha})_{pq}\rangle\, dp\, dq$$

$$= \iiiint \sigma(p,q) W(f,g)(\xi, x) W\phi\big(\langle p \rangle^{-\alpha/2}(\xi - p),\, \langle p \rangle^{\alpha/2}(x - q)\big)\, d\xi\, dx\, dp\, dq$$

$$= \iint T\sigma(\xi, x) W(f,g)(\xi, x)\, d\xi\, dx = \langle (T\sigma)(D, X)f,\, g \rangle. \quad \blacksquare$$

(3.10) Theorem. *Suppose $\sigma \in S_{\rho,\delta}^m$ with $\rho > \delta$.*
(a) *Given an even $\Phi \in S(\mathbf{R}^{2n})$ with $\int \Phi = 1$, let*

$$(3.11) \quad T\sigma(\xi, x) = \iint \sigma(p, q)\Phi(\langle p \rangle^{-(\rho+\delta)/2}(\xi - p), \langle p \rangle^{(\rho+\delta)/2}(x - q)) \, dp \, dq.$$

Then $T\sigma - \sigma \in S_{\rho,\delta}^{m-(\rho-\delta)}$.
(b) *If $\phi \in S(\mathbf{R}^n)$ is even, $\|\phi\|_2 = 1$, and Q_ϕ^α is defined by (3.7), then*

$$\sigma(D, X) - (Q_\phi^{\rho+\delta})^* M_\sigma Q_\phi^{\rho+\delta} \in OPS_{\rho,\delta}^{m-(\rho-\delta)}.$$

Proof: When $\alpha = \rho + \delta$ and $\Phi = W\phi$, the definitions of $T\sigma$ in (3.9) and (3.11) coincide. Hence, in view of Lemma (3.8), (b) is an immediate corollary of (a). In proving (a), we shall employ the notation

$$N(p) = \langle p \rangle^{(\rho+\delta)/2}.$$

We begin by observing that the operator T is a convolution as far as the second variable is concerned, and hence $D_x^\alpha T\sigma = T(D_x^\alpha \sigma)$. As for ξ-derivatives, we have

$$(D_{\xi_j} + D_{p_j})\Phi(N(p)^{-1}(\xi-p), N(p)(x-q)) = \Phi'(N(p)^{-1}(\xi-p), N(p)(x-q))N_j'(p)$$

where

$$\Phi'(\xi, x) = \frac{d}{dt}\Phi(t^{-1}\xi, tx)|_{t=1} = x \cdot \nabla_x \Phi - \xi \cdot \nabla_\xi \Phi,$$

$$N_j' = N^{-1}D_{p_j} N.$$

We regard $N_j'(p)$ as a function of p and q, independent of q; as such, it is in $S_{1,0}^{-1}$. Moreover, Φ' is even and $\int \Phi' = 0$. Thus,

$$D_{\xi_j} T\sigma = T(D_{\xi_j}\sigma) + T'(N_j'\sigma)$$

where T' is defined like T but with Φ' replacing Φ, and the arguments $D_{\xi_j}\sigma$ and $N_j'\sigma$ are in $S_{\rho,\delta}^{m-\rho}$. An inductive argument then shows that $D_\xi^\alpha D_x^\beta(T\sigma - \sigma)$ is a sum of terms of the form

$$(3.12) \quad B(\xi, x) = \iint \Psi(N(p)^{-1}(\xi-p), N(p)(x-q))\tau(p, q) \, dp \, dq$$

$$-\tau(p, q) \iint \Psi(p, q) \, dp \, dq$$

in which $\Psi \in S$ is even and $\tau \in S_{\rho,\delta}^{m-\rho|\alpha|+\delta|\beta|}$. (The second term is nonzero only when $\Psi = \Phi$ and $\tau = D_\xi^\alpha D_x^\beta \sigma$.) So it will suffice to prove that

$$\tau \in S_{\rho,\delta}^\mu \implies |B(\xi,x)| \leq C\langle \xi \rangle^{\mu-(\rho-\delta)}.$$

We break up the first integral in (3.12) into two pieces, I_1 and I_2, given by integrating over the regions $|p - \xi| > \frac{1}{2}\langle \xi \rangle$ and $|p - \xi| \leq \frac{1}{2}\langle \xi \rangle$. To estimate I_1, we observe that the factor of Ψ in the integrand is dominated by

$$(3.13) \qquad \left(1 + N(p)^{-1}|\xi - p| + N(p)|x - q|\right)^{-2n-1}\left(1 + N(p)^{-1}|\xi - p|\right)^{-2K}$$

for any K. Since $\langle \xi \rangle/\langle p \rangle$ and $\langle p \rangle/\langle \xi \rangle$ are bounded by $2\langle \xi - p \rangle$, the first factor of (3.13) is at most

$$C\left(1 + N(\xi)^{-1}|\xi - p| + N(\xi)|x - q|\right)^{-2n-1}\langle \xi - p \rangle^{(2n+1)(\rho+\delta)/2}.$$

Also, if $|p - \xi| \geq \frac{1}{2}\langle \xi \rangle$ then

$$\langle p \rangle \leq \langle \xi \rangle + |\xi - p| \leq 3\langle p - \xi \rangle$$

and hence $N(p) \leq C\langle \xi - p \rangle^{(\rho+\delta)/2}$, so the second factor of (3.13) is dominated by $\langle \xi - p \rangle^{-K(2-\rho-\delta)}$. Moreover,

$$|\tau(p,q)| \leq C\langle p \rangle^\mu \leq C'\langle \xi - p \rangle^{|\mu|},$$

so the whole integrand in I_1 is dominated by

$$\left(1 + N(p)^{-1}|\xi - p| + N(p)|x - q|\right)^{-2n-1}\langle \xi - p \rangle^{(n+1)(\rho+\delta)+|\mu|-K(2-\rho-\delta)}.$$

Since $\rho + \delta < 2$ we can choose K so large that the exponent of the second term is negative, in fact less than $-M$ for any M, and we can then replace $\langle \xi - p \rangle$ by $\langle \xi \rangle$ in this factor since we are assuming that $\langle \xi \rangle \leq 2\langle \xi - p \rangle$. Since

$$\iint \left(1 + N(\xi)^{-1}|\xi - p| + N(\xi)|x - q|\right)^{-2n-1} dp\, dq$$

is finite and independent of ξ and x, we conclude that

$$(3.14) \qquad \left| \iint_{|\xi-p|>\langle\xi\rangle/2} \Psi\left(N(p)^{-1}(\xi-p), N(p)(x-q)\right)\tau(p,q)\, dp\, dq \right| \leq C_M \langle \xi \rangle^{-M}$$

for any M.

To estimate I_2, we apply Taylor's formula to τ:

$$(3.15) \quad \tau(p,q) = \tau(\xi,x) + (p-\xi)\cdot\nabla_\xi\tau(\xi,x) + (q-x)\cdot\nabla_x\tau(\xi,x) + R(\xi,x,p,q)$$

where, for some point (ξ', x') on the line segment from (ξ, x) to (p, q),

$$R(\xi,x,p,q) = \sum_{|\alpha|+|\beta|=2} \frac{1}{\alpha!\beta!}(p-\xi)^\alpha(x-q)^\beta\partial_\xi^\alpha\partial_x^\beta\tau(\xi',x').$$

If $|p-\xi| \le \frac{1}{2}\langle\xi\rangle$ then $\langle\xi'\rangle$ and $\langle\xi\rangle$ are comparable, so since $\tau \in S_{\rho,\delta}^\mu$,

$$|R(\xi,x,p,q)| \le C\big[|\xi-p|^2\langle\xi\rangle^{\mu-2\rho} + |\xi-p||x-q|\langle\xi\rangle^{\mu-\rho+\delta} + |x-q|^2\langle\xi\rangle^{\mu+2\delta}\big]$$
$$\le C\big[N(\xi)^{-1}|\xi-p| + N(\xi)|x-q|\big]^2\langle\xi\rangle^{\mu-(\rho-\delta)}.$$

Hence the contribution of R to I_2 is easily estimated:

$$(3.16) \quad \iint_{|p-\xi|\le\langle\xi\rangle/2} |R(\xi,x,p,q)|\,\big|\Psi\big(N(p)^{-1}(\xi-p), N(p)(x-q)\big)\big|\,dp\,dq$$
$$\le C_K\langle\xi\rangle^{\mu-(\rho-\delta)} \iint \big(1+N(\xi)^{-1}(\xi-p) + N(\xi)(x-q)\big)^{2-K}dp\,dq$$
$$\le C\langle\xi\rangle^{\mu-(\rho-\delta)}.$$

To handle the contributions of the other terms in (3.15), we make some preliminary calculations on Ψ. If

$$\Psi'(\xi,x) = x\cdot\nabla_x\Psi - \xi\cdot\nabla_\xi\Psi$$

as above, Taylor's theorem applied to the function $t \to \Psi(t^{-1}\eta, ty)$ yields

$$(3.17) \quad |\Psi(t^{-1}\eta, ty) - \Psi(\eta,y) - (t-1)\Psi'(\eta,y)| \le C_K(1+|\eta|+|y|)^{-K}(t-1)^2$$

when (say) $t \in [10^{-1}, 10]$. If $|p-\xi| \le \frac{1}{2}\langle\xi\rangle$ this is the case if we take $t = N(p)/N(\xi)$, and moreover

$$t-1 = \frac{N(p)-N(\xi)}{N(\xi)} = \frac{p-\xi}{N(\xi)}\cdot\nabla N(\xi) + O\left(\frac{|p-\xi|^2}{N(\xi)}|\nabla\nabla N(\xi)|\right),$$
$$(t-1)^2 = \frac{[N(p)-N(\xi)]^2}{N(\xi)^2} = O\left(\frac{|p-\xi|^2}{N(\xi)^2}|\nabla N(\xi)|^2\right).$$

Now

$$|\nabla N(\xi)| \le C\langle\xi\rangle^{-1+(\rho+\delta)/2}, \qquad |\nabla\nabla N(\xi)| \le C\langle\xi\rangle^{-2+(\rho+\delta)/2} = C\frac{\langle\xi\rangle^{-2+\rho+\delta}}{N(\xi)},$$

so if we replace η, y, and t in (3.17) by $N(\xi)^{-1}(\xi - p)$, $N(\xi)(x - q)$, and $N(p)/N(\xi)$, we get

$$\left| \Psi\big(N(p)^{-1}(\xi-p), N(p)(x-q)\big) - \Psi\big(N(\xi)^{-1}(\xi-p), N(\xi)(x-q)\big) \right.$$
$$\left. - N(\xi)^{-1}(p-\xi)\cdot\nabla N(\xi)\Psi'\big(N(\xi)^{-1}(\xi-p), N(\xi)(x-q)\big) \right|$$
$$\leq C_k\big(1 + N(\xi)^{-1}|\xi - p| + N(\xi)|x - q|\big)^{2-K}\langle\xi\rangle^{-2+\rho+\delta}.$$

Hence, taking $K > 2n + 2$, it follows that

$$\iint_{|p-\xi|\leq\frac{1}{2}\langle\xi\rangle} \Psi\big(N(p)^{-1}(\xi-p), N(p)(x-q)\big)\, dp\, dq$$

$$= \iint_{|p-\xi|\leq\frac{1}{2}\langle\xi\rangle} \left[\Psi\big(N(\xi)^{-1}(\xi-p), N(\xi)(x-q)\big) \right.$$
$$\left. + (N(\xi)^{-1}(p-\xi)\cdot\nabla N(\xi)\Psi'\big(N(\xi)^{-1}(\xi-p), N(\xi)(x-q)\big) \right] dp\, dq + O(\langle\xi\rangle^{-2+\rho+\delta}).$$

In the integral on the right, the second term integrates to zero since Ψ' is even, and the first term is

$$\left(\iint_{\mathbf{R}^{2n}} - \iint_{|\eta|>\langle\xi\rangle/2N(\xi)} \right) \Psi(\eta, y)\, d\eta\, dy.$$

The second term here is rapidly decreasing in ξ since $\langle\xi\rangle/N(\xi) = \langle\xi\rangle^{1-(\rho+\delta)/2}$ and $\rho + \delta < 2$, so

$$(3.18) \quad \iint_{|p-\xi|\leq\langle\xi\rangle/2} \Psi\big(N(p)^{-1}(\xi-p), N(p)(x-q)\big)\, dp\, dq - \iint \Psi(p, q)\, dp\, dq$$
$$= O(\langle\xi\rangle^{-2+\rho+\delta}).$$

In the same way,

$$\iint_{|p-\xi|\leq\langle\xi\rangle/2} (p-\xi)^\alpha(q-x)^\beta \Psi\big(N(p)^{-1}(\xi-p), N(p)(x - q)\big)\, dp\, dq$$

$$= \iint_{|p-\xi|\leq\langle\xi\rangle/2} \left[(p-\xi)^\alpha(q-x)^\beta \Psi\big(N(\xi)^{-1}(\xi-p), N(\xi)(x - q)\big) \right.$$
$$- N(\xi)^{-1}(p-\xi)\cdot\nabla N(\xi)\Psi'\big(N(\xi)^{-1}(\xi-p), N(\xi)(x - q)\big)$$
$$\left. + O\big(1+N(\xi)^{-1}|\xi-p|+N(\xi)|x-q|\big)^{2-K}\langle\xi\rangle^{-2+\rho+\delta} \right] dp\, dq$$

$$= \iint_{|\eta|\leq\langle\xi\rangle/2N(\xi)} \left[\eta^\alpha y^\beta \Psi(\eta, y) - \eta\cdot\nabla N(\xi)\Psi'(\eta, y) \right.$$
$$\left. + O\big(1 + |\eta| + |y|\big)^{2-K}\langle\xi\rangle^{-2+\rho+\delta} \right] N(\xi)^{|\alpha|-|\beta|} d\eta\, dy.$$

When $|\alpha| + |\beta| = 1$ the integral of $\Psi(\eta, y)\eta^\alpha y^\beta$ vanishes since Ψ is even. The remaining terms give, for $|\alpha| = 1$ and $\beta = 0$,

$$O\big((|\nabla N(\xi)| + \langle\xi\rangle^{-2+\rho+\delta})N(\xi)\big) = O\big(\langle\xi\rangle^{-1+\rho+\delta} + \langle\xi\rangle^{-2+3(\rho+\delta)/2}\big)$$
$$= O\big(\langle\xi\rangle^{-1+\rho+\delta}\big),$$

and for $\alpha = 0$, $|\beta| = 1$,

$$O\big((|\nabla N(\xi)| + \langle\xi\rangle^{-2+\rho+\delta})N(\xi)^{-1}\big) = O\big(\langle\xi\rangle^{-1} + \langle\xi\rangle^{-2+(\rho+\delta)/2}\big) = O\big(\langle\xi\rangle^{-1}\big).$$

Therefore, if we plug the expansion (3.15) for $\tau(p,q)$ into I_2 and use (3.16), (3.18), and the estimates just above, we obtain

$$\left| \iint_{|p-\xi|\le\langle\xi\rangle/2} \Psi\big(N(p)^{-1}(\xi-p),\, N(p)(x-q)\big)\tau(p,q)dp\,dq - \tau(\xi,x)\iint \Psi(p,q)dp\,dq \right|$$
$$\le C\big[|\tau(\xi,x)|\langle\xi\rangle^{-2+\rho+\delta} + |\nabla_\xi\tau(\xi,x)|\langle\xi\rangle^{-1+\rho+\delta} + |\nabla_x\tau(\xi,x)|\langle\xi\rangle^{-1} + \langle\xi\rangle^{\mu-\rho+\delta}\big]$$
$$\le C\big[\langle\xi\rangle^{\mu-2+\rho+\delta} + \langle\xi\rangle^{\mu-1+\delta} + \langle\xi\rangle^{\mu-1+\delta} + \langle\xi\rangle^{\mu-\rho+\delta}\big]$$
$$\le C\langle\xi\rangle^{\mu-(\rho-\delta)}$$

since $\rho \le 1$. Combining this with (3.14), we are done. ∎

(3.19) Corollary. *For any $\phi \in \mathcal{S}$ with $\|\phi\|_2 = 1$ and any $\alpha \in (0,2)$,*

$$I - (Q_\phi^\alpha)^* Q_\phi^\alpha \in OPS_{1,0}^{m-(\rho-\delta)}$$

where $m = \min(\alpha, 2 - \alpha)$, $\rho = \min(\alpha, 1)$ and $\delta = \max(0, \alpha - 1)$.

This corollary is just the special case of Theorem (3.10) with $\sigma = 1$. Actually, it probably can be improved when $\alpha < 1$; after all, $(Q_\phi^0)^* Q_\phi^0 = P_\phi^* P_\phi = I$ with no error term. For $\alpha = 1$ it is sharp.

The sharp Gårding inequality is also an immediate corollary of Theorem (3.10), for if $\sigma \ge 0$ then M_σ is obviously a positive operator and hence so is $(Q_\phi^\alpha)^* M_\sigma Q_\phi^\alpha$.

In the case where $\rho = 1$, $\delta = 0$, and ϕ is a Gaussian, Theorem (3.10) is essentially due to Córdoba and Fefferman [33], although they only show that the error is in $S_{1,1/2}^{m-1}$. We included Theorem (3.10) mainly in order to encompass the work of Córdoba and Fefferman; for most purposes, Theorem (3.6) works at least as well, and its proof is simpler. Some related results, in which wave packet transforms are used that involve higher-order Hermite functions as well as the basic Gaussians, can be found in Unterberger [140], [141].

Open question: What happens if we use a $\phi \in \mathcal{S}$ that is not even? The proofs of Theorems (3.6) and (3.10) both break down, but are the theorems themselves still valid?

2. A Characterization of Wave Front Sets

In this section we show how the wave front set of a tempered distribution f can be characterized in terms of the wave packet transforms of f. Given a nonzero even function $\phi \in \mathcal{S}(\mathbf{R}^n)$, we set

$$\phi^t(x) = t^{n/4}\phi(t^{1/2}x)$$

as before, and define

$$P_\phi^t f(\xi, x) = \langle f, (\phi^t)_{\xi x} \rangle = \langle f, \rho(-x, \xi)\phi^t \rangle$$

for $f \in \mathcal{S}'$. (The reasons for writing (ξ, x) instead of (p, q) here are purely cosmetic; in the present context, (ξ, x) is the "natural" notation, as will soon become clear.) Clearly $P_\phi^t f$ is a tempered C^∞ function on $\mathbf{R}^{2n}$ for any $f \in \mathcal{S}'$ and $t > 0$, and we have

$$(3.20) \qquad \begin{aligned} P_\phi^t f(\xi, x) &= \int f(y) e^{\pi i \xi x - 2\pi i \xi y} \overline{\phi^t(y - x)} \, dy \\ &= e^{\pi i \xi x}(f\psi_{x,t})\widehat{\ }(\xi) = e^{\pi i \xi x}\widehat{f} * \zeta_{x,t}(\xi) \end{aligned}$$

where

$$\psi_{x,t}(y) = \overline{\phi^t(y - x)} \quad \text{and} \quad \zeta_{x,t}(\xi) = \widehat{\psi}_{x,t}(\xi) = e^{-2\pi i x \xi} t^{-n/4} \overline{\widehat{\phi}(-t^{-1/2}\xi)}.$$

We now define the **wave front set of f with respect to ϕ**, denoted $WF_\phi(f)$, for a tempered distribution f. This is a closed conic subset of $\mathbf{R}_*^{2n} = \{(\xi, x) : \xi \neq 0\}$, the complement of the set on which $P_\phi^t f(\xi, x)$ is rapidly decreasing when t and $|\xi|$ go to infinity at the same rate. Precisely:

$(\xi_0, x_0) \notin WF_\phi(f)$ if and only if there is a conic neighborhood V of (ξ_0, x_0) such that for all $a, N \geq 1$,

$$|P_\phi^t(t\xi, x)| \leq C_{a,N} t^{-N} \text{ for } t \geq 1, \ a^{-1} \leq |\xi| \leq a, \text{ and } (\xi, x) \in V.$$

This definition, when ϕ is a Gaussian, is due to Córdoba and Fefferman [33]. We begin our investigation of $WF_\phi(f)$ by showing that it depends only on the local properties of f.

(3.21) Lemma. *If $f \in S'$ and $f = 0$ on an open set $U \subset \mathbf{R}^n$, then $(\xi_0, x_0) \notin WF_\phi(f)$ for all $x_0 \in U$ and all $\xi_0 \in \mathbf{R}^n \backslash \{0\}$.*

Proof: If $x_0 \in U$, pick $\epsilon > 0$ small enough so that $x \in U$ for $|x - x_0| < 3\epsilon$, and pick $\psi \in C^\infty(\mathbf{R}^n)$ with $\psi(x) = 1$ for $|x| \geq 2\epsilon$ and $\psi(x) = 0$ for $|x| \leq \epsilon$. Then if $|x - x_0| \leq \epsilon$ we have

$$P_\phi^t f(t\xi, x) = \langle f, \rho(-x, t\xi)(\psi\phi^t) \rangle,$$

and we want to show that this decreases rapidly as $t \to \infty$, uniformly for $|x - x_0| \leq \epsilon$ and ξ in any compact subset of $\mathbf{R}^n \backslash \{0\}$. But since f is a continuous linear functional on S, there exists $k \geq 0$ such that

$$|\langle f, \zeta \rangle| \leq C \|\zeta\|_{[k]}, \quad \text{where} \quad \|\zeta\|_{[k]} = \sum_{|\alpha| + |\beta| \leq k} \sup_y |y^\alpha D^\beta \zeta(y)|.$$

It is easily verified that

$$\|\rho(-x, \xi)\zeta\|_{[k]} \leq C \langle x \rangle^k \langle \xi \rangle^k \|\zeta\|_{[k]},$$

so that if x and ξ remain bounded,

$$\|\rho(-x, t\xi)\zeta\|_{[k]} \leq C' t^k \|\zeta\|_{[k]}.$$

Moreover, for any N, since ψ vanishes near the origin and $\phi \in S$,

$$\|\psi\phi^t\|_{[k]} \leq C_N t^{(n/4) - (N/2)} \|\phi\|_{[k+N]}.$$

Combining these estimates, we have the desired result:

$$|P_\phi^t f(t\xi, x)| \leq C_N' t^{k + (n/4) - (N/2)}. \quad \blacksquare$$

This lemma implies that if $f, g \in S'$ and $f = g$ on an open set U, then

$$\{(\xi, x) \in WF_\phi(f) : x \in U\} = \{(\xi, x) \in WF_\phi(g) : x \in U\}.$$

In particular, in investigating the portion of $WF_\phi(f)$ lying over U we may multiply f by a cutoff function that equals 1 on U and hence assume that f has compact support. With this in mind, we come to the main result of this section.

(3.22) Theorem. *If $\phi \in S$ is even and nonzero then $WF_\phi(f) = WF(f)$ for all $f \in S'$.*

Proof: For each $(\xi_0, x_0) \in \mathbf{R}_*^{2n}$, we shall show that $(\xi_0, x_0) \notin WF_\phi(f)$ if and only if $(\xi_0, x_0) \notin WF(f)$. We begin with the following lemma.

(3.23) Lemma. *Suppose $f \in \mathcal{S}'$ has compact support, $\widehat{f}(\xi)$ is rapidly decreasing on a conic neighborhood U of ξ_0, and V is a conic neighborhood of ξ_0 with $\overline{V} \subset U$. Then there exists $K \geq 0$ such that*

$$|(\zeta f)\widehat{\ }(\xi)| \leq C_N \langle \xi \rangle^{-N} \int \langle \eta \rangle^{N+K} |\widehat{\zeta}(\eta)| \, d\eta$$

for all $\xi \in V$, $\zeta \in \mathcal{S}$, and $N \geq 0$.

Proof: Choose $\epsilon > 0$ so small that $\xi - \eta \in U$ when $\xi \in V$ and $|\eta| < \epsilon|\xi|$, and write

$$(\zeta f)\widehat{\ }(\xi) = \int_{|\eta| < \epsilon|\xi|} \widehat{\zeta}(\eta)\widehat{f}(\xi - \eta) \, d\eta + \int_{|\eta| \geq \epsilon|\xi|} \widehat{\zeta}(\eta)\widehat{f}(\xi - \eta) \, d\eta$$
$$= I_1 + I_2.$$

When $|\eta| < \epsilon|\xi|$ we have $\langle \xi - \eta \rangle \geq C\langle \xi \rangle$, so for $\xi \in V$,

$$|I_1| \leq C_N \int_{|\eta| < \epsilon|\xi|} |\widehat{\zeta}(\eta)| \langle \xi - \eta \rangle^{-N} d\eta \leq C_N' \langle \xi \rangle^{-N} \int |\widehat{\zeta}(\eta)| \, d\eta.$$

On the other hand, $|\widehat{f}(\eta)| \leq C\langle \eta \rangle^K$ for some K, and when $|\eta| \geq \epsilon|\xi|$ we have $\langle \xi - \eta \rangle \leq C\langle \eta \rangle$, so

$$|I_2| \leq C\langle \xi \rangle^{-N} \int_{|\eta| \geq \epsilon|\xi|} |\widehat{\zeta}(\eta)| \left(\epsilon^{-1}\langle \eta \rangle \right)^N \langle \xi - \eta \rangle^K d\eta$$
$$\leq C'\langle \xi \rangle^{-N} \int |\widehat{\zeta}(\eta)| \langle \eta \rangle^{N+K} d\eta. \blacksquare$$

Now suppose $(\xi_0, x_0) \notin WF(f)$. By Theorem (2.72), there exists $\psi \in C_c^\infty(\mathbf{R}^n)$ with $\psi = 1$ near x_0 and $\widehat{\psi f}$ rapidly decreasing on a conic neighborhood of ξ_0. We write

$$P_\phi^t f = P_\phi^t(\psi f) + P_\phi^t((1 - \psi)f).$$

By Lemma (3.23), with f replaced by ψf and $\zeta(y) = \overline{\phi^t(y - x)}$,

$$|P_\phi^t(\psi f)(t\xi, x)| \leq C_N \langle t\xi \rangle^{-N} \int |t^{-n/4}\widehat{\phi}(-t^{-1/2}\eta)| \langle \eta \rangle^{N+K} d\eta$$

if (ξ, x) is in a sufficiently small conic neighborhood of (ξ_0, x_0). But for $t \geq 1$,

$$\int |\widehat{\phi}(-t^{-1/2}\eta)| \langle \eta \rangle^{N+K} d\eta = t^{n/2} \int |\widehat{\phi}(\eta)| \langle t^{1/2}\eta \rangle^{N+K} d\eta$$
$$\leq t^{(n+N+K)/2} \int |\widehat{\phi}(\eta)| \langle \eta \rangle^{N+K} d\eta,$$

and hence, if $a^{-1} \leq |\xi| \leq a$,

$$|P_\phi^t(\psi f)(t\xi, x)| \leq C_N' t^{(n/4)+(N+K)/2} \langle t\xi \rangle^{-N} \leq C_{N,a} t^{(n+2K-2N)/4}.$$

On the other hand, if x is close enough to x_0 so that $\psi = 1$ near x, by Lemma (3.21) we have

$$|P_\phi^t((1-\psi)f)(t\xi, x)| \leq C_N t^{-N}$$

for all ξ. In short, $(\xi_0, x_0) \notin WF_\phi(f)$.

For the converse, we need another lemma.

(3.24) Lemma. *Suppose* $\sigma \in S_{1,0}^0$ *and* $S = \sigma(D, X)$. *There exists* $T \in OPS_{1,0}^{-1}$ *such that if* $f \in S'$ *and* $WF_\phi(f) \cap \mu \operatorname{supp} S = \emptyset$ *(i.e., if* $\sigma(\xi, x)$ *is rapidly decreasing in* ξ *on* $WF_\phi(f)$*), then* $Sf - Tf \in C^\infty$.

Proof: We use Theorem (3.6) with $\rho = 1$, $\delta = 0$. Writing $\sigma = \sum_{-1}^\infty \sigma_j$ as in Theorem (3.6), with σ_j supported in the set where $2^{j-1} \leq |\xi| \leq 2^{j+2}$ for $j \geq 0$, we set

$$S_j = (\sigma * W\phi^{2^j})(D, X), \qquad T = S - \sum_0^\infty S_j.$$

Then $T \in S_{1,0}^{-1}$ and $S_j = (P_\phi^{2^j})^* M_{\sigma_j} P_\phi^{2^j}$, that is,

$$(3.25) \qquad S_j f(x) = \iint \sigma_j(\eta, y) P_\phi^{2^j} f(\eta, y) e^{-\pi i \eta y + 2\pi i \eta x} \phi^{2^j}(x - y) \, dy \, d\eta.$$

(Actually, (3.25) was established in Proposition (3.5) only for $f \in S$. But since $\sigma_j(\eta, y)$ has compact support in η, $\phi^{2^j}(x - y)$ is rapidly decreasing in y, and $P_\phi^{2^j} f$ is a tempered function, the integral converges for any $f \in S'$, and a routine limiting argument shows that (3.25) remains valid in general.) Now, if $WF_\phi(f) \cap \mu \operatorname{supp} S = \emptyset$ then

$$|\sigma_j(\eta, y) P_\phi^{2^j} f(\eta, y)| \leq C_N 2^{-Nj} \quad \text{for } 2^{j-1} \leq |\eta| \leq 2^{j+2},$$
$$= 0 \quad \text{otherwise.}$$

It therefore follows easily from (3.25) that the series $\sum D^\alpha S_j f$ converges absolutely and uniformly for any α, and hence $Sf - Tf = \sum S_j f \in C^\infty$. (Exercise for the reader: rewrite this argument using Theorem (3.10) instead of Theorem (3.6).) ∎

Now suppose $(\xi_0, x_0) \notin WF_\phi(f)$. Let U and V be conic neighborhoods of (ξ_0, x_0) with $\overline{V} \subset U$ and $U \cap WF_\phi(f) = \emptyset$. Pick $\sigma_0^0 \in S_{1,0}^0$ supported in U and equal to 1 on a conic neighborhood of $\overline{V}$ [except near $\xi = 0$]; then by recursion pick a sequence $\{\sigma_k^0\} \subset S_{1,0}^0$ with σ_k^0 supported in the set where $\sigma_{k-1}^0 = 1$ and $\sigma_k^0 = 1$ on a conic neighborhood of $\overline{V}$; finally, pick $\sigma_\infty^0 \in S_{1,0}^0$ supported in V and equal to 1 on a conic neighborhood of (ξ_0, x_0). Fix $\psi = \psi(x) \in C_c^\infty$ with $\psi = 1$ near x_0, and set $\sigma_k = \psi \sharp \sigma_k^0$ and $S_k = \sigma_k(D, X)$. Then:

$$(3.26) \qquad \mu \operatorname{supp} S_k \cap \mu \operatorname{supp}(I - S_j) = \emptyset \text{ for } 0 \le j < k \le \infty,$$

$$(3.27) \qquad \operatorname{supp} S_k f \subset \operatorname{supp} \psi \text{ for } 0 \le k \le \infty \text{ and } f \in \mathcal{S}'.$$

By (3.27), $S_0 f$ has compact support, so $S_0 f \in H_s$ for some s. We shall prove by induction that $S_{2k} f \in H_{s+k}$. Indeed, by Lemma (3.24) we have $S_{2k} f = T_k f + g_k$ where $T_k \in OPS_{1,0}^{-1}$ and $g_k \in C^\infty$. Hence

$$
\begin{aligned}
S_{2k} f &= S_{2k-1} S_{2k} f + (I - S_{2k-1}) S_{2k} f \\
&= S_{2k-1} T_k f + S_{2k-1} g_k + (I - S_{2k-1}) S_{2k} f \\
&= S_{2k-1} T_k S_{2k-2} f + S_{2k-1} T_k (I - S_{2k-2}) f + S_{2k-1} g_k + (I - S_{2k-1}) S_{2k} f.
\end{aligned}
$$

In this last sum, the first term is in H_{s+k} since $S_{2k-2} f \in H_{s+k-1}$ by inductive hypothesis, and the other terms are all in C_c^∞ by (3.26), (3.27), Lemma (2.67), and Theorem (2.70). But then, in the same way,

$$
\begin{aligned}
S_\infty f &= S_\infty S_{2k} f + S_\infty (I - S_{2k}) f \\
&\in H_{s+k} + C_c^\infty = H_{s+k}
\end{aligned}
$$

for all k, so $S_\infty f \in C^\infty$. Since $(\xi_0, x_0) \notin \operatorname{char} S_\infty$, we conclude that $(\xi_0, x_0) \notin WF(f)$, and the proof of Theorem (3.22) is complete. ∎

As Weinstein [152] has observed, this theorem readily yields another characterization of $WF(f)$ in terms of the Wigner distribution Wf.

(3.28) Corollary. *Suppose $\phi \in \mathcal{S}$ is even and nonzero, and $f \in \mathcal{S}'$. Then $(\xi_0, x_0) \notin WF(f)$ if and only if there is a conic neighborhood U of (ξ_0, x_0) such that*

$$|(Wf * W\phi^t)(t\xi, x)| \le C_{a,N} t^{-N}$$

for all $t \ge 1$ and all $(\xi, x) \in U$ with $a^{-1} \le |\xi| \le a$.

Proof: We have merely to observe that by (1.93),

$$
\begin{aligned}
|P\phi^t f(x)|^2 &= |\langle f, \rho(-x, \xi)\phi^t \rangle|^2 = \langle Wf, W[\rho(-x, \xi)\phi^t] \rangle \\
&= \iint Wf(\eta, y) W\phi^t(\eta - \xi, y - x) \, d\eta \, dy = Wf * W\phi^t(\xi, x)
\end{aligned}
$$

since ϕ^t, and hence $W\phi^t$, is even. ∎

This corollary has a pleasing geometric interpretation. The discussion in Section 1.8 shows, on the one hand, that Wf is a sort of "portrait in phase space" of f, so the microlocal behavior of f at (ξ, x) should be reflected by the behavior of Wf along the ray $(t\xi, x)$ as $t \to \infty$. On the other hand, useful information about f is obtained not from the pointwise values $Wf(\xi, x)$ but from the smeared-out values $Wf * \Phi(\xi, x)$ where Φ "occupies" a rectangle

$$\left\{ (\xi, x) : |x| \le a, \ |\xi| \le a^{-1} \right\}$$

for some $a > 0$. Moreover, a suitable Φ for this purpose is $\Phi = W\phi^{a^2}$ where $\phi \in \mathcal{S}$. Thus we want to consider $Wf * W\phi^{a^2}(t\xi, x)$ where a is small (so as to localize at x) and t is large. But we also want a^{-1} to be small in comparison to t, so that $Wf * W\phi^{a^2}(t\xi, x)$ still involves only the behavior of Wf in the high-frequency region. We achieve both these aims by taking $a = \sqrt{t}$.

According to this line of thought, we actually should be able to take $a = t^\alpha$ for any $\alpha \in (0, 1)$. This is indeed correct:

(3.29) Theorem. *Theorem (3.22) and Corollary (3.28) remain valid if, in the definition of $WF_\phi f$ and the statement of Corollary (3.28), we redefine $\phi^t(x)$ to be $t^{\alpha n/2}\phi(t^\alpha x)$ where α is any number in $(0, 1)$.*

We leave it to the reader to verify that our proof that $WF_\phi(f) \subset WF(f)$ goes through with no essential change, as does our proof that $WF(f) \subset WF_\phi(f)$ once one observes that $S^0_{1,0} \subset S^0_{\rho,\delta}$ and that $\alpha = (\rho + \delta)/2$ is the appropriate exponent for dealing with $S^0_{\rho,\delta}$ in Theorem (3.6).

3. Analyticity and the FBI Transform

The first use of wave packet transforms to study the microlocal regularity properties of distributions, due to Iagolnitzer and Bros (see [78]), was in the analytic category rather than the C^∞ category. (By "analytic" we always mean "real analytic"; for "complex analytic" we shall use the term "holomorphic.") A complete discussion of these matters is beyond the scope of this monograph; we shall content ourselves with illustrating the ideas by proving—in dimension one, to minimize technicalities—that local analyticity of a distribution can be characterized by the exponential decay of a suitable wave packet transform.

The wave packet transform to be used is essentially the same as the P^t_ϕ of the previous section in which ϕ is a Gaussian, but since the factors of $t^{n/4}$ and $e^{\pi i x \xi}$ are irrelevant, we shall omit them and define, for $f \in \mathcal{S}'$,

$$(3.30) \qquad P^t f(\xi, x) = \int e^{-2\pi i y \xi - \pi t (y - x)^2} f(y) \, dy.$$

$P^t f$ is called the **FBI transform** of f; the initials are for Fourier, Bros, and Iagolnitzer. More generally, the function $(y - x)^2$ in the exponent could be replaced by a function $\psi(y - x)$ where $\psi(x)$ is analytic and real near $x = 0$, $\psi(0) = 0$, and ψ has a strict minimum at 0 with positive definite Hessian. However, we shall stick with the FBI transform defined by (3.30). The exponential decay condition on $P^t f$ in which we are interested is the following.

Condition $A(x_0)$: There exist a neighborhood U of x_0 and positive constants C, ϵ, K such that for all $t > 0$,

$$|P^t f(t\xi, x)| \leq Ce^{-\epsilon t} \text{ for } |\xi| \geq K \text{ and } x \in U.$$

(3.31) Theorem. *Suppose $f \in S'(\mathbf{R})$. Then f is analytic near $x_0 \in \mathbf{R}$ if and only if f satisfies Condition $A(x_0)$.*

Proof: Let us first observe that the satisfaction of Condition $A(x_0)$, like analyticity at x_0, depends only on the behavior of f in a neighborhood of x_0. Indeed, the proof of Lemma (3.21), with $\phi(x) = e^{-\pi x^2}$, shows that if $g = f$ near x_0 then $f - g$ automatically satisfies condition $A(x_0)$. Since analyticity at x_0 and Condition $A(x_0)$ both imply that f is C^∞ near x_0 (the latter by Theorem (3.22)), we may as well replace f by ψf where ψ is a suitable cutoff function that equals 1 near x_0, and assume that $f \in C_c^\infty$.

Suppose then that $f \in C_c^\infty(\mathbf{R})$ and f is analytic at x_0. This means that f can be extended holomorphically into some region

$$\{ y + iv : |y - x_0| < 2\delta, \ |v| < 2\delta \} \subset \mathbf{C}.$$

Choose $\chi \in C_c^\infty(\mathbf{R})$ supported in $\{|y - x_0| < 2\delta\}$ with $\chi = 1$ on $\{|y - x_0| \leq \delta\}$ and $0 \leq \chi \leq 1$ everywhere. Then, given $\xi \neq 0$, by Cauchy's theorem we can replace the axis of integration in (3.30) by the contour

$$\gamma(y) = y - i\delta'\chi(y), \qquad \delta' = \delta \operatorname{sgn} \xi,$$

obtaining

$$P^t f(x) =$$
$$\int e^{-2\pi i \xi y - 2\pi t \delta \chi(y)|\xi| - \pi t(y-x)^2 + 2\pi i t \delta' \chi(y) + \pi t \delta^2 \chi(y)^2} f(y - i\delta'\chi(y))(1 - i\delta'\chi'(y)) \, dy,$$

whence

$$|P^t f(t\xi, x)| \leq C \sup_y \exp\{-\pi t[2\delta\psi(y)|\xi| - \delta^2\psi(y)^2 + (y - x)^2]\}.$$

If $|\xi| \geq \delta$ and $|x - x_0| < \delta/2$, the expression in square brackets is greater than δ^2 when $|y - x_0| \leq \delta$ and greater than $(\delta/2)^2$ when $|y - x_0| > \delta$, so

$$|P^t f(t\xi, x)| \leq Ce^{-\delta^2 t/4} \text{ for } |\xi| \geq \delta \text{ and } |x - x_0| < \delta/2,$$

and Condition $A(x_0)$ is satisfied.

To prove the converse, as in the proof of Theorem (3.22), we need a way of recovering f from its FBI transform. Heuristically, we start with the formula

$$\delta(x) = \int e^{2\pi i x \xi} d\xi$$

and replace the contour of integration (the real axis) by the v-shaped contour

$$\gamma(\xi) = \xi + iax|\xi| \qquad (a > 0)$$

to obtain

$$\delta(x) = \int e^{2\pi i x \xi - 2\pi a x^2 |\xi|}(1 + iax \operatorname{sgn} \xi) \, d\xi.$$

To make this rigorous, observe that if $g \in \mathcal{S}$ then

$$\int e^{2\pi i x \xi - \pi b x^2} g(x) \, dx = (b^{-1/2} e^{-\pi \xi^2 / b}) * g(\xi),$$

so

$$\int e^{2\pi i x \xi - 2\pi a |\xi| x^2} g(x) \, dx = \int (2a|\xi|)^{-1/2} e^{-\pi(\xi - \eta)^2 / 2a|\xi|} \widehat{g}(\eta) \, d\eta.$$

This is easily seen to be rapidly decreasing in ξ, hence integrable with respect to ξ. Thus, it makes sense to define a tempered distribution $D = D_a$ by

(3.32) $$\langle f, D \rangle = \iint e^{2\pi i x \xi - 2\pi a x^2 |\xi|}(1 + iax \operatorname{sgn} \xi) f(x) \, dx \, d\xi.$$

(3.33) Lemma. *If D is defined by (3.32) with $a > 0$ then $D = \delta$.*

Proof: First, for any $x \neq 0$ we have

(3.34) $$\int e^{2\pi i x \xi - 2\pi a x^2 |\xi|}(1 + iax \operatorname{sgn} \xi) \, d\xi$$

$$= (1 - iax) \int_{-\infty}^{0} e^{2\pi i(x - iax^2)\xi} d\xi + (1 + iax) \int_{0}^{\infty} e^{2\pi i(x + iax^2)\xi} d\xi$$

$$= \frac{1 - iax}{2\pi i(x - iax^2)} - \frac{1 + iax}{2\pi i(x + iax^2)} = 0.$$

Suppose $f \in \mathcal{S}$ vanishes to second order at $x = 0$. Replacing the integral in (3.32) by its absolute value, we get

$$\iint e^{-2\pi a x^2 |\xi|}(1 + a^2 x^2)^{1/2} |f(x)| \, dx \, d\xi = \int (1 + a^2 x^2)^{1/2} \frac{|f(x)|}{2\pi a x^2} \, dx < \infty,$$

so we can apply Fubini's theorem and (3.34) to conclude that $\langle f, D \rangle = 0$. Next, if $f \in S$ vanishes to first order at $x = 0$, we can write $f = f_1 + f_2$ where f_1 is odd and f_2 vanishes to second order at 0. We have just seen that $\langle f_2, D \rangle = 0$, and substituting $(-\xi, -x)$ for (ξ, x) in (3.32) shows that $\langle f_1, D \rangle = 0$. In short, if $f(0) = 0$ then $\langle f, D \rangle = 0$, so it must be that $D = c\delta$ for some $c \in \mathbf{C}$. Pick $f \in C_c^\infty$ with $f = 1$ near the origin, and set $f_k(x) = f(x/k)$. Then $c = \langle f_k, D \rangle$ for all k, so by letting $k \to \infty$ we obtain

$$c = \iint e^{2\pi i x\xi - 2\pi ax^2 |\xi|}(1 + iax \operatorname{sgn}\xi)\, dx\, d\xi.$$

Therefore, since

$$(e^{-\pi bx^2}\Upsilon(\xi) = b^{-1/2}e^{-\pi\xi^2/b} \quad \text{and} \quad (xe^{-\pi bx^2}\Upsilon(\xi) = b^{-3/2}i\xi e^{-\pi\xi^2/b},$$

we find that

$$c = \int\left[(2a|\xi|)^{-1/2}e^{-\pi|\xi|/2a} + (ia\operatorname{sgn}\xi)(2a|\xi|)^{-3/2}i\xi e^{-\pi|\xi|/2a}\right]d\xi$$

$$= (8a)^{-1/2}\int |\xi|^{-1/2}e^{-\pi|\xi|/2a}d\xi = \pi^{-1/2}\int_0^\infty \eta^{-1/2}e^{-\eta}d\eta = 1. \quad \blacksquare$$

(3.35) Lemma. If $f, g \in C_c^\infty$ satisfy Condition $A(x_0)$, then so does fg.

Proof: According to hypothesis, there exist C, ϵ, K, U such that

$$|P^t f(t\xi, x)| \le Ce^{-\epsilon t} \quad \text{and} \quad |P^t g(t\xi, x)| \le Ce^{-\epsilon t} \quad \text{for } |\xi| \ge K \text{ and } x \in U.$$

Henceforth we assume that $|\xi| \ge K$ and $x \in U$. Setting

$$h_1(y) = e^{-\pi t(y-x)^2/2}f(y), \qquad h_2(y) = e^{-\pi t(y-x)^2/2}g(y),$$

we have

$$\widehat{h_1}(\xi) = P^{t/2}f(\xi, x), \qquad \widehat{h_2}(\xi) = P^{t/2}g(\xi, x),$$
$$e^{-\pi t(y-x)^2}f(y)g(y) = h_1(y)h_2(y),$$

and hence

$$P^t(fg)(t\xi, x) = (h_1 h_2 \Upsilon(t\xi) = \widehat{h_1} * \widehat{h_2}(t\xi)$$
$$= \int P^{t/2}f(t\xi - \eta, x)P^{t/2}g(\eta, x)d\eta = I_1 + I_2$$

where I_1 and I_2 are the integrals over the regions $|\eta| \leq \frac{1}{2}t|\xi|$ and $|\eta| > \frac{1}{2}t|\xi|$. If $|\eta| \leq \frac{1}{2}t|\xi|$ then $|\xi - t^{-1}\eta| \geq \frac{1}{2}|\xi|$, so

$$|P^{t/2}f(t\xi - \eta, x)| = |P^{t/2}f((t/2)(2\xi - 2t^{-1}\eta), x)| \leq Ce^{-\epsilon t/2}.$$

On the other hand,

$$\int |P^{t/2}g(\eta, x)| \, d\eta = \|\hat{h}_2\|_1 \leq C \sum_{|\alpha| \leq 2} \|D^\alpha h_2\|_1 \leq C'(t+1)^2,$$

and hence

$$|I_1| \leq C(t+1)^2 e^{-\epsilon t/2} \leq C' e^{-\epsilon t/3}.$$

Similarly, if $|\eta| > \frac{1}{2}t|\xi|$ then

$$|P^{t/2}g(\eta, x)| = |P^{t/2}g((t/2)(2t^{-1}\eta), x)| \leq Ce^{-\epsilon t/2},$$

from which it follows that $|I_2| \leq C' e^{-\epsilon t/3}$. Combining these estimates, we are done. ∎

Now we can complete the proof of Theorem (3.31). Suppose that $f \in C_c^\infty$ and f satisfies Condition $A(x_0)$. By Lemma (3.33), for any $a > 0$ we have

$$(3.36) \qquad f(x) = \iint e^{2\pi i(x-y)\xi - 2\pi a(x-y)^2|\xi|}(1 + ia(x-y)\operatorname{sgn}\xi)f(y) \, dy \, d\xi.$$

Let $h(y) = yf(y)$: then $h = fg$ where $g \in C_c^\infty$ and $g(y) = y$ on a neighborhood of $\{x_0\} \cup \operatorname{supp} f$. Since g is analytic at x_0 it satisfies Condition $A(x_0)$, and hence, by Lemma (3.35), so does h. Thus, for some positive C, K, ϵ, and δ,

$$|P^t f(\eta, x)| \leq Ce^{-\epsilon t} \quad \text{and} \quad |P^t h(\eta, x)| \leq Ce^{-\epsilon t} \quad \text{for } |\eta| \geq K \text{ and } |x - x_0| < \delta.$$

But if $z = x + iv$ with x, v real, we have

$$\int e^{2\pi i(z-y)\xi - 2\pi a(z-y)^2|\xi|}(1 + ia(z-y)\operatorname{sgn}\xi)f(y) \, dy$$

$$= e^{2\pi iz\xi - 4\pi iaxv|\xi| + 2\pi av^2|\xi|} \times$$

$$\left[(1 + iaz\operatorname{sgn}\xi)P^{2a|\xi|}f(\xi - 2av|\xi|, x) - ia\operatorname{sgn}\xi P^{2a|\xi|}h(\xi - 2av|\xi|, x)\right].$$

Therefore, if we choose $a = (4K)^{-1}$ and require that

$$(3.37) \qquad z = x + iv, \qquad |x - x_0| < \delta, \qquad |v| < \min\left(\frac{\epsilon}{16\pi K}, \frac{1}{2}\sqrt{\frac{\epsilon}{\pi}}, K\right),$$

we obtain

$$|e^{2\pi iz\xi - 4\pi iaxv|\xi| + 2\pi av^2|\xi|}| \leq e^{\epsilon a|\xi|} \quad \text{and} \quad |\xi - 2av|\xi|| \geq 2a|\xi|K,$$

and hence

$$\left|\int e^{2\pi i(z-y)\xi - 2\pi a(z-y)^2|\xi|}(1 + ia(z-y)\operatorname{sgn}\xi)f(y) \, dy\right| \leq C' e^{-\epsilon a|\xi|}.$$

But then we can substitute z for x in (3.36), thereby obtaining a holomorphic extension of f to the complex neighborhood (3.37) of x_0. f is therefore analytic at x_0. ∎

This argument can be adapted to higher dimensions: see Iagolnitzer [78] or Sjöstrand [129]. It can also be localized in ξ-space. The microlocal analogue of Condition $A(x_0)$ is

Condition $A(\xi_0, x_0)$: There exist a conic neighborhood U of (ξ_0, x_0) and positive constants C, ϵ, K such that for all $t > 0$,

$$|P^t f(t\xi, x)| \le C e^{-\epsilon t} \text{ for } |\xi| \ge K \text{ and } (\xi, x) \in U.$$

One can then prove the following extension of Theorem (3.31).

(3.38) Theorem. $f \in \mathcal{S}'(\mathbf{R}^n)$ satisfies Condition $A(\xi_0, x_0)$ if and only if there exist open convex cones $\Gamma_1, \dots, \Gamma_k \subset \{y \in \mathbf{R}^n : y\xi_0 < 0\}$, a complex neighborhood W of x_0, and holomorphic functions g_j on $W \cap (\mathbf{R}^n + i\Gamma_j)$, $1 \le j \le k$, such that $f|(W \cap \mathbf{R}^n)$ is the sum of the distribution boundary values of the g_j's.

There are several ways to define the "analytic wave front set" or "essential support" of a distribution f, which is intuitively the set of (ξ, x) such that f is not analytic at x in the direction ξ. One of these exploits the analogy with the characterization of $WF(f)$ in Theorem (3.22):

$$WF_a(f) = \big\{ (\xi, x) : \text{ Condition } A(\xi, x) \text{ is not satisfied} \big\}.$$

The characterization of $WF_a(f)$ given by Theorem (3.38) is essentially the definition of Sato; one can also impose growth restrictions on the functions g_j. There is yet another definition of $WF_a(f)$, due to Hörmander, in terms of decay conditions on $(\chi_j f)\hat{}$ for a suitable sequence $\{\chi_j\}$ of cutoff functions. For the proof of Theorem (3.38), the equivalence of all these definitions, and related matters, see Iagolnitzer [78], [79], Sjöstrand [129], Bony [24], and Hörmander [71].

4. Gabor Expansions

In this section we work in dimension $n = 1$.

In an influential paper [53] of 1946 on the theory of signal communication, D. Gabor (who later won the Nobel prize for his work in laser holography) stressed the importance of analysis in the time-frequency plane—what we have been calling phase space—and the utility of the Gaussian wave packets

$$(3.39) \qquad \phi_{ab}^c(x) = 2^{1/4} c^{1/2} e^{2\pi i a x} e^{-\pi c^2 (x-b)^2} \qquad (a, b \in \mathbf{R}, \ c > 0)$$

as basic building blocks for signals. (In this context, x represents time.) ϕ_{ab}^c is "maximally localized" in phase space according to the uncertainty principle

(1.35): ϕ_{ab}^c and its Fourier transform are essentially supported in intervals of length c^{-1} and c centered at b and a respectively. Since rectangles of area 1 are in some sense the smallest meaningful units of phase space, Gabor suggested considering phase space as the union of the rectangles of side length c and c^{-1} centered at

$$(a_j, b_k) = (jc, kc^{-1}) \qquad (j, k \in \mathbf{Z}),$$

associating to each rectangle the corresponding wave packet $\phi_{a_j b_k}^c$, and expanding an arbitrary $f \in L^2$ as a superposition of these functions. Gabor made no attempt at a rigorous analysis of such expansions, but it seems plausible that his idea should work well. The $\phi_{a_j b_k}^c$ are not mutually orthogonal, but they are nearly so in the sense that $\langle \phi_{a_j b_k}^c, \phi_{a_l b_m}^c \rangle$ tends rapidly to zero as $|j-l|+|k-m|$ increases. Expanding f with respect to the $\phi_{a_j b_k}^c$ should be a smoothed-out version of expanding f in a Fourier series on each interval $[(k-\frac{1}{2})c^{-1}, (k+\frac{1}{2})c^{-1}]$.

By substituting $c^{-1}x$, ca, and $c^{-1}b$ for x, a, and b in (3.39), we can reduce to the case $c = 1$. Thus, we shall work with the functions

$$\phi(x) = 2^{1/4} e^{-\pi x^2},$$

$$\phi_{ab}(x) = 2^{1/4} e^{2\pi i a x} e^{-\pi(x-b)^2} = \rho^{\text{pol}}(-b, a)\phi(x),$$

where the notation ρ^{pol} is as in (1.106). We are particularly interested in the set

$$\{\,\phi_{jk} : j, k \in \mathbf{Z}\,\}.$$

The first question to be settled is whether the set $\{\phi_{jk}\}$ is *complete*, that is, whether the equations $\langle f, \phi_{jk} \rangle = 0$ for all j, k imply that $f = 0$. In fact, the completeness of $\{\phi_{jk}\}$ was asserted without proof in 1932 by von Neumann [147, Section 5.4]; but it was not until 1971 that a proof—two independent ones, in fact, due to Bargmann et al. [14] and Perelomov [117]—appeared in the literature.

Following Bargmann et al. [14], we shall use the Weil-Brezin transform

$$T_0 f(p, q) = \sum_{-\infty}^{\infty} f(p + k) e^{2\pi i k q}$$

as the cornerstone of our arguments. We recall some facts from from Section 1.10. First, T_0 is a unitary map from $L^2(\mathbf{R})$ to the space $\mathcal{H}^1$ of functions on $\mathbf{R}^2$ that are square-integrable on the unit square and satisfy the quasi-periodicity condition

$$g(p + j, q + k) = e^{-2\pi i j q} g(p, q) \qquad (j, k \in \mathbf{Z}).$$

Second, if we denote the Weil-Brezin transform of ϕ by Θ,

$$\Theta(p, q) = T_0 \phi(p, q) = 2^{1/4} \sum e^{2\pi i k q} e^{-\pi(p+k)^2},$$

we have

$$\Theta(p, q) = 2^{1/4} e^{-\pi p^2} \vartheta_3 \big(\pi(q + ip), \, e^{-\pi} \big)$$

where ϑ_3 is a Jacobi theta function. Finally,

$$T_0 \phi_{jk}(p, q) = e^{2\pi i(jp+kq)} T_0 \phi(p, q) = e^{2\pi i(jp+kq)} \Theta(p, q).$$

If we abbreviate $\vartheta_3(z, e^{-\pi})$ by $\vartheta_3(z)$, the quasiperiodicity of Θ means that

$$\vartheta_3(z + \pi) = \vartheta_3(z), \qquad \vartheta_3(z + i\pi) = -e^\pi e^{-2iz} \vartheta_3(z).$$

Since ϑ_3 is holomorphic, it follows by a simple application of the argument principle that ϑ_3 has precisely one zero in each square of side π, and hence that Θ has exactly one zero in each square of side 1. But

$$\Theta(\tfrac{1}{2}, \tfrac{1}{2}) = 2^{1/4} \sum (-1)^k e^{-\pi(2^{-1}+k)^2}$$
$$= 2^{1/4} \sum (-1)^{-j-1} e^{-\pi(-2^{-1}-j)^2} = -\Theta(\tfrac{1}{2}, \tfrac{1}{2}),$$

so we have proved:

(3.40) Lemma. $\Theta(p, q)$ *has simple zeros at the points* $(\tfrac{1}{2}+j, \, \tfrac{1}{2}+k)$ *for* $j, k \in \mathbf{Z}$, *and no other zeros.*

Now we can establish completeness, and in fact a bit more.

(3.41) Theorem. *The set* $\{\phi_{jk} : j, k \in \mathbf{Z}\}$ *is complete in* $L^2(\mathbf{R})$. *It remains complete if any one element is deleted, but not if two elements are deleted.*

Proof: Suppose $f \in L^2(\mathbf{R})$ and $\langle f, \phi_{jk} \rangle = 0$ for all (j, k) except (a, b). Then for $(j, k) \neq (a, b)$ we have

$$0 = \langle f, \phi_{jk} \rangle = \langle T_0 f, \, T_0 \phi_{jk} \rangle_{\mathcal{H}^1} = \int_0^1 \int_0^1 T_0 f(p, q) \overline{\Theta(p, q)} e^{-2\pi i(jp+kq)} \, dp \, dq.$$

But this last integral is the (j, k)-th Fourier coefficient of $(T_0 f)\overline{\Theta}$, so it follows that $(T_0 f)\overline{\Theta}$ is a scalar multiple of $e^{2\pi i(ap+bq)}$ on the unit square, that is,

$$T_0 f(p, q) = C e^{2\pi i(ap+bq)} \Theta(p, q)^{-1} \quad \text{for } p, q \in [0, 1).$$

But since Θ^{-1} has a pole at $(\tfrac{1}{2}, \tfrac{1}{2})$ it is not L^2 on the unit square, whereas $T_0 f$ is; hence $C = 0$ and $f = 0$. On the other hand, if (a, b) and (c, d) are distinct pairs of integers, we can choose nonzero constants C_1 and C_2 such that $C_1 e^{\pi i(a+b)} + C_2 e^{\pi i(c+d)} = 0$. Then the function G on the unit square defined by

$$G(p,q) = (C_1 e^{2\pi i(ap+bq)} + C_2 e^{2\pi i(cp+dq)})\Theta(p,q)^{-1}$$

is bounded, hence L^2. We regard G as an element of $\mathcal{H}^1$ and set $f = T_0^{-1}G$; then the above argument shows that $\langle f, \phi_{jk} \rangle = 0$ for all (j,k) except (a,b) and (c,d). ∎

More generally, one can ask whether the set $\{\phi_{ab} : (a,b) \in \Lambda\}$ is complete in $L^2(\mathbf{R})$ where Λ is an arbitrary lattice in $\mathbf{R}^2$. The answer depends on the area $|\Lambda|$ of a fundamental parallelogram for Λ (i.e., $|\Lambda| = |a_1 b_2 - a_2 b_1|$ where (a_1, b_1) and (a_2, b_2) are a set of generators for Λ). Here are the results.

Case I: $|\Lambda| = 1$. In this case there exists $\mathcal{A} \in SL(2, \mathbf{R}) = Sp(1, \mathbf{R})$ which maps Λ onto the integer lattice $\mathbf{Z}^2$, and we can essentially reduce the problem to the special case studied above by invoking the results of Chapter 4 below. Specifically, there is a unitary operator $\mu(\mathcal{A})$ on $L^2(\mathbf{R})$, the metaplectic operator corresponding to $\mathcal{A}$, which transforms $\{\phi_{ab} : (a,b) \in \Lambda\}$ into $\{C_{jk}\rho^{\text{pol}}(j,k)\psi : (j,k) \in \mathbf{Z}^2\}$, where $C_{jk} \in \mathbf{C}$ and $\psi(x) = e^{\pi i \tau x^2}$ for some $\tau \in \mathbf{C}$ with $\text{Im}\,\tau > 0$. The preceding arguments apply equally well to the latter set of functions; one has merely to replace the variables $\pi(q + ip)$ and $e^{-\pi}$ in ϑ_3 by $\pi(q + \tau p)$ and $e^{\pi i \tau}$. Conclusion: $\{\phi_{ab} : (a,b) \in \Lambda\}$ is complete and remains so after deletion of any one element.

Case II: $|\Lambda| < 1$. We observe that

$$\langle f, \phi_{ab} \rangle = e^{-(\pi/2)(p^2+q^2)} e^{\pi ipq} Bf(-q + ip)$$

where B is the Bargmann transform. $Bf(z)$ is an entire function which grows no faster than $e^{(\pi/2)|z|^2}$. One can therefore invoke theorems relating the order of growth of an entire function to upper bounds for the density of its zeros to conclude that if $|\Lambda| < 1$, $\{\phi_{ab} : (a,b) \in \Lambda\}$ is complete and remains so after deletion of any finite set (or even any sufficiently sparse infinite set).

Case III: $|\Lambda| > 1$. In this case one can construct an explicit $F \in \mathcal{F}_n$ such that $F(-b + ia) = 0$ for all $(a,b) \in \Lambda$. Taking $f = B^{-1}F$, we have $\langle f, \phi_{ab} \rangle = 0$ for all $(a,b) \in \Lambda$, and hence $\{\phi_{ab} : (a,b) \in \Lambda\}$ is not complete.

For the details of the arguments in Cases II and III, see Bargmann et al. [14] or Perelomov [117].

Incidentally, although $\{\phi_{jk} : j, k \in \mathbf{Z}\}$ is complete in L^2, it is not complete in $\mathcal{S}$ (in the topology of $\mathcal{S}$). Indeed, every f in the closed linear span of $\{\phi_{jk}\}$ in $\mathcal{S}$ is annihilated by the linear functional $f \to T_0 f(\frac{1}{2}, \frac{1}{2})$. For a more detailed study of tempered distributions that annihilate ϕ_{ab} for all (a,b) in some lattice, see Janssen [83].

Once we know that $\{\phi_{jk}\}$ is complete in L^2, the question remains whether every $f \in L^2$ has an L^2-convergent expansion $f = \sum c_{jk}\phi_{jk}$. Here the situation is surprisingly unpleasant. A warning of troubles is already provided by the remarks of the preceding paragraph, which show that an $f \in \mathcal{S}$ will usually not have an $\mathcal{S}$-convergent expansion. More generally, if

$$f = \sum_{j,k \in \mathbf{Z}} c_{jk} \phi_{jk} \qquad \text{(convergence in } L^2\text{)},$$

then

$$T_0 f(p,q) = \sum c_{jk} T_0 \phi_{jk}(p,q) = \Theta(p,q) \sum c_{jk} e^{2\pi i(jp+kq)},$$

or

$$(\Theta^{-1} T_0 f)(p,q) = \sum c_{jk} e^{2\pi i(jp+kq)}.$$

$\Theta^{-1} T_0 f$ is a doubly periodic function on $\mathbf{R}^2$, and the series on the right must be its Fourier series. But since Θ^{-1} has poles, $\Theta^{-1} T_0 f$ will usually not be in L^2, so the coefficients c_{jk} will not be square-summable. In fact, there exist $f \in L^2$ such that $\Theta^{-1} T_0 f \notin L^1$, in which case it is not even obvious how to define c_{jk} ($\Theta^{-1} T_0 f$ must be interpreted as a distribution). Even if $T_0 f$ is smooth, unless $T_0 f(\frac{1}{2}, \frac{1}{2}) = 0$ the best that can be expected is that $|c_{jk}| \leq C(|j| + |k|)^{-1}$. These arguments do not immediately prove the nonexistence of the expansion $f = \sum c_{jk} \phi_{jk}$, for since the ϕ_{jk} are not orthogonal, square-summability of $\{c_{jk}\}$ may not be necessary for the convergence of $\sum c_{jk} \phi_{jk}$. Nonetheless, with some more work one can show that (a) there exists $f \in L^2$ which cannot be written as $\sum c_{jk} \phi_{jk}$ with convergence in L^2, but (b) any $f \in L^2$ can be written as $\sum c_{jk} \phi_{jk}$ with convergence in $\mathcal{S}'$. For these and related matters, see Janssen [84], [86].

Our arguments do show, however, that in spite of Theorem (3.41), the set $\{\phi_{jk}\}$ with one element (say ϕ_{00}) deleted is not a basis for L^2. Specifically, one cannot write

$$\phi_{00} = \sum_{(j,k) \neq (0,0)} c_{jk} \phi_{jk}$$

with convergence in L^2, for on applying the Weil-Brezin transform to both sides one would get

$$1 = \sum_{(j,k) \neq (0,0)} c_{jk} e^{2\pi i(jp+kq)},$$

which is impossible even if convergence on the right is only in the sense of distributions.

The situation is not improved if one replaces the Gaussian $\phi(x) = 2^{1/4} e^{-\pi x^2}$ by some other Schwartz class function ψ, for by Proposition (1.114), $T_0 \psi$ still has zeros, and these are what cause the pathology.

In retrospect, it is perhaps not surprising that $\{\phi_{jk}\}$ does not work well as a basis for L^2, for the following reason. The wave packets ϕ_{jk} are all of the same shape; but for analyzing high frequencies, where rapid variations take place over small regions, it is better to use wave packets that are more tightly localized in position space, such as the functions ϕ_{ab}^c in (3.39) with c large. We have already seen this phenomenon in Section 3.1, where we were forced to dilate the wave packets appropriately in order to handle pseudodifferential operators.

A much more successful sort of discrete wave packet expansion has recently been constructed by Y. Meyer and his collaborators. They work in $\mathbf{R}^n$, and rather than starting with a single ϕ and subjecting it to a discrete set of translations in phase space, they start with a finite set $\{\phi^m : 1 \leq m \leq 2^n - 1\} \subset \mathcal{S}(\mathbf{R}^n)$ of functions whose Fourier transforms are all supported in the annulus $\{(2\pi/3) \leq |\xi| \leq (8\pi/3)\}$ and subject them to a discrete set of translations and dilations in position space. In fact, they show that the ϕ^m's can be chosen (cleverly!) in such a way that

$$\left\{ \phi_{jK}^m : 1 \leq m \leq 2^n - 1, \, j \in \mathbf{Z}, \, K \in \mathbf{Z}^n \right\}, \text{ with } \phi_{jK}^m(x) = 2^{nj/2}\phi^m\left(2^j(x+K)\right),$$

is an orthonormal basis for $L^2(\mathbf{R}^n)$. (Note that $\widehat{\phi}_{jk}^m$ is supported in an annulus where $|\xi| \sim 2^j$, and these annuli fill up all of momentum space as j varies.) Moreover, $\{\phi_{jk}^m\}$ is also an unconditional basis for most of the other classical function spaces (L^p, Sobolev, etc.) and has other interesting properties. See Meyer [104].

CHAPTER 4.
THE METAPLECTIC REPRESENTATION

The metaplectic representation—also called the oscillator representation, harmonic representation, or Segal-Shale-Weil representation—is a particularly fascinating double-valued unitary representation of the symplectic group $Sp(n, \mathbf{R})$ (or, if one prefers, a unitary representation of the double cover of $Sp(n, \mathbf{R})$) on $L^2(\mathbf{R}^n)$. It appears implicitly in a number of contexts going back at least as far as Fresnel's work on optics around 1820. (See Guillemin-Sternberg [67] for a detailed discussion of symplectic geometry in optics.) However, it was first rigorously constructed on the Lie algebra level—as a representation of $\mathsf{sp}(n, \mathbf{R})$ by essentially skew-adjoint operators on a common invariant domain—by van Hove [143] in 1950, and on the group level by Segal [126] and Shale [128] a decade later. These authors were motivated by quantum mechanics; at about the same time, Weil [151] developed analogues of the metaplectic representation over arbitrary local fields, with a view to applications in number theory. Since then the metaplectic representation has attracted the attention of many people.

The first section of this chapter consists of a collection of useful facts about the symplectic group and its Lie algebra. These are all well known, in some cases as instances of theorems about semi-simple Lie groups, but it seems worthwhile to give an elementary and self-contained presentation. After that, we proceed to construct the metaplectic representation and to examine it from several viewpoints.

1. Symplectic Linear Algebra

Throughout this chapter we shall be working with $2n \times 2n$ matrices, which we shall generally denote by capital calligraphic letters. We shall frequently write them in block form:

$$\mathcal{A} = \begin{pmatrix} A & B \\ C & D \end{pmatrix},$$

where A, B, C, and D are $n \times n$ matrices. Of fundamental importance is the matrix

$$\mathcal{J} = \begin{pmatrix} 0 & I \\ -I & 0 \end{pmatrix},$$

which describes the symplectic form on $\mathbf{R}^{2n}$:

$$[w_1, w_2] = w_1 \mathcal{J} w_2.$$

We observe that

$$\mathcal{J}^* = -\mathcal{J} = \mathcal{J}^{-1}.$$

The **symplectic group** $Sp(n, \mathbf{R})$ is the group of all $2n \times 2n$ real matrices which, as operators on $\mathbf{R}^{2n}$, preserve the symplectic form:

$$\mathcal{A} \in Sp(n, \mathbf{R}) \iff [\mathcal{A}w_1, \mathcal{A}w_2] = [w_1, w_2] \text{ for all } w_1, w_2 \in \mathbf{R}^{2n}.$$

(4.1) Proposition. If $\mathcal{A} = \begin{pmatrix} A & B \\ C & D \end{pmatrix} \in GL(2n, \mathbf{R})$, the following are equivalent:

(a) $\qquad \mathcal{A} \in Sp(n, \mathbf{R})$.

(b) $\qquad \mathcal{A}^* \mathcal{J} \mathcal{A} = \mathcal{J}$.

(c) $\qquad \mathcal{A}^{-1} = \mathcal{J} \mathcal{A}^* \mathcal{J}^{-1} = \begin{pmatrix} D^* & -B^* \\ -C^* & A^* \end{pmatrix}$.

(d) $\qquad \mathcal{A}^* \in Sp(n, \mathbf{R})$.

(e) $\qquad A^*C = C^*A, \; B^*D = D^*B, \text{ and } A^*D - C^*B = I$.

(f) $\qquad AB^* = BA^*, \; CD^* = DC^*, \text{ and } AD^* - BC^* = I$.

Proof: We have

$$[\mathcal{A}w_1, \mathcal{A}w_2] = (\mathcal{A}w_1)\mathcal{J}(\mathcal{A}w_2) = w_1 \mathcal{A}^* \mathcal{J} \mathcal{A} w_2,$$

so (a) and (b) are equivalent. (b) and (c) are clearly equivalent. Taking the inverse transpose of (b) we get $\mathcal{A}^{-1} \mathcal{J} \mathcal{A}^{*-1} = \mathcal{J}$, and replacing $\mathcal{A}$ by $\mathcal{A}^{-1}$ we get (d). (e) is merely (b) written out in block form, and (f) is (b) written out in block form with $\mathcal{A}$ replaced by $\mathcal{A}^*$. ∎

The **symplectic Lie algebra** $sp(n, \mathbf{R})$ is the set of all $\mathcal{A} \in M_{2n}(\mathbf{R})$ such that $e^{t\mathcal{A}} \in Sp(n, \mathbf{R})$ for all $t \in \mathbf{R}$.

(4.2) Proposition. If $\mathcal{A} = \begin{pmatrix} A & B \\ C & D \end{pmatrix} \in M_{2n}(\mathbf{R})$, the following are equivalent.

(a) $\qquad \mathcal{A} \in sp(n, \mathbf{R})$.

(b) $\qquad \mathcal{J} \mathcal{A} + \mathcal{A}^* \mathcal{J} = 0$.

(c) $\qquad \mathcal{J} \mathcal{A} = (\mathcal{J} \mathcal{A})^*$.

(d) $\qquad D = -A^*, \; B = B^*, \text{ and } C = C^*$.

(e) $\qquad \mathcal{J} \mathcal{A}^* + \mathcal{A} \mathcal{J} = 0$.

(f) $\qquad \mathcal{A} \mathcal{J} = (\mathcal{A} \mathcal{J})^*$.

Proof: We have

$$\frac{d}{dt}(e^{tA})^*\mathcal{J}e^{tA} = (e^{tA})^*(A^*\mathcal{J} + \mathcal{J}A)e^{tA}.$$

Thus $(e^{tA})\mathcal{J}e^{tA}$ is independent of t, and hence always equal to its value $\mathcal{J}$ at $t = 0$, precisely when $A^*\mathcal{J} + \mathcal{J}A = 0$. So (a) is equivalent to (b); (c) is a restatement of (b) in view of the fact that $\mathcal{J}^* = -\mathcal{J}$; and (d) is (c) written out in block form. From (d) it is clear that $\mathsf{sp}(n, \mathbf{R})$ is closed under adjoints, and (e) and (f) are (b) and (c) with A replaced by A^*. ∎

From (d) it is easy to check that $\mathsf{sp}(n, \mathbf{R})$ is indeed a Lie algebra, that is, if $A, B \in \mathsf{sp}(n, \mathbf{R})$ then $[A, B] = AB - BA \in \mathsf{sp}(n, \mathbf{R})$.

Henceforth, when the dimension n is understood, we shall abbreviate:

$$Sp = Sp(n, \mathbf{R}), \qquad \mathsf{sp} = \mathsf{sp}(n, \mathbf{R}).$$

We proceed to derive some simple properties of Sp, beginning with the polar decomposition.

(4.3) Proposition. *Any $A \in GL(2n, \mathbf{R})$ can be written uniquely as $A = \mathcal{U}\mathcal{P}$ where $\mathcal{U}$ is orthogonal and $\mathcal{P}$ is positive definite.*

Proof: If $A = \mathcal{U}\mathcal{P}$ where $\mathcal{U} \in O(2n)$ and $\mathcal{P} > 0$ then $A^*A = \mathcal{P}\mathcal{U}^*\mathcal{U}\mathcal{P} = \mathcal{P}^2$, so $\mathcal{P}$ must be the positive square root of A^*A and $\mathcal{U}$ must be $A\mathcal{P}^{-1}$. But $\mathcal{P} = (A^*A)^{1/2}$ is indeed positive definite, and

$$(A\mathcal{P}^{-1})^*A\mathcal{P}^{-1} = (A^*A)^{1/2}A^*A(A^*A)^{-1/2} = I,$$

so $\mathcal{U} = A\mathcal{P}^{-1}$ is orthogonal. ∎

(4.4) Proposition. *If $A \in Sp$ and $\mathcal{U}, \mathcal{P}$ are as in Proposition (4.3), then $\mathcal{U}$ and $\mathcal{P}$ are also in Sp.*

Proof: By Proposition (4.1) we have

$$\mathcal{U}^{*-1}\mathcal{P}^{*-1} = A^{*-1} = \mathcal{J}A\mathcal{J}^{-1} = (\mathcal{J}\mathcal{U}\mathcal{J}^{-1})(\mathcal{J}\mathcal{P}\mathcal{J}^{-1}).$$

Since $\mathcal{J}$ is orthogonal, $\mathcal{J}\mathcal{U}\mathcal{J}^{-1} \in O(2n)$ and $\mathcal{J}\mathcal{P}\mathcal{J}^{-1} > 0$, so by uniqueness of the polar decomposition, $\mathcal{J}\mathcal{U}\mathcal{J}^{-1} = \mathcal{U}^{*-1}$ and $\mathcal{J}\mathcal{P}\mathcal{J}^{-1} = \mathcal{P}^{*-1}$. In other words, $\mathcal{U} \in Sp$ and $\mathcal{P} \in Sp$. ∎

(4.5) Proposition. *$Sp \cap O(2n)$ is a maximal compact subgroup of Sp.*

Proof: If K is a subgroup of Sp that contains $Sp \cap O(2n)$ and $A \in K$, let $A = \mathcal{U}\mathcal{P}$ as in Proposition (4.3). Then by Proposition (4.4), $\mathcal{U} \in K$ and hence $\mathcal{P} = \mathcal{U}^{-1}A \in K$. But $\mathcal{P}$ is positive definite and $\det \mathcal{P} = 1$ (since $\mathcal{P}^*\mathcal{J}\mathcal{P} = \mathcal{J}$), so either $\mathcal{P} = I$ or some eigenvalue of $\mathcal{P}$ is greater than 1. In the former case, $A = \mathcal{U} \in O(2n)$; in the latter case, $\mathcal{P}^j \in K$ for all j and $\|\mathcal{P}^j\| \to \infty$, so K is not compact. ∎

(4.6) Proposition. *If we identify $\mathbf{R}^{2n}$ with $\mathbf{C}^n$ via $(p, q) \leftrightarrow p + iq$, then* $Sp \cap O(2n) = U(n)$.

Proof: If $A \in Sp \cap O(2n)$ then $A\mathcal{J} = \mathcal{J}A$, and $\mathcal{J}$ is multiplication by $-i$ when $\mathbf{R}^{2n}$ is identified with $\mathbf{C}^n$, so A is complex linear. The assertion is now clear since the real and imaginary parts of the Hermitian inner product on $\mathbf{C}^n$ are the inner product and symplectic form on $\mathbf{R}^{2n}$. ∎

(4.7) Proposition. *Suppose $\mathcal{P} \in Sp$. Then $\mathcal{P}$ is positive definite if and only if $\mathcal{P} = e^{\mathcal{B}}$ where $\mathcal{B} \in sp$ and $\mathcal{B} = \mathcal{B}^*$.*

Proof: By the spectral theorem, the positive definite matrices are precisely the exponentials of the symmetric matrices. If $\mathcal{B} = \mathcal{B}^*$ and $\mathcal{P} = e^{\mathcal{B}} \in Sp$, we have $e^{-\mathcal{B}} = \mathcal{P}^{-1} = \mathcal{J}\mathcal{P}\mathcal{J}^{-1} = e^{\mathcal{J}\mathcal{B}\mathcal{J}^{-1}}$, so $\mathcal{J}\mathcal{B}\mathcal{J}^{-1} = -\mathcal{B} = -\mathcal{B}^*$ and hence $\mathcal{B} \in sp$. ∎

(4.8) Proposition. *Sp is connected, and its fundamental group is $\mathbf{Z}$.*

Proof: By the preceding propositions, Sp is topologically the product of $U(n)$ with the vector space $\{\mathcal{B} \in sp : \mathcal{B} = \mathcal{B}^*\}$, so the assertions follow from the corresponding well-known properties of $U(n)$. (These may easily be deduced from the exact homotopy sequence of the fibration $U(n-1) \to U(n) \to S^{2n-1}$, or proved directly by the argument sketched in Varadarajan [144, p.135]. The fundamental group is generated by the map $e^{i\theta} \to e^{i\theta}I$.) ∎

(4.9) Proposition. *The subsets*

$$N = \left\{ \begin{pmatrix} I & A \\ 0 & I \end{pmatrix} : A = A^* \right\}, \qquad \overline{N} = \left\{ \begin{pmatrix} I & 0 \\ A & I \end{pmatrix} : A = A^* \right\},$$

$$D = \left\{ \begin{pmatrix} A & 0 \\ 0 & A^{*-1} \end{pmatrix} : A \in GL(n, \mathbf{R}) \right\}$$

of $GL(2n, \mathbf{R})$ are subgroups of Sp. Moreover,

$$\overline{N}DN = \left\{ \begin{pmatrix} A & B \\ C & D \end{pmatrix} \in Sp : \det A \neq 0 \right\}.$$

Proof: The verification of the first assertion is an easy exercise which we leave to the reader. If

$$\begin{pmatrix} I & 0 \\ G & I \end{pmatrix} \in \overline{N}, \qquad \begin{pmatrix} E & 0 \\ 0 & E^{*-1} \end{pmatrix} \in D, \qquad \begin{pmatrix} I & F \\ 0 & I \end{pmatrix} \in N,$$

then

$$\begin{pmatrix} I & 0 \\ G & I \end{pmatrix} \begin{pmatrix} E & 0 \\ 0 & E^{*-1} \end{pmatrix} \begin{pmatrix} I & F \\ 0 & I \end{pmatrix} = \begin{pmatrix} E & EF \\ GE & GEF + E^{*-1} \end{pmatrix},$$

so if $\left(\begin{smallmatrix} A & B \\ C & D \end{smallmatrix}\right) \in Sp$ and $\det A \neq 0$, we take $E = A$, $F = A^{-1}B$, and $G = CA^{-1}$, and we must verify that

$$F = F^*, \qquad G = G^*, \qquad D = GEF + E^{*-1} = CA^{-1}B + A^{*-1}.$$

But this follows easily from Proposition (4.1e,f). ∎

(4.10) Proposition. *With notation as above, Sp is generated by* $\mathsf{D} \cup \overline{\mathsf{N}} \cup \{\mathcal{J}\}$ *and also by* $\mathsf{D} \cup \mathsf{N} \cup \{\mathcal{J}\}$.

Proof: A simple calculation shows that $\mathcal{J}\overline{\mathsf{N}}\mathcal{J}^{-1} = \mathsf{N}$, and $\mathcal{J}^{-1} = \mathcal{J}(-I)$ is in $\mathcal{J}\mathsf{D}$. Hence the subgroup G generated by $\mathsf{D} \cup \overline{\mathsf{N}} \cup \{\mathcal{J}\}$, or by $\mathsf{D} \cup \mathsf{N} \cup \{\mathcal{J}\}$, contains $\overline{\mathsf{N}}\mathsf{D}\mathsf{N}$, which is an open neighborhood of the identity by Proposition (4.9). If follows that G is open and hence also closed (since its complement, being the union of cosets of G, is open). By Proposition (4.8), $\mathsf{G} = Sp$. ∎

In connection with the Fock model we shall wish to use complex coordinates $z = p + iq$ corresponding to the real coordinates p, q on $\mathbf{R}^{2n}$. Since the action of Sp on $\mathbf{R}^{2n}$ is not complex linear, however, it will be more appropriate to map $\mathbf{R}^{2n}$ into $\mathbf{C}^{2n}$ by

$$W_0(p, q) = (p + iq, \, p - iq).$$

Under this mapping, any linear transformation T on $\mathbf{R}^{2n}$ turns into the linear transformation $T_c = W_0 T W_0^{-1}$ on $\mathbf{C}^{2n}$, so on the level of matrices we have the map

$$\mathcal{A} \in M_{2n}(\mathbf{R}) \longrightarrow \mathcal{A}_c \in M_{2n}(\mathbf{C})$$

given by

$$(4.11) \qquad \mathcal{A}_c = W_0 \mathcal{A} W_0^{-1} = W \mathcal{A} W^{-1},$$

where

$$(4.12) \qquad W_0 = \begin{pmatrix} I & iI \\ I & -iI \end{pmatrix}, \qquad W = \frac{1}{\sqrt{2}} W_0 = \frac{1}{\sqrt{2}} \begin{pmatrix} I & iI \\ I & -iI \end{pmatrix}.$$

(The factor of $\sqrt{2}$ in W is irrelevant for present purposes, but it makes W unitary and simplifies some formulas later on.) The following properties of W are easily verified.

$$(4.13) \qquad W^{-1} = W^* = \frac{1}{\sqrt{2}} \begin{pmatrix} I & I \\ -iI & iI \end{pmatrix},$$

$$(4.14) \qquad \overline{W}W^{-1} = W\overline{W}^{-1} = \begin{pmatrix} 0 & I \\ I & 0 \end{pmatrix},$$

$$(4.15) \qquad W\mathcal{J}W^{-1} = \begin{pmatrix} -iI & 0 \\ 0 & iI \end{pmatrix}.$$

We denote the images of $M_{2n}(\mathbf{R})$ and Sp under the map (4.11) by $M_{2n}(\mathbf{R})_c$ and Sp_c:

$$M_{2n}(\mathbf{R})_c = \{\, \mathcal{A}_c : \mathcal{A} \in M_{2n}(\mathbf{R}) \,\}, \qquad Sp_c = \{\, \mathcal{A}_c : \mathcal{A} \in Sp \,\}.$$

(4.16) Proposition. $M_{2n}(\mathbf{R})_c = \left\{ \left(\begin{smallmatrix} P & Q \\ \overline{Q} & \overline{P} \end{smallmatrix} \right) : P, Q \in M_n(\mathbf{C}) \right\}.$

Proof: Suppose $\mathcal{A} \in M_{2n}(\mathbf{C})$. Then we have $\mathcal{A} \in M_{2n}(\mathbf{R})$ iff $\mathcal{A} = \overline{\mathcal{A}}$ iff $\mathcal{W}^{-1}\mathcal{A}_c\mathcal{W} = \overline{\mathcal{W}}^{-1}\overline{\mathcal{A}_c}\overline{\mathcal{W}}$ iff $\overline{\mathcal{A}_c} = \overline{\mathcal{W}}\mathcal{W}^{-1}\mathcal{A}_c\mathcal{W}\overline{\mathcal{W}}^{-1}$. By (4.14), if $\mathcal{A} = \left(\begin{smallmatrix} P & Q \\ R & S \end{smallmatrix} \right)$ this says that

$$\begin{pmatrix} \overline{P} & \overline{Q} \\ \overline{R} & \overline{S} \end{pmatrix} = \begin{pmatrix} 0 & I \\ I & 0 \end{pmatrix} \begin{pmatrix} P & Q \\ R & S \end{pmatrix} \begin{pmatrix} 0 & I \\ I & 0 \end{pmatrix} = \begin{pmatrix} S & R \\ Q & P \end{pmatrix},$$

that is, $S = \overline{P}$ and $R = \overline{Q}$. ∎

(4.17) Proposition. If $\mathcal{A}_c = \left(\begin{smallmatrix} P & Q \\ \overline{Q} & \overline{P} \end{smallmatrix} \right) \in M_{2n}(\mathbf{R})_c$, the following are equivalent.

(a) $\qquad\qquad\qquad \mathcal{A}_c \in Sp_c.$

(b) $\qquad\qquad\qquad \mathcal{A}_c^{-1} = \mathcal{K}\mathcal{A}_c^*\mathcal{K}$ where $\mathcal{K} = \begin{pmatrix} I & 0 \\ 0 & -I \end{pmatrix}.$

(c) $\qquad\qquad\qquad PP^* - QQ^* = I$ and $PQ^\dagger = QP^\dagger.$

(d) $\qquad\qquad\qquad P^*P - Q^\dagger\overline{Q} = I$ and $P^\dagger\overline{Q} = Q^*P.$

Proof: $\mathcal{A}_c \in Sp$ iff $\mathcal{A} = \mathcal{W}^{-1}\mathcal{A}_c\mathcal{W}$ satisfies $\mathcal{A}^{-1} = \mathcal{J}\mathcal{A}^*\mathcal{J}^{-1}$; in other words, since $\mathcal{W}^* = \mathcal{W}^{-1}$,

$$\mathcal{W}^{-1}\mathcal{A}_c^{-1}\mathcal{W} = \mathcal{J}\mathcal{W}^*\mathcal{A}_c^*\mathcal{W}^{*-1}\mathcal{J}^{-1}, \text{ or } \mathcal{A}_c^{-1} = \mathcal{W}\mathcal{J}\mathcal{W}^{-1}\mathcal{A}_c\mathcal{W}\mathcal{J}^{-1}\mathcal{W}^{-1}.$$

But by (4.15),
$$\mathcal{W}\mathcal{J}^{-1}\mathcal{W}^{-1} = -\mathcal{W}\mathcal{J}\mathcal{W}^{-1} = i\mathcal{K},$$

whence (a) is equivalent to (b). In block form, (b) says that

$$\begin{pmatrix} P & Q \\ \overline{Q} & \overline{P} \end{pmatrix}^{-1} = \begin{pmatrix} P^* & -Q^\dagger \\ -Q^* & P^\dagger \end{pmatrix}.$$

(c) and (d) are then obtained by writing the equations $\mathcal{A}_c\mathcal{A}_c^{-1} = \mathcal{A}_c^{-1}\mathcal{A}_c = I$ in block form. ∎

Incidentally, Proposition 4.17(b) can be rewritten as $\mathcal{A}_c^* \mathcal{K} \mathcal{A} = \mathcal{K}$, which shows that

$$Sp_c = M_{2n}(\mathbf{R})_c \cap U(n, n).$$

(4.18) Corollary. If $\left(\begin{smallmatrix} P & Q \\ \overline{Q} & \overline{P} \end{smallmatrix} \right) \in Sp_c$ then:

(a) P is invertible and $\|P\| \geq 1$.

(b) $\|P^{-1}Q\|^2 = \|\overline{Q}P^{-1}\|^2 = 1 - \|P\|^{-2}$.

(c) $P^{-1}Q = (P^{-1}Q)^\dagger$ and $\overline{Q}P^{-1} = (\overline{Q}P^{-1})^\dagger$.

Proof: The equation $P^*P - Q^\dagger \overline{Q} = I$ yields

$$|Pu|^2 = \langle P^*Pu, u \rangle = |u|^2 + |Qu|^2,$$

which proves (a). If we set $u = P^{-1}v$, the same equation shows that $\|\overline{Q}P^{-1}\|^2 = 1 - \|P\|^{-2}$, and the equation $\|P^{-1}Q\|^2 = 1 - \|P\|^{-2}$ follows in the same way from $PP^* - QQ^* = I$. (c) is a restatement of the equations $PQ^\dagger = QP^\dagger$ and $P^\dagger \overline{Q} = Q^*P$. ∎

(4.19) Proposition. *The image of $Sp \cap O(2n)$ in Sp_c is* $\left\{ \left(\begin{smallmatrix} P & 0 \\ 0 & \overline{P} \end{smallmatrix} \right) : P \in U(n) \right\}$.

Proof: If we identify $\mathbf{R}^{2n}$ with $\mathbf{C}^n$ via $(p, q) \leftrightarrow p + iq$, the map $(p, q) \to \mathcal{W}_0(p, q)$ given by (4.12) becomes simply $z \to (z, \overline{z})$. The result now follows from Proposition (4.6). ∎

We conclude this section with a few miscellaneous results that will be needed later.

(4.20) Proposition. $[\mathsf{sp}, \mathsf{sp}] = \mathsf{sp}$.

Proof: Let e_{ij} denote the $n \times n$ matrix with 1 in the ij-th entry and 0's elsewhere, and let

$$E_{ij} = \begin{pmatrix} e_{ij} & 0 \\ 0 & -e_{ji} \end{pmatrix}, \qquad F_{ij} = \begin{pmatrix} 0 & e_{ij} + e_{ji} \\ 0 & 0 \end{pmatrix}, \qquad G_{ij} = \begin{pmatrix} 0 & 0 \\ e_{ij} + e_{ji} & 0 \end{pmatrix}.$$

These matrices span sp by Proposition (4.2d), and so the result follows from the following easily verified facts:

$$[E_{ii}, E_{ij}] = (1 - \delta_{ij})E_{ij}, \qquad [E_{ii}, F_{ij}] = (1 + \delta_{ij})F_{ij},$$
$$[E_{ii}, G_{ij}] = -(1 + \delta_{ij})G_{ij}, \qquad [F_{ii}, G_{ii}] = 2E_{ii}. \quad ∎$$

(4.21) Proposition. *If G is Sp or any of its connected covering groups, there are no nontrivial homomorphisms from G into the circle group $T = \{z \in \mathbf{C} : |z| = 1\}$.*

Proof: If $\pi : G \to T$ is a homomorphism, its differential

$$d\pi(\mathcal{A}) = \frac{d}{dt}\pi(\exp(t\mathcal{A}))|_{t=0}, \qquad \mathcal{A} \in \mathsf{sp},$$

is a Lie algebra homomorphism from sp to the Abelian Lie algebra $i\mathbf{R}$. By Proposition (4.20), $d\pi = 0$. Hence $\pi(g) = I$ for all $g = \exp(t\mathcal{A})$ with $\mathcal{A} \in \mathsf{sp}$, and in particular for all g in a neighborhood of the identity in G. Since G is connected, $\pi = I$. ∎

(4.22) Proposition. *If $\mathcal{A} \in GL(2n, \mathbf{R})$ is positive definite, there exists $\mathcal{S} \in Sp$ such that $\mathcal{S}^* \mathcal{A} \mathcal{S} = \begin{pmatrix} D & 0 \\ 0 & D \end{pmatrix}$ where D is diagonal.*

Proof: Define the positive definite bilinear form Q on $\mathbf{R}^{2n}$ by $Q(u, v) = u\mathcal{A}v$. Since Q and the symplectic form are nondegenerate, there is a unique $T \in GL(2n, \mathbf{R})$ such that $Q(u, Tv) = [u, v]$ for all $u, v \in \mathbf{R}^{2n}$. T is skew-symmetric with respect to Q, so its eigenvalues are of the form $\pm i\lambda$ with $\lambda > 0$. Its eigenvectors in $\mathbf{C}^{2n}$ occur in conjugate pairs, so by taking their real and imaginary parts we obtain a Q-orthonormal basis $\{E_j, F_k\}_{j,k=1}^n$ of $\mathbf{R}^{2n}$ such that $TE_j = -\lambda_j F_j$ and $TF_j = \lambda_j E_j$ where $\lambda_j > 0$. Let $e_j = \lambda_j^{-1/2} E_j$ and $f_j = \lambda_j^{-1/2} F_j$, and let $\mathcal{S}$ be the linear transformation that maps the standard basis for $\mathbf{R}^{2n}$ to the basis $\{e_j, f_k\}$. The Q-orthonormality of $\{E_j, F_k\}$ means that $\mathcal{S}^* \mathcal{A} \mathcal{S} = \begin{pmatrix} D & 0 \\ 0 & D \end{pmatrix}$ where D is the diagonal matrix with eigenvalues $\lambda_1^{-1}, \dots, \lambda_n^{-1}$, and together with the definition of T it implies that

$$[e_j, e_k] = [f_j, f_k] = 0, \qquad [e_j, f_k] = \delta_{jk},$$

so that $\mathcal{S} \in Sp$. ∎

2. Construction of the Metaplectic Representation

We begin our study of the metaplectic representation by presenting it abstractly as a group of intertwining operators.

Let $\mathcal{T}$ denote the group of automorphisms of $\mathbf{H}_n$ that leave the center pointwise fixed. From Theorem (1.22) we know that $\mathcal{T}$ is generated by the automorphisms

$$T_{(a,b,c)}(p, q, t) = (a, b, c)(p, q, t)(a, b, c)^{-1} = (p, q, t + aq - bp), \qquad (a, b, c) \in \mathbf{H}_n,$$

and

$$T_{\mathcal{A}}(p, q, t) = (\mathcal{A}(p, q), t), \qquad \mathcal{A} \in Sp.$$

If $T \in \mathcal{T}$, we can compose the Schrödinger representation ρ with T to obtain a new representation $\rho \circ T$ of $\mathbf{H}_n$ on $L^2(\mathbf{R}^n)$ such that $\rho \circ T(0, 0, t) = e^{2\pi i t} I$. By the Stone–von Neumann theorem (1.50), ρ and $\rho \circ T$ are equivalent: there exists a unitary operator $\mu(T)$ on $L^2(\mathbf{R}^n)$ such that

$$\rho \circ T(X) = \mu(T)\rho(X)\mu(T)^{-1}, \qquad X \in \mathbf{H}_n.$$

Moreover, by Schur's lemma (cf. Appendix B), $\mu(T)$ is determined up to a phase factor, i.e., a scalar factor of modulus one. It follows that

$$\mu(TS) = c_{T,S}\mu(T)\mu(S), \qquad |c_{T,S}| = 1,$$

so that μ defines a projective unitary representation of T, that is, a homomorphism from T into the quotient of the unitary group by its center $\{cI : |c| = 1\}$.
If $T = T_Y$ as above where $Y = (a, b, c) \in \mathbf{H}_n$, we can take $\mu(T)$ to be $\rho(Y)$:

$$\rho(YXY^{-1}) = \rho(Y)\rho(X)\rho(Y)^{-1},$$

so in this case there is nothing new. We therefore restrict attention to the symplectic automorphisms $T_\mathcal{A}$ and write $\mu(\mathcal{A})$ for $\mu(T_\mathcal{A})$, so $\mu(\mathcal{A})$ is determined up to a phase factor by the relation

(4.23) $$\rho(\mathcal{A}(p,q)) = \mu(\mathcal{A})\rho(p,q)\mu(\mathcal{A})^{-1}.$$

We shall prove below that the phase factors can be chosen in one and only one way up to factors of ± 1 so that μ becomes a double-valued unitary representation of Sp, i.e.,

$$\mu(\mathcal{AB}) = \pm\mu(\mathcal{A})\mu(\mathcal{B}).$$

With this choice of phase factors, which we henceforth assume to have been made, μ is called the **metaplectic representation** of Sp.

Of course, the fastidious way to deal with the double-valuedness is to pass to the double covering group Sp_2 (sometimes called Mp) of Sp. (In view of Proposition (4.8), Sp has a unique connected double cover.) μ then defines a unitary representation of Sp_2. As a matter of fact, this representation is faithful, so we can regard its range as a model for Sp_2; and then μ^{-1} is just the 2-to-1 covering map from Sp_2 to Sp. However, for our purposes it is more trouble than it is worth to deal with Sp_2. We prefer to think of μ as a homomorphism from Sp to the group of unitary operators modulo $\{\pm I\}$; and if $\mathcal{A} \in Sp$, we shall think of $\mu(\mathcal{A})$ either as a pair of unitary operators that differ from each other by a factor of -1, or as one of the members of this pair. In explicit formulas, the ambiguity of ± 1 will usually appear as the ambiguity in sign of a square root.

As we shall see, the explicit calculation of $\mu(\mathcal{A})$ for a general $\mathcal{A} \in Sp$ is rather complicated, but it is easy to use (4.23) to find $\mu(\mathcal{A})$ up to a phase factor when $\mathcal{A}$ belongs to certain subgroups of Sp. Let us now do this.

(i) Let $\mathcal{A} = \begin{pmatrix} A & 0 \\ 0 & A^{*-1} \end{pmatrix}$. Then

$$[\rho \circ \mathcal{A}(p,q)]f(x) = \rho(Ap, A^{*-1}q)f(x) = e^{2\pi i A^{*-1}q \cdot x + \pi i Ap \cdot A^{*-1}q}f(x + Ap)$$
$$= e^{2\pi i q A^{-1}x + \pi i pq}(f \circ A)(A^{-1}x + p) = U\rho(p,q)U^{-1}f(x)$$

where
$$U f(x) = |\det A|^{-1/2} f(A^{-1} x).$$

(The factor of $|\det A|^{-1/2}$ is inserted to make U unitary.)

(ii) Let $\mathcal{A} = \begin{pmatrix} I & 0 \\ C & I \end{pmatrix}$ with $C = C^*$. Then

$$[\rho \circ \mathcal{A}(p,q)] f(x) = \rho(p, Cp + q) f(x) = e^{2\pi i (Cp+q)x + \pi i p(Cp+q)} f(x+p)$$
$$= e^{-\pi i x C x} e^{2\pi i q x + \pi i p q} e^{\pi i (x+p) C(x+p)} f(x+p) = U \rho(p,q) U^{-1} f(x)$$

where
$$U f(x) = e^{-\pi i x C x} f(x).$$

(iii) Let $\mathcal{A} = \mathcal{J} = \begin{pmatrix} 0 & I \\ -I & 0 \end{pmatrix}$. Then $\rho \circ \mathcal{A}(p,q) = \rho(q, -p)$, so we want an operator that intertwines translation and multiplication by exponentials. The inverse Fourier transform does the job:

$$\rho(p,q) \mathcal{F} f(x) = e^{2\pi i q x + \pi i p q} \int e^{-2\pi i (x+p) y} f(y)\, dy$$

$$= e^{2\pi i q x + \pi i p q} \int e^{-2\pi i (x+p)(y+q)} f(y+q)\, dy$$

$$= \int e^{-2\pi i x y} e^{-2\pi i p y - \pi i p q} f(y+q)\, dy = \mathcal{F} \rho(q, -p) f(x).$$

Thus we have identified $\mu(\mathcal{A})$ up to a phase factor for $\mathcal{A}$ of the above three types. At this point we anticipate the results of the next section and reveal the correct phase factors:

$$(4.24) \qquad \mu\left[\begin{pmatrix} A & 0 \\ 0 & A^{*-1} \end{pmatrix}\right] f(x) = (\det^{-1/2} A) f(A^{-1} x),$$

$$(4.25) \qquad \mu\left[\begin{pmatrix} I & 0 \\ C & I \end{pmatrix}\right] f(x) = \pm e^{-\pi i x C x} f(x),$$

$$(4.26) \qquad \mu(\mathcal{J}) = i^{n/2} \mathcal{F}^{-1}.$$

In (4.24) and (4.26) the sign of the square root is deliberately left undetermined.

By Proposition (4.10), Sp is generated by matrices of these three types, so we have in some sense computed the metaplectic representation up to phase factors. This computation, although not very explicit, is enough to allow us to deduce the following fundamental fact, which we have already used in Chapter 2:

(4.27) Proposition. *If $\mathcal{A} \in Sp$, the operator $\mu(\mathcal{A})$ maps $\mathcal{S}$ isomorphically onto itself and extends by continuity to an isomorphism on $\mathcal{S}'$.*

Proof: The operators (4.24), (4.25), and (4.26) clearly have this property. Since any $\mu(\mathcal{A})$ is a product of such operators by Proposition (4.10), the result follows. ∎

At this point we record the effect of the metaplectic representation on Fourier-Wigner and Wigner transforms. Since V and W are sesquilinear, in these formulas the phase factor in $\mu(\mathcal{A})$ is irrelevant as long as the same factor is used in both arguments.

(4.28) Proposition. If $\mathcal{A} \in Sp$ and $f, g \in \mathcal{S}'$,

$$V(\mu(\mathcal{A})f, \, \mu(\mathcal{A})g) = V(f,g) \circ \mathcal{A}^{-1} \quad \text{and} \quad W(\mu(\mathcal{A})f, \, \mu(\mathcal{A})g) = W(f,g) \circ \mathcal{A}^*.$$

Proof: It suffices to assume $f, g \in L^2$, in which case we have

$$V(\mu(\mathcal{A})f, \, \mu(\mathcal{A})g)(p, q) = \langle \rho(p,q)\mu(\mathcal{A})f, \, \mu(\mathcal{A})g \rangle = \langle \mu(\mathcal{A})^{-1}\rho(p,q)\mu(\mathcal{A})f, \, g \rangle$$
$$= \langle \rho \circ \mathcal{A}^{-1}(p,q)f, \, g \rangle = V(f,g)(\mathcal{A}^{-1}(p,q)).$$

This takes care of V, and since $\det \mathcal{A} = 1$,

$$W(\mu(\mathcal{A})f, \, \mu(\mathcal{A})g) = V(\mu(\mathcal{A})f, \, \mu(\mathcal{A})g)\widehat{} = [V(f,g) \circ \mathcal{A}^{-1}]\widehat{} = W(f,g) \circ \mathcal{A}^*. \quad \blacksquare$$

Remark. The proof that the ambiguity of phase factors can be reduced to ± 1 will require a fair amount of work. However, without doing any calculations at all, one can deduce a weaker result from the following general theorems:

(i) Let G be a simply connected Lie group with Lie algebra $\mathbf{g}$. If the cohomology group $H^2(\mathbf{g}, \mathbf{R})$ (whose definition need not concern us here) is trivial, then every projective unitary representation of G comes from a genuine unitary representation of G.

(ii) If $\mathbf{g}$ is semisimple then $H^2(\mathbf{g}, \mathbf{R})$ is trivial.

(iii) $\mathbf{sp}$ is semisimple.

(i) is a deep result of Bargmann [10]; (ii) is a rather easy calculation which can also be found in [10]; (iii) is common knowledge (see Varadarajan [144]). Combining them, we see that the projective representation μ of Sp can be made into a genuine representation of the universal cover of Sp. By Proposition (4.8), the universal cover of Sp has infinitely many sheets, so it is still not obvious that this representation factors through the double cover.

The Fock Model. The easiest way to construct the metaplectic representation globally so that the phase factors can be sorted out is to move it over to Fock space. We recall that on Fock space $\mathcal{F}_n$ we have the representation β of $\mathbf{H}_n$ obtained from the Schrödinger representation ρ by conjugation with the Bargmann transform B:

$$\beta(p + iq) = B\rho(p,q)B^{-1}, \qquad \beta(w)F(z) = e^{-(\pi/2)|w|^2 - \pi z \overline{w}}F(z + w).$$

Since we use the complex coordinates $w = p + iq$ for describing β, we also use the complex form Sp_c of Sp as described in Section 4.1. We adopt the following notational convention: if $\mathcal{A} \in Sp_c$ and $w \in \mathbf{C}^n$, we define $\mathcal{A}w$ to be the vector $z \in \mathbf{C}^n$ such that $\mathcal{A}(w, \overline{w}) = (z, \overline{z})$. That is,

$$\mathcal{A}w = z \iff z = Pw + Q\overline{w} \text{ where } \mathcal{A} = \begin{pmatrix} P & Q \\ \overline{Q} & \overline{P} \end{pmatrix}.$$

The Fock-metaplectic representation ν of Sp_c is then defined up to phase factors by the condition

$$(4.29) \qquad \beta(\mathcal{A}w) = \nu(\mathcal{A})\beta(w)\nu(\mathcal{A})^{-1}.$$

The advantage of Fock space is that the operators $\nu(\mathcal{A})$ are integral operators whose kernels are rather easily computable. Indeed, by Proposition (1.68), we have

$$(4.30) \qquad \nu(\mathcal{A})F(z) = \int K_{\mathcal{A}}(z, \overline{w})F(w)e^{-\pi|w|^2} \, dw$$

$$\text{where} \quad K_{\mathcal{A}}(z, \overline{w}) = \langle \nu(\mathcal{A})E_w, E_z \rangle_{\mathcal{F}}.$$

We proceed to calculate $K_{\mathcal{A}}$, following Bargmann [13].

(4.31) Proposition. *If $\mathcal{A} = \begin{pmatrix} P & Q \\ \overline{Q} & \overline{P} \end{pmatrix} \in Sp_c$ and $\nu(\mathcal{A})$ is a unitary operator on $\mathcal{F}_n$ given by (4.30) and satisfying (4.29), then*

$$(4.32) \qquad K_{\mathcal{A}}(z, \overline{w}) = C_{\mathcal{A}} \exp\left\{ \tfrac{1}{2}\pi(z\overline{Q}P^{-1}z + 2\overline{w}P^{-1}z - \overline{w}P^{-1}Q\overline{w}) \right\}$$

for some $C_{\mathcal{A}} \in \mathbf{C}$.

Proof: To begin with, we recall that

$$E_w(z) = e^{\pi z\overline{w}} = e^{(\pi/2)|w|^2}\beta(-w)E_0(z),$$

and hence

$$(4.33) \qquad K_{\mathcal{A}}(z, \overline{w}) = e^{(\pi/2)(|w|^2 + |z|^2)}\langle \nu(\mathcal{A})\beta(-w)E_0, \, \beta(-z)E_0 \rangle_{\mathcal{F}}.$$

Suppose for the moment that $z = \mathcal{A}w$, i.e.,

$$z = Pw + Q\overline{w}.$$

Then $w = \mathcal{A}^{-1}z$, so by Proposition (4.17b),

$$\overline{w} = P^\dagger \overline{z} - Q^* z.$$

Solving the first equation for w and the second for $\overline{z}$, and using Proposition (4.17d), we find that

$$w = P^{-1}z - P^{-1}Q\overline{w}, \qquad \overline{z} = P^{\dagger-1}\overline{w} + P^{\dagger-1}Q^*z = P^{\dagger-1}\overline{w} + \overline{Q}P^{-1}z.$$

Therefore,

$$(4.34) \quad \begin{aligned} |w|^2 + |z|^2 &= \overline{w}w + z\overline{z} = \overline{w}P^{-1}z - \overline{w}P^{-1}Q\overline{w} + zP^{\dagger-1}\overline{w} + z\overline{Q}P^{-1}z \\ &= z\overline{Q}P^{-1}z + 2\overline{w}P^{-1}z - \overline{w}P^{-1}Q\overline{w}. \end{aligned}$$

On the other hand, when $z = \mathcal{A}w$,

$$(4.35) \quad \begin{aligned} \langle \nu(\mathcal{A})\beta(-w)E_0, \beta(-z)E_0 \rangle_{\mathcal{F}} &= \langle \nu(\mathcal{A})\beta(-w)E_0, \nu(\mathcal{A})\beta(-w)\nu(\mathcal{A})^{-1}E_0 \rangle_{\mathcal{F}} \\ &= \langle E_0, \nu(\mathcal{A})^{-1}E_0 \rangle_{\mathcal{F}} \end{aligned}$$

since $\nu(\mathcal{A})$ and $\beta(-w)$ are unitary. Combining (4.33), (4.34), and (4.35), and taking $C_{\mathcal{A}} = \langle E_0, \nu(\mathcal{A})^{-1}E_0 \rangle_{\mathcal{F}}$, we have proved (4.32) in the case $z = \mathcal{A}w$. But if we set

$$u = \tfrac{1}{2}(\overline{w} + P^{-1}z - P^{-1}Q\overline{w}), \qquad v = \tfrac{1}{2}i(\overline{w} - P^{-1}z + P^{-1}Q\overline{w}),$$

so that

$$z = P(u + iv) + Q(u - iv), \qquad \overline{w} = u - iv,$$

then $K_{\mathcal{A}}(z, \overline{w})$ becomes an entire function of u and v, so it is completely determined by its values for u and v real. But u and v are real precisely when $w = u + iv$, i.e., $w = P^{-1}z - P^{-1}Q\overline{w}$, and this means that $z = Pw + Q\overline{w}$, i.e., $z = \mathcal{A}w$. So $K_{\mathcal{A}}(z, \overline{w})$ is determined by its values for $z = \mathcal{A}w$, and since the right side of (4.32) is clearly an entire function of z and $\overline{w}$, it follows that (4.32) is valid for arbitrary z and $\overline{w}$. ∎

The phase factor for $\nu(\mathcal{A})$, that is, the argument of the constant $C_{\mathcal{A}}$ in (4.32), has not yet been determined, but the absolute value of $C_{\mathcal{A}}$ is fixed by the requirement that $\nu(\mathcal{A})$ be unitary. Indeed, we have

$$(4.36) \qquad \nu(\mathcal{A})E_0(z) = \langle \nu(\mathcal{A})E_0, E_z \rangle_{\mathcal{F}} = K_{\mathcal{A}}(z, 0),$$

and $\|E_0\|_{\mathcal{F}} = 1$, so

$$\begin{aligned} 1 = \|\nu(\mathcal{A})E_0\|_{\mathcal{F}}^2 &= \int |K_{\mathcal{A}}(z, 0)|^2 e^{-\pi|z|^2} dz \\ &= |C_{\mathcal{A}}|^2 \int e^{(\pi/2)(z\overline{Q}P^{-1}z + \overline{z}Q\overline{P}^{-1}\overline{z})} e^{-\pi|z|^2} dz \\ &= |C_{\mathcal{A}}|^2 \det{}^{-1/2}(I - Q\overline{P}^{-1}\overline{Q}P^{-1}) \end{aligned}$$

by Theorem 3 of Appendix A. (This result is applicable since $\overline{Q}P^{-1}$ is symmetric and has norm < 1, by Corollary (4.18).) But by Proposition (4.17d),

$$I - Q\overline{P}^{-1}\overline{Q}P^{-1} = I - P^{*-1}Q^\dagger\overline{Q}P^{-1} = P^{*-1}(P^*P - Q^\dagger\overline{Q})P^{-1} = (PP^*)^{-1},$$

and hence

$$|C_{\mathcal{A}}| = \det{}^{1/4}(PP^*)^{-1} = |\det{}^{-1/2} P|.$$

At long last, we specify the phase factor up to ± 1 by decreeing that

$$C_{\mathcal{A}} = \det{}^{-1/2} P,$$

the sign of the square root being left undetermined. We now come to the final result.

(4.37) Theorem. (Bargmann [13], Itzykson [81]) For $\mathcal{A} = \left(\begin{smallmatrix} P & Q \\ Q & P \end{smallmatrix}\right) \in Sp_c$, define the operator $\nu(\mathcal{A})$ (modulo ± 1) on $\mathcal{F}_n$ by

$$\nu(\mathcal{A})F(z) = \int K_{\mathcal{A}}(z, \overline{w}) F(w) e^{-\pi |w|^2} \, dw,$$

(4.38)

$$K_{\mathcal{A}}(z, \overline{w}) = (\det{}^{-1/2} P) \exp\left\{ \tfrac{1}{2}\pi(z\overline{Q}P^{-1}z + 2\overline{w}P^{-1}z - \overline{w}P^{-1}Q\overline{w}) \right\}.$$

Then for all $\mathcal{A}, \mathcal{B} \in Sp_c$ and $w \in \mathbf{C}^n$,

(a) $\nu(\mathcal{A})$ is unitary,

(b) $\nu(\mathcal{A})\beta(w)\nu(\mathcal{A})^{-1} = \beta(\mathcal{A}w),$

(c) $\nu(\mathcal{A})\nu(\mathcal{B}) = \pm\nu(\mathcal{A}\mathcal{B}).$

Moreover, if $\{\nu'(\mathcal{A}) : \mathcal{A} \in Sp\}$ is any family of operators satisfying these three conditions, then $\nu'(\mathcal{A}) = \pm\nu(\mathcal{A})$ for all $\mathcal{A} \in Sp$.

Proof: The Stone–von Neumann theorem (1.50) guarantees the existence of operators satisfying (a) and (b), and the calculations above show that they must be given by (4.38) up to phase factors. Moreover, for any $\mathcal{A}_1, \mathcal{A}_2 \in Sp$ we must have

$$\nu(\mathcal{A}_1)\nu(\mathcal{A}_2) = \chi(\mathcal{A}_1, \mathcal{A}_2)\nu(\mathcal{A}_1\mathcal{A}_2)$$

where $|\chi(\mathcal{A}_1, \mathcal{A}_2)| = 1$. Now, by (4.36) and Theorem 3 of Appendix A, if $\mathcal{A}_j = \left(\begin{smallmatrix} P_j & Q_j \\ \overline{Q}_j & \overline{P}_j \end{smallmatrix}\right)$,

$$\chi(\mathcal{A}_1, \mathcal{A}_2)\langle \nu(\mathcal{A}_1\mathcal{A}_2)E_0, E_0\rangle_{\mathcal{F}} = \langle \nu(\mathcal{A}_1)\nu(\mathcal{A}_2)E_0, E_0\rangle_{\mathcal{F}}$$

$$= \langle \nu(\mathcal{A}_2)E_0, \nu(\mathcal{A}_1)^* E_0\rangle_{\mathcal{F}}$$

$$= \int K_{\mathcal{A}_2}(z, 0) K_{\mathcal{A}_1}(0, \overline{z}) e^{-\pi |z|^2} \, dz$$

$$= (\det{}^{-1/2} P_1)(\det{}^{-1/2} P_2) \int \exp\left\{ \tfrac{1}{2}\pi(z\overline{Q}_2 P_2^{-1}z - \overline{z}P_1^{-1}Q_1\overline{z}) \right\} e^{-\pi |z|^2} \, dz$$

$$= (\det{}^{-1/2} P_1)(\det{}^{-1/2}[I + P_1^{-1}Q_1\overline{Q}_2 P_2^{-1}])(\det{}^{-1/2} P_2)$$

$$= \det{}^{-1/2}(P_1 P_2 + Q_1\overline{Q}_2).$$

But $P_1 P_2 + Q_1 \overline{Q}_2$ is the upper left entry of $\mathcal{A}_1 \mathcal{A}_2$, so

$$\det{}^{-1/2}(P_1 P_2 + Q_1 \overline{Q}_2) = K_{\mathcal{A}_1 \mathcal{A}_2}(0,0) = \langle \nu(\mathcal{A}_1 \mathcal{A}_2) E_0, \, E_0 \rangle_{\mathcal{F}}.$$

Therefore $\chi(\mathcal{A}_1, \mathcal{A}_2) = \pm 1$ (the above calculation is only valid up to the sign in the square root), so (c) is proved.

If $\nu'(\mathcal{A})$ also satisfies (a) and (b) we must have $\nu'(\mathcal{A}) = \alpha(\mathcal{A})\nu(\mathcal{A})$ where $|\alpha(\mathcal{A})| = 1$, and if $\nu'(\mathcal{A})$ also satisfies (c), α must be a homomorphism from Sp_c into the circle group modulo ± 1, which is isomorphic to the circle group itself. By Proposition (4.21), α is trivial, so $\nu'(\mathcal{A}) = \pm\nu(\mathcal{A})$. ∎

Remark. The uniqueness part of this theorem shows, in particular, that ν cannot be made into a single-valued representation of Sp_c, as it is impossible to define a continuous single-valued branch of $\det^{-1/2} P$ on Sp_c.

We saw above that the operators $\mu(\mathcal{A})$ on $L^2(\mathbf{R}^n)$ are very simple for certain types of $\mathcal{A} \in Sp$. The subgroup of Sp_c which the Fock model ν represents in a simple fashion is the unitary group $U(n)$, which is imbedded in Sp_c by

$$U(n) \ni P \to \begin{pmatrix} P & 0 \\ 0 & \overline{P} \end{pmatrix}.$$

(4.39) Proposition. If $P \in U(n)$ then

$$\nu\left[\begin{pmatrix} P & 0 \\ 0 & \overline{P} \end{pmatrix}\right] F(z) = (\det{}^{-1/2} P) F(P^{-1} z).$$

Proof: By (4.38) we have

$$\nu\left[\begin{pmatrix} P & 0 \\ 0 & \overline{P} \end{pmatrix}\right] F(z) = (\det{}^{-1/2} P) \int e^{\pi \overline{w} P^{-1} z} F(w) e^{-\pi |w|^2} dw$$

$$= (\det{}^{-1/2} P) \langle F, E_{P^{-1} z} \rangle_{\mathcal{F}} = (\det{}^{-1/2} P) F(P^{-1} z). \quad ∎$$

This proposition raises a rather paradoxical point. Since the non-simple connectivity of the symplectic group comes entirely from the subgroup $U(n)$ (cf. the proof of Proposition (4.8)), one would think that the obstruction to making ν single-valued must come from $U(n)$. But it is clear from Proposition (4.39) that the restriction of ν to $U(n)$ can be made into a single-valued unitary representation of $U(n)$ by discarding the factor of $\det^{-1/2} P$. However, this single-valued representation cannot be extended to the whole symplectic group: if one discards the factor $\det^{-1/2} P$ in (4.38), one loses unitarity as well as the property that $\nu(\mathcal{A})\nu(\mathcal{B}) = \pm\nu(\mathcal{A}\mathcal{B})$ when $\mathcal{A}$ or $\mathcal{B}$ is not in $U(n)$.

We can now go back to the Schrödinger picture and define the metaplectic representation μ, phase factors and all, by

$$\mu(\mathcal{A}) = B^{-1}\nu(\mathcal{A}_c)B, \qquad \mathcal{A}_c \text{ as in (4.11)}.$$

Theorem (4.37) translates into a corresponding result for μ, whose precise statement we leave to the reader. Actually, the uniqueness assertion can be stated in the following somewhat stronger form.

(4.40) Proposition. *Let G denote the universal covering group of Sp, $\pi :$ $G \to Sp$ the projection, and μ_0 the representation of G determined by the condition $\mu(\pi(g)) = \pm\mu_0(g)$, $g \in G$. Then μ_0 is the only unitary representation of G on $L^2(\mathbf{R}^n)$ such that*

$$\mu_0(g)\rho(p,q)\mu_0(g)^{-1} = \rho\big(\pi(g)(p,q)\big), \qquad (p,q) \in \mathbf{R}^{2n}.$$

Proof: If μ_1 is another such representation, then by the uniqueness of intertwining operators, for each $g \in G$ we have $\mu_1(g) = c_g\mu_0(g)$ for some constant c_g of modulus 1. Clearly $g \to c_g$ is a homomorphism from G to the circle group, so $c_g = 1$ for all g by Proposition (4.21). ∎

3. The Infinitesimal Representation

We have seen in Proposition (4.2) that the symplectic Lie algebra sp is precisely the set of $\mathcal{A} \in M_{2n}(\mathbf{R})$ such that $\mathcal{AJ}$ is symmetric. Symmetric matrices are naturally associated with homogeneous quadratic polynomials, and the latter form a Lie algebra under the Poisson bracket. This Lie algebra turns out to be isomorphic to sp.

More precisely, let $\mathcal{Q}$ denote the space of real homogeneous quadratic polynomials on $\mathbf{R}^{2n}$, equipped with the Poisson bracket. Each $P \in \mathcal{Q}$ has the form $P(w) = -\frac{1}{2}w\mathcal{B}w$ for a unique symmetric matrix $\mathcal{B}$, the factor of $-\frac{1}{2}$ being introduced for ulterior purposes. Accordingly, to each $\mathcal{A} \in sp$ we associate the polynomial $P_\mathcal{A} \in \mathcal{Q}$ defined by

$$P_\mathcal{A}(w) = -\tfrac{1}{2}w\mathcal{AJ}w.$$

According to Proposition (4.2d), if $\mathcal{A} \in sp$ we have

$$\mathcal{A} = \begin{pmatrix} A & B \\ C & -A^* \end{pmatrix} \quad \text{and} \quad \mathcal{AJ} = \begin{pmatrix} -B & A \\ A^* & C \end{pmatrix}$$

where $B = B^*$ and $C = C^*$, and hence

(4.41) $$P_\mathcal{A}(\xi, x) = \tfrac{1}{2}\xi B\xi - \xi Ax - \tfrac{1}{2}xCx.$$

A mechanical calculation which we leave to the reader then shows that:

(4.42) Proposition. *The map $\mathcal{A} \to P_\mathcal{A}$ is a Lie algebra isomorphism from sp to $\mathcal{Q}$.*

Moreover, by Corollary (2.51) (or another direct calculation), we know that the Weyl correspondence $P \to P(D, X)$ satisfies

$$[P(D,X), Q(D,X)] = \frac{1}{2\pi i}\{P, Q\}(D, X) \quad \text{for} \quad P, Q \in \mathcal{Q}.$$

There follows immediately:

(4.43) Proposition. *The map*

$$\mathcal{A} \longrightarrow 2\pi i P_{\mathcal{A}}(D, X)$$

is a faithful representation of sp *by skew-Hermitian operators on* $\mathcal{S}$.

Explicitly, if $\mathcal{A}$ is written in block form as above, we have

$$(4.44) \qquad 2\pi i P_{\mathcal{A}}(D, X) = \frac{1}{4\pi i} \sum B_{jk} \frac{\partial^2}{\partial x_j \partial x_k} - \sum A_{jk} x_k \frac{\partial}{\partial x_j}$$
$$- \tfrac{1}{2} \operatorname{tr} A - \pi i \sum C_{jk} x_j x_k.$$

The term $\frac{1}{2} \operatorname{tr} A$ comes from the fact that

> if $P(\xi, x) = \xi_j x_k$ then $P(D, X) = \frac{1}{2}(D_j X_k + X_k D_j) = X_k D_j + \frac{1}{2} \delta_{jk}$.

We claim that this representation of sp is nothing but the infinitesimal version $d\mu$ of the metaplectic representation, defined by

$$d\mu(\mathcal{A}) = \frac{d}{dt} \mu(e^{t\mathcal{A}})|_{t=0}, \qquad \mathcal{A} \in \text{sp}.$$

Here and in what follows, the sign of $\mu(e^{t\mathcal{A}})$ is determined by the requirement that $\mu(e^{t\mathcal{A}})$ be continuous in t and equal to I at $t = 0$. The domain of $d\mu(\mathcal{A})$ is the space of all $f \in L^2$ such that $t^{-1}[\mu(e^{t\mathcal{A}}) - I]f$ converges in the L^2 norm as $t \to 0$, but we shall consider $d\mu(\mathcal{A})$ as an operator on the space of C^∞ vectors for μ, namely the set of all f that belong to the domain of $\prod_1^k d\mu(\mathcal{A}_j)$ for all $\mathcal{A}_j \in$ sp and all positive integers k.

(4.45) Theorem. *The Schwartz space* $\mathcal{S}(\mathbf{R}^n)$ *consists of* C^∞ *vectors for* μ, *and for* $\mathcal{A} \in$ sp *and* $f \in \mathcal{S}(\mathbf{R}^n)$ *we have*

$$d\mu(\mathcal{A})f = 2\pi i P_{\mathcal{A}}(D, X)f.$$

Proof: We shall use formulas (4.24)–(4.26). So far these formulas are known to be correct only up to phase factors, but this will be enough. To begin with, if

$$\mathcal{A} = \begin{pmatrix} A & 0 \\ 0 & -A^* \end{pmatrix}, \quad \begin{pmatrix} 0 & 0 \\ C & 0 \end{pmatrix}, \quad \text{or} \quad \begin{pmatrix} 0 & B \\ 0 & 0 \end{pmatrix},$$

then

$$e^{t\mathcal{A}} = \begin{pmatrix} e^{tA} & 0 \\ 0 & e^{-tA^*} \end{pmatrix}, \quad \begin{pmatrix} I & 0 \\ tC & I \end{pmatrix}, \quad \text{or} \quad \begin{pmatrix} I & tB \\ 0 & I \end{pmatrix},$$

respectively. Moreover,

$$\begin{pmatrix} I & tB \\ 0 & I \end{pmatrix} = \begin{pmatrix} 0 & I \\ -I & 0 \end{pmatrix} \begin{pmatrix} I & 0 \\ -tB & I \end{pmatrix} \begin{pmatrix} 0 & -I \\ I & 0 \end{pmatrix}.$$

Hence, from formulas (4.24)–(4.26), with a phase factor c_{tA} thrown in, we have:

if $A = \begin{pmatrix} A & 0 \\ 0 & -A^* \end{pmatrix}$ then $\mu(e^{tA})f(x) = c_{tA}(\det^{-1/2} e^{-tA})f(e^{-tA}x)$,

if $A = \begin{pmatrix} 0 & 0 \\ C & 0 \end{pmatrix}$ then $\mu(e^{tA})f(x) = c_{tA}e^{-\pi i t x C x}f(x)$,

if $A = \begin{pmatrix} 0 & B \\ 0 & 0 \end{pmatrix}$ then $\mu(e^{tA})f = c_{tA}\mathcal{F}^{-1}\mu(e^{-tA^*})\mathcal{F}f = \mathcal{F}^{-1}(e^{\pi i t \xi B \xi}\widehat{f})$.

Here the phase factor c_{tA} depends smoothly on t and $\det^{-1/2}$ denotes the positive branch of the square root. On the other hand, if $f \in \mathcal{S}$,

$$\frac{d}{dt}\left[(\det^{-1/2} e^{-tA})f(e^{-tA}x)\right]_{t=0} = \frac{d}{dt}\left[e^{-(t/2)\operatorname{tr} A}f(e^{-tA}x)\right]_{t=0}$$
$$= -\tfrac{1}{2}(\operatorname{tr} A)f(x) - \sum A_{jk}x_k\frac{\partial f}{\partial x_j}(x),$$

and also

$$\frac{d}{dt}\left[e^{-\pi i t x C x}f(x)\right]_{t=0} = -\pi i \sum C_{jk}x_j x_k f(x),$$
$$\frac{d}{dt}\left[\mathcal{F}^{-1}(e^{\pi i t \xi B \xi}\widehat{f})\right]_{t=0} = \mathcal{F}^{-1}(\pi i \xi B \xi \widehat{f}) = \pi i \sum B_{jk}D_j D_k f.$$

Combining these results with (4.44), since $d\mu(A)$ depends linearly on A we conclude that $\mathcal{S}$ consists of C^∞ vectors for μ and that

$$d\mu(A) = 2\pi i P_A(D, X) + c'_A I, \qquad A \in \operatorname{sp}.$$

Here c'_A is a scalar (the derivative of c_{tA} at $t = 0$ when A is one of the above types). Since $A \to d\mu(A)$ and $A \to 2\pi i P_A(D, X)$ are Lie algebra homomorphisms, so is $A \to c'_A$. But then

$$c'_{[A,B]} = [c'_A, c'_B] = 0,$$

so $c'_A = 0$ by Proposition (4.20), and the proof is complete. ∎

At this point we can dispose of some unfinished business from Section 4.2.

(4.46) Proposition. *Formulas (4.24), (4.25), and (4.26) are correct.*

Proof: We have actually established (4.24) and (4.25) in the course of the preceding proof. Indeed, it is clear that $c_{(t+s)\mathcal{A}} = c_{t\mathcal{A}}c_{s\mathcal{A}}$, and it follows that $c_{t\mathcal{A}} = \exp(tc'_{\mathcal{A}}) = 1$. As for (4.26), the trick is to embed $\mathcal{J}$ into a one-parameter group. In fact, we have

$$e^{t\mathcal{J}} = \begin{pmatrix} (\cos t)I & (\sin t)I \\ -(\sin t)I & (\cos t)I \end{pmatrix}, \quad \text{in particular,} \quad \mathcal{J} = e^{(\pi/2)\mathcal{J}}.$$

Now, $P_{\mathcal{J}}(\xi, x) = \frac{1}{2}(\xi^2 + x^2)$ and hence

$$d\mu(\mathcal{J}) = 2\pi i P_{\mathcal{J}}(D, X) = \pi i (D^2 + X^2).$$

This is $\frac{1}{2}i$ times the Hermite operator, so according to (1.83b), its action on the Hermite functions is given by

$$\pi i (D^2 + X^2)h_\alpha = \frac{1}{2}i(2|\alpha| + n)h_\alpha.$$

It follows that

(4.47) $$\mu(e^{t\mathcal{J}})h_\alpha = e^{it(2|\alpha|+n)/2}h_\alpha,$$

and taking $t = \frac{1}{2}\pi$,

$$\mu(\mathcal{J})h_\alpha = i^{|\alpha|+(n/2)}h_\alpha.$$

On the other hand, from (1.84) we also have $\mathcal{F}^{-1}h_\alpha = i^{|\alpha|}h_\alpha$. Since the h_α's are a complete orthonormal set, this establishes (4.26). ∎

This last argument clearly exhibits the double-valued nature of μ, at least when n is odd. Indeed, we have $e^{2\pi\mathcal{J}} = I$, but $\mu(e^{2\pi\mathcal{J}}) = (-1)^n$ according to (4.47). Thus when n is odd, the path $t \to e^{t\mathcal{J}}$, $0 \le t \le 4\pi$, is a doubly-traversed circle in Sp that unwinds to a simple closed curve in $\mu(Sp)$. We can get the same effect for arbitrary n by replacing $\mathcal{J}$ by $\mathcal{J}_1$, which is "$\mathcal{J}$ in the first pair of variables":

$$\mathcal{J}_1(\xi, x) = (x_1, \xi_2, \ldots, \xi_n, -\xi_1, x_2, \ldots, x_n).$$

We have

$$2\pi i P_{\mathcal{J}_1}(D, X) = \pi i (D_1^2 + X_1^2), \qquad \mu(e^{t\mathcal{J}_1})h_\alpha = e^{it(2\alpha_1+n)/2}h_\alpha,$$

so the double-valuedness in dimension one persists in dimension n.

The isomorphism $\mathcal{A} \to P_{\mathcal{A}}$ from **sp** to $\mathcal{Q}$ can be interpreted geometrically as follows. For any $\mathcal{A} \in M_{2n}(\mathbf{R})$, $\{e^{t\mathcal{A}^*}\}$ is a one-parameter group of linear diffeomorphisms of $\mathbf{R}^{2n}$, and its infinitesimal generator is the vector field $X_{\mathcal{A}}$ defined by

$$X_{\mathcal{A}}f(x) = \frac{d}{dt}f(e^{t\mathcal{A}^*}w)|_{t=0} = \mathcal{A}^*w \cdot \nabla f(w).$$

The correspondence $\mathcal{A} \to X_{\mathcal{A}}$ is easily seen to be a Lie algebra homomorphism. (If we had used $e^{t\mathcal{A}}$ instead of $e^{t\mathcal{A}^*}$ it would have been an antihomomorphism.) Now, if $\mathcal{A} \in$ sp then $\{e^{t\mathcal{A}^*}\}$ is a group of canonical transformations, so $X_{\mathcal{A}}$ is a Hamiltonian vector field. In fact, it is the Hamiltonian vector field associated to the polynomial $P_{\mathcal{A}}$, for if $\mathcal{A} = \left(\begin{smallmatrix} A & B \\ C & -A^* \end{smallmatrix}\right)$ and $w = (\xi, x)$ then by (4.41),

$$
\begin{aligned}
(4.48) \qquad X_{\mathcal{A}}f(w) &= \mathcal{A}^*w \cdot \nabla f(w) \\
&= (A^*\xi + Cx) \cdot \nabla_\xi f(w) + (B\xi - Ax) \cdot \nabla_x f(w) \\
&= [\nabla_\xi P_{\mathcal{A}} \cdot \nabla_x f - \nabla_x P_{\mathcal{A}} \cdot \nabla_\xi f](w) = \{P_{\mathcal{A}}, f\}(w).
\end{aligned}
$$

In the proof above of Theorem (4.45) we established the formula $d\mu(\mathcal{A}) = 2\pi i P_{\mathcal{A}}(D, X)$ piecemeal. However, once it is granted that $\mathcal{S}$ consists of C^∞ vectors for μ, we can use (4.48) to rederive this formula in a more direct, global way. Namely, suppose σ is in some symbol class $S^m_{1,0}$. By Theorem (2.15),

$$\mu(e^{t\mathcal{A}})\sigma(D, X)\mu(e^{-t\mathcal{A}}) = \sigma \circ e^{t\mathcal{A}^*}(D, X), \qquad \mathcal{A} \in \text{sp}.$$

Considering both sides as operators on $\mathcal{S}$, we differentiate in t and set $t = 0$. By (4.48) and Corollary (2.51), we get

$$[d\mu(\mathcal{A}), \sigma(D, X)] = \{P_{\mathcal{A}}, \sigma\}(D, X) = 2\pi i[P_{\mathcal{A}}(D, X), \sigma(D, X)].$$

Since σ is quite arbitrary, we must have

$$d\mu(\mathcal{A}) = 2\pi i P_{\mathcal{A}}(D, X) + C_{\mathcal{A}}I$$

where $C_{\mathcal{A}}$ is a scalar. As in the proof of Theorem (4.45), $C_{\mathcal{A}} = 0$.

We have defined the metaplectic representation as a group of intertwining operators and then shown that its infinitesimal version is the Weyl quantization of $\mathcal{Q}$. One can also proceed in the opposite direction, as follows. One first constructs the representation

$$R(\mathcal{A}) = 2\pi i P_{\mathcal{A}}(D, X)$$

of sp by skew-Hermitian operators on $\mathcal{S}$, as we did at the beginning of this section. The next step is to show that this representation can be exponentiated to a representation of Sp or a covering group thereof. The fact that the operators $R(\mathcal{A})$ are all essentially skew-adjoint on $\mathcal{S}$ is not sufficient to ensure this; it is necessary to invoke the theory of analytic vectors. We recall that if $T_1, \ldots, T_k$ are (unbounded) operators on a Hilbert space $\mathcal{H}$, a vector $v \in \mathcal{H}$ is called **analytic** for $\{T_j\}_1^k$ if v is in the domain of all products of the T_j's and there exist $C, K > 0$ such that

$$\|T_{j_1} \cdots T_{j_N} v\|_{\mathcal{H}} \leq C K^N N! \text{ for all } j_i \in \{1, \ldots, k\} \text{ and } N \in \mathbf{Z}^+.$$

We shall invoke without proof the following theorem of Nelson [112] (see also Taylor [137]):

> Suppose G is a simply connected Lie group with Lie algebra **g**, and let $X_1, \ldots, X_k$ be a basis for **g**. Suppose R is a representation of **g** by (unbounded) skew-Hermitian operators on a Hilbert space $\mathcal{H}$. If there is a dense set of analytic vectors for $\{R(X_1), \ldots, R(X_k)\}$, then there is a unique unitary representation π of G such that $d\pi = R$.

In our case, with $Z_j = X_j + iD_j$ and $Z_j^* = X_j - iD_j$, we have:

(4.49) Proposition. *The finite linear combinations of the Hermite functions are analytic vectors for*

$$\left\{ Z_j Z_k, \; Z_j^* Z_k, \; Z_j Z_k^*, \; Z_j^* Z_k^* : 1 \leq j, k \leq n \right\}.$$

Proof: Obviously it suffices to show that each Hermite function is an analytic vector. We recall from (1.82) that

$$\sqrt{\pi}\, Z_j^* h_\alpha = \sqrt{\alpha_j + 1}\, h_{\alpha + 1_j}, \qquad \sqrt{\pi}\, Z_j h_\alpha = \sqrt{\alpha_j}\, h_{\alpha - 1_j}.$$

Suppose $W = \prod_1^{2N} W_k$ where each W_k is either a Z_j or a Z_j^*. Since $\|h_\alpha\|_2 = 1$ for all α, it follows easily from the above equations that

$$\|W h_\alpha\|_2 \leq \sqrt{(|\alpha| + 1) \cdots (|\alpha| + 2N)} = \sqrt{\frac{(|\alpha| + 2N)!}{|\alpha|!}}.$$

By the binomial theorem,

$$\frac{(|\alpha| + 2N)!}{|\alpha|!\,(2N)!} \leq 2^{|\alpha| + 2N},$$

and thus

$$\|W h_\alpha\|_2 \leq 2^{N + (|\alpha|/2)} \sqrt{(2N)!} \leq 2^{2N + (|\alpha|/2)} N!. \quad \blacksquare$$

The operators $Z_j Z_k$, etc., are not in the range of the representation $\mathcal{A} \to 2\pi i P_{\mathcal{A}}(D, X)$ of **sp** but only of its complexification. But this is sufficient: if $\{\mathcal{A}_1, \ldots, \mathcal{A}_N\}$ is any basis for **sp**, the operators $2\pi i P_{\mathcal{A}_j}(D, X)$ are linear combinations with complex coefficients of the former operators, and one concludes without difficulty that the Hermite functions are analytic vectors for them. By Nelson's theorem, therefore, there is a unitary representation μ' of the universal cover G of Sp such that $d\mu'(\mathcal{A}) = 2\pi i P_{\mathcal{A}}(D, X)$ for $\mathcal{A} \in \mathfrak{sp}$.

When $\mathcal{A}$ is of the form

$$\begin{pmatrix} A & 0 \\ 0 & -A^* \end{pmatrix}, \quad \begin{pmatrix} 0 & B \\ 0 & 0 \end{pmatrix}, \quad \begin{pmatrix} 0 & 0 \\ C & 0 \end{pmatrix},$$

it is easy enough to exponentiate $2\pi i P_{\mathcal{A}}(D, X)$ explicitly to obtain

$$\mu' \left[\exp \begin{pmatrix} A & 0 \\ 0 & -A^* \end{pmatrix} \right] f(x) = (\det^{-1/2} A) f(A^{-1}x),$$

$$\mathcal{F}\mu' \left[\exp \begin{pmatrix} 0 & B \\ 0 & 0 \end{pmatrix} \right] f(\xi) = e^{\pi i \xi B \xi} \widehat{f}(\xi).$$

$$\mu' \left[\exp \begin{pmatrix} 0 & 0 \\ C & 0 \end{pmatrix} \right] f(x) = e^{-\pi i x C x} f(x),$$

By the simple calculations we performed in Section 4.1, these operators $\mu'(g)$ (where $g = \exp \mathcal{A}$ as above) all have the intertwining property

$$(4.50) \qquad \mu'(g)\rho(p, q)\mu'(g)^{-1} = \rho\big(\pi(g)(p, q)\big),$$

where $\pi : G \to Sp$ is the projection. But such g's generate G, so (4.50) holds in general. Thus we have recovered the characterization of the metaplectic representation in terms of intertwining operators. Of course, in this approach there is still some work to be done to show that μ' actually lives on the double cover of Sp, and passing to the Fock model remains the most perspicuous way to do this.

4. Other Aspects of the Metaplectic Representation

Integral Formulas.   The formulas (4.24)–(4.26) describe the restriction of the metaplectic representation to certain subgroups of Sp. We can combine them to obtain oscillatory integral formulas for the metaplectic representation that are valid on dense open sets in Sp. Here is the first one.

(4.51) Theorem. If $\mathcal{A} = \begin{pmatrix} A & B \\ C & D \end{pmatrix} \in Sp$ and $\det A \neq 0$, then

$$(4.52) \qquad \mu(\mathcal{A})f(x) = (\det^{-1/2} A) \int e^{2\pi i S(x, \xi)} \widehat{f}(\xi) \, d\xi,$$

$$\text{where} \quad S(x, \xi) = -\tfrac{1}{2}xCA^{-1}x + \xi A^{-1}x + \tfrac{1}{2}\xi A^{-1}B\xi.$$

Proof: From Proposition (4.9) and its proof, if $\det A \neq 0$ we have

$$\begin{pmatrix} A & B \\ C & D \end{pmatrix} = \begin{pmatrix} I & 0 \\ CA^{-1} & I \end{pmatrix} \begin{pmatrix} A & 0 \\ 0 & A^{*-1} \end{pmatrix} \begin{pmatrix} I & A^{-1}B \\ 0 & I \end{pmatrix}.$$

Moreover,

$$\begin{pmatrix} I & A^{-1}B \\ 0 & I \end{pmatrix} = \begin{pmatrix} 0 & I \\ -I & 0 \end{pmatrix} \begin{pmatrix} I & 0 \\ -A^{-1}B & I \end{pmatrix} \begin{pmatrix} 0 & -I \\ I & 0 \end{pmatrix}.$$

Hence, by (4.24), (4.25), and (4.26) (i.e., by Proposition (4.46)),

$$\mu(\mathcal{A})f(x) = e^{-\pi i x C A^{-1} x} (\det^{-1/2} A) \mathcal{F}^{-1}[e^{\pi i \xi A^{-1}B\xi} \widehat{f}](A^{-1}x)$$

$$= (\det^{-1/2} A) \int e^{-\pi i x C A^{-1} x + 2\pi i A^{-1} x + \pi i \xi A^{-1}B\xi} \widehat{f}(\xi) \, d\xi. \quad \blacksquare$$

Remark 1. The integral in (4.52) converges absolutely if $\widehat{f} \in L^1$, and in particular if $f \in S$. For more general f, the reader may provide suitable interpretations of (4.52) ad libitum. The same goes for the following results.

Remark 2. If $S(x, \xi)$ is as in (4.52), we have

$$\nabla_x S(x, \xi) = -CA^{-1}x + A^{*-1}\xi, \qquad \nabla_\xi S(x, \xi) = A^{-1}x + A^{-1}B\xi,$$

and hence in view of the fact that $D = CA^{-1}B + A^{*-1}$ (Proposition (4.1e)),

$$\begin{pmatrix} \nabla_x S \\ x \end{pmatrix} = \begin{pmatrix} D & -C \\ -B & A \end{pmatrix} \begin{pmatrix} \xi \\ \nabla_\xi S \end{pmatrix}.$$

Moreover,

$$\begin{pmatrix} D & -C \\ -B & A \end{pmatrix} = \mathcal{J}\mathcal{A}\mathcal{J}^{-1} = \mathcal{A}^{*-1}.$$

In other words, S is a "generating function" for the canonical transformation $\mathcal{A}^{*-1}$ in the sense used in classical mechanics. (The fact that we get $\mathcal{A}^{*-1}$ instead of $\mathcal{A}$ is an artifact of the way we set up the Schrödinger representation. If we had used ρ' instead of ρ as in (1.27), $\mu(\mathcal{A})$ would turn into $\mu(\mathcal{A}^{*-1})$.) Thus formula (4.52) is in the spirit of the prescription of Egorov [42] for associating a Fourier integral operator to a canonical transformation, although Egorov's homogeneity condition is satisfied in (4.52) only in the trivial case $B = C = 0$.

Theorem (4.51) gives $\mu(\mathcal{A})f$ as an integral transform of $\widehat{f}$; our next result gives $\mu(\mathcal{A})f$ as a similar transform of f itself.

(4.53) Theorem. If $\mathcal{A} = \begin{pmatrix} A & B \\ C & D \end{pmatrix} \in Sp$ and $\det B \neq 0$ then

(4.54)
$$\mu(\mathcal{A})f(x) = i^{n/2}(\det^{-1/2} B) \int e^{2\pi i S'(x,y)} f(y)\, dy,$$

$$\text{where} \quad S'(x,y) = -\tfrac{1}{2}x DB^{-1}x + y B^{-1}x - \tfrac{1}{2}y B^{-1}Ay.$$

Proof: We have $\mathcal{A}\mathcal{J}^{-1} = \begin{pmatrix} B & -A \\ D & -C \end{pmatrix}$, so by Theorem (4.51),

$$\mu(\mathcal{A}\mathcal{J}^{-1})f(x) = (\det^{-1/2} B) \int e^{\pi i(-x DB^{-1}x + 2y B^{-1}x - y B^{-1}Ay)} \widehat{f}(y)\, dy.$$

But $\mu(\mathcal{A}) = \mu(\mathcal{A}\mathcal{J}^{-1})\mu(\mathcal{J})$ and $\mu(\mathcal{J}) = i^{n/2}\mathcal{F}^{-1}$, so the result follows. ∎

As an important special case, we obtain formulas for the one-parameter group $\mu(e^{t\mathcal{J}})$, whose action on the Hermite functions we computed in the previous section.

(4.55) Corollary. Let

$$A_\theta = e^{\theta \mathcal{J}} = \begin{pmatrix} (\cos\theta)I & (\sin\theta)I \\ (-\sin\theta)I & (\cos\theta)I \end{pmatrix}, \qquad \theta \in \mathbf{R}.$$

If $\cos\theta \neq 0$, then

$$\mu(A_\theta)f(x) = (\sec\theta)^{n/2} \int e^{-\pi i(\tan\theta)(x^2 + \xi^2) + 2\pi i(\sec\theta)x\xi} \widehat{f}(\xi)\, d\xi.$$

If $\sin\theta \neq 0$, then

$$\mu(A_\theta)f(x) = (i\csc\theta)^{n/2} \int e^{-\pi i(\cot\theta)(x^2 + y^2) + 2\pi i(\csc\theta)xy} f(y)\, dy.$$

The latter of these formulas is closely related to Mehler's formula (1.87). We shall say more about this in Section 5.2.

Exercise. Show directly that if $\mathcal{A} = \begin{pmatrix} A & B \\ C & D \end{pmatrix} \in Sp$, $\det A \neq 0$, and $\det B \neq 0$, then formulas (4.52) and (4.54) agree. (This amounts to showing that $(e^{2\pi i S})\widehat{} = e^{2\pi i S'}$. The essential tools are Theorem 2 of Appendix A and Proposition (4.1e,f).)

Theorems (4.51) and (4.53) together still do not give explicit formulas for $\mu(\mathcal{A})$ for arbitrary $\mathcal{A} \in Sp$. (Exercise: find a 4×4 symplectic matrix whose four 2×2 blocks are all singular.) It seems to be a fact of life that there is no simple description of the operator $\mu(\mathcal{A})$ that is valid for all $\mathcal{A} \in Sp$. (The Fock model ν is of course superior in this respect.)

Irreducible Subspaces.   The metaplectic representation is reducible: a glance at any of the formulas we have given will show that the spaces L^2_{even} and L^2_{odd} of even and odd functions in $L^2(\mathbf{R}^n)$ are invariant. (Another way of looking at this: the nontrivial element of the center of Sp is $-I$, so the eigenspaces of $\mu(-I)$ are invariant. But $\mu(-I)f(x) = \pm i^n f(-x)$.) However, these are the only nontrivial closed invariant subspaces.

(4.56) Theorem. *The subrepresentations of μ on L^2_{even} and L^2_{odd} are irreducible and inequivalent.*

Proof: Consider the Hermite operator $H = 2\pi(D^2 + X^2)$ and its ground state $\phi_0(x) = e^{-\pi x^2}$, which is the unique eigenvector of H (up to scalar multiples) with eigenvalue n. If $L^2_{\text{even}} = \mathcal{M} \oplus \mathcal{N}$ with $\mathcal{M}$ and $\mathcal{N}$ both invariant under μ, we can write $\phi_0 = \phi_{\mathcal{M}} + \phi_{\mathcal{N}}$ where $\phi_{\mathcal{M}} \in \mathcal{M}$, $\phi_{\mathcal{N}} \in \mathcal{N}$, and

$$e^{int}\phi_{\mathcal{M}} + e^{int}\phi_{\mathcal{N}} = e^{int}\phi_0 = e^{itH}\phi_0 = e^{itH}\phi_{\mathcal{M}} + e^{itH}\phi_{\mathcal{N}}.$$

But $e^{itH} = \mu(e^{2tJ})$, as we observed in proving Proposition (4.46), so $e^{itH}\phi_{\mathcal{M}} \in \mathcal{M}$ and $e^{itH}\phi_{\mathcal{N}} \in \mathcal{N}$ since $\mathcal{M}$ and $\mathcal{N}$ are invariant. It follows that

$$e^{itH}\phi_{\mathcal{M}} = e^{int}\phi_{\mathcal{M}}, \qquad e^{itH}\phi_{\mathcal{N}} = e^{int}\phi_{\mathcal{N}}.$$

Differentiating in t, we see that $H\phi_{\mathcal{M}} = n\phi_{\mathcal{M}}$ and $H\phi_{\mathcal{N}} = n\phi_{\mathcal{N}}$. But this means that $\phi_{\mathcal{M}}$ and $\phi_{\mathcal{N}}$ are multiples of ϕ_0, and since $\phi_{\mathcal{M}} \perp \phi_{\mathcal{N}}$, one of them— say, $\phi_{\mathcal{N}}$—is zero.

We now have $\phi_0 \in \mathcal{M}$, and ϕ_0 is a C^∞ vector for μ by Theorem (4.45), so $B\phi_0 \in \mathcal{M}$ where B is any linear combination of products of the operators $d\mu(\mathcal{A})$, $\mathcal{A} \in \mathsf{sp}$. For suitable $\mathcal{A}$ we have $d\mu(\mathcal{A})f(x) = x_j x_k f(x)$, so $P\phi_0 \in \mathcal{M}$ for all even polynomials P, and in particular for all even Hermite polynomials. But these functions span L^2_{even}; hence $\mathcal{M} = L^2_{\text{even}}$ and $\mathcal{N} = \{0\}$.

The argument for L^2_{odd} is similar. Let $\phi_1(x) = x_1 e^{-\pi x^2}$; then ϕ_1 is the unique vector (up to scalar multiples) satisfying $H\phi_1 = (n+2)\phi_1$ and $H_1\phi_1 = 3\phi_1$ where $H_1 = 2\pi(D_1^2 + X_1^2)$. The preceding argument shows that if $L^2_{\text{odd}} = \mathcal{M} \oplus \mathcal{N}$ where $\mathcal{M}$ and $\mathcal{N}$ are μ-invariant, then $\phi_1 \in \mathcal{M}$ or $\phi_1 \in \mathcal{N}$. Say $\phi_1 \in \mathcal{M}$: then $\phi_1 \circ U \in \mathcal{M}$ for any $U \in SO(n)$, by (4.24). In other words, $l\phi_0 \in \mathcal{M}$ for any linear functional l on $\mathbf{R}^n$. Applying the operators $d\mu(\mathcal{A})$ as above, then, we see that $P\phi_0 \in \mathcal{M}$ for all odd polynomials P, and hence $\mathcal{M} = L^2_{\text{odd}}$ and $\mathcal{N} = \{0\}$.

Finally, that the subrepresentations of μ on L^2_{even} and L^2_{odd} are inequivalent may be seen by observing that L^2_{even} contains a vector that is invariant under $\mu(\mathcal{A})$ for all $\mathcal{A} \in Sp \cap O(2n)$ (namely ϕ_0), whereas L^2_{odd} does not. (This is perhaps clearer in the Fock model, where $Sp \cap O(2n)$ acts by rotations of $\mathbf{C}^n$, according to Proposition (4.39). Here it is obvious that the constant functions are the only elements of $\mathcal{F}_n$ that are invariant under this action.) ∎

Dependence on Planck's Constant.   For each $h \in \mathbf{R}\backslash\{0\}$ we have the Schrödinger representation ρ_h and hence a corresponding metaplectic representation μ_h determined by the condition

$$\mu_h(\mathcal{A})\rho_h(p,q)\mu_h(\mathcal{A})^{-1} = \rho_h\big(\mathcal{A}(p,q)\big).$$

Although the representations ρ_h are inequivalent, the same is not true of the μ_h.

(4.57) Theorem. μ_h and $\mu_{h'}$ are equivalent if and only if h and h' have the same sign.

Proof: If $h > 0$ we have

$$\rho_h(p,q) = \rho(hp,q) = (\rho \circ \delta_h \circ \mathcal{B}_h)(p,q),$$

where

$$\delta_h(p,q) = (h^{1/2}p,\, h^{1/2}q) \quad \text{and} \quad \mathcal{B}_h(p,q) = (h^{1/2}p,\, h^{-1/2}q).$$

Clearly $\mathcal{B}_h \in Sp$, and δ_h is a multiple of I and so commutes with all elements of Sp. Therefore, if $\mathcal{A} \in Sp$,

$$\begin{aligned}
\rho_h\big(\mathcal{A}(p,q)\big) &= (\rho \circ \delta_h \circ \mathcal{B}_h\mathcal{A})(p,q) = (\rho \circ \mathcal{B}_h\mathcal{A})\big(\delta_h(p,q)\big) \\
&= \mu(\mathcal{B}_h\mathcal{A})\rho\big(\delta_h(p,q)\big)\mu(\mathcal{B}_h\mathcal{A})^{-1} \\
&= \mu(\mathcal{B}_h)\mu(\mathcal{A})\mu(\mathcal{B}_h)^{-1}\rho_h(p,q)[\mu(\mathcal{B}_h)\mu(\mathcal{A})\mu(\mathcal{B}_h)^{-1}]^{-1}.
\end{aligned}$$

From the uniqueness of μ_h (the generalization of Proposition (4.40) to arbitrary h), it follows that

$$\mu_h(\mathcal{A}) = \mu(\mathcal{B}_h)\mu(\mathcal{A})\mu(\mathcal{B}_h)^{-1},$$

so μ_h and $\mu = \mu_1$ are equivalent. The same argument shows that μ_h is equivalent to μ_{-1} when $h < 0$.

The quickest way to see that μ_1 and μ_{-1} are inequivalent is to pass to the Fock model. If we denote by ν_h the Fock-metaplectic representation of Sp_c corresponding to the representation β_h of $\mathbf{H}_n$, it is clear from the construction of β_h in Section 1.6 that

$$\nu_{-1}(\mathcal{A})F(\overline{z}) = \overline{\nu_1(\mathcal{A})F(z)}.$$

If we imbed the circle group $U(1)$ into Sp_c by

$$e^{i\theta} \longrightarrow A_\theta = \begin{pmatrix} e^{i\theta}I & 0 \\ 0 & e^{-i\theta}I \end{pmatrix}$$

and take $F(z) \equiv 1$, by Proposition (4.39) we have

$$\nu_1(A_\theta)F = e^{-in\theta/2}F \quad \text{and} \quad \nu_{-1}(A_\theta)F = e^{in\theta/2}F.$$

These are clearly inequivalent one-dimensional representations of (the double cover of) the circle group, so a fortiori ν_1 and ν_{-1} are inequivalent. ∎

The Extended Metaplectic Representation. Since Sp acts on $\mathbf{H}_n$ by automorphisms, we can form the semidirect product of these two groups, which we denote by HSp. The underlying manifold of HSp is $\mathbf{H}_n \times Sp$, and the group law is

$$(X, A)(X', A') = (X(AX'),\ AA').$$

Here $X = (p, q, t)$ and $AX = (A(p, q),\ t)$. The Schrödinger representation ρ of $\mathbf{H}_n$ and the metaplectic representation μ of Sp fit together to form a double-valued unitary representation of HSp which we denote by ω:

$$\omega(X, A) = \rho(X)\mu(A).$$

Let us check that this really is a representation:

$$\omega(X, A)\omega(X', A') = \rho(X)\mu(A)\rho(X')[\mu(A)^{-1}\mu(A)]\mu(A')$$
$$= \rho(X)\rho(AX')\mu(AA') = \omega(X(AX'),\ AA').$$

ω is irreducible since ρ is; and since μ is faithful, the kernel of ω is the set of elements $(X, I) \in HSp$ such that $X = (0, 0, k)$, $k \in \mathbf{Z}$.

(4.58) Proposition. ω *is the unique unitary representation of HSp (or rather of its universal cover) whose restriction to $\mathbf{H}_n$ is ρ.*

Proof: If ω' is another such, we must have

$$\rho(AX) = \omega'(AX, I) = \omega'(0, A)\omega'(X, I)\omega'(0, A^{-1}) = \omega'(0, A)\rho(X)\omega'(0, A)^{-1},$$

so $\omega'(0, A) = \mu(A)$ by Proposition (4.40) and thus

$$\omega'(X, A) = \omega'(X, I)\omega'(0, A) = \rho(X)\mu(A) = \omega(X, A). \quad \blacksquare$$

The Lie algebra $\mathfrak{h}_n$ is isomorphic to the algebra of first-degree polynomials on $\mathbf{R}^{2n}$ with the Poisson bracket, and under this identification, the infinitesimal representation $d\rho$ is just Weyl quantization times $2\pi i$:

$$(a, b, c) \in \mathfrak{h}_n \longleftrightarrow \sigma(\xi, x) = a\xi + bx + c,$$
$$d\rho(a, b, c) = 2\pi i(aD + bX + cI) = 2\pi i\sigma(D, X).$$

With this and Theorem (4.45) in mind, it is easy to see that the Lie algebra of HSp is isomorphic to the algebra $\mathfrak{p}_2$ of polynomials of degree ≤ 2 on $\mathbf{R}^{2n}$ with the Poisson bracket, and that the infinitesimal representation $d\omega$ is $2\pi i$ times the Weyl quantization of such polynomials.

Another way one might think of extending the metaplectic representation is to return to the full group of automorphisms of $\mathbf{H}_n$ that leave the center pointwise fixed, namely the group generated by Sp and the inner automorphisms. This group is isomorphic to the inhomogeneous symplectic group ISp, that is, the semidirect product of $\mathbf{R}^{2n}$ with Sp defined by

$$(w, \mathcal{A})(w', \mathcal{A}') = (w + \mathcal{A}w', \ \mathcal{A}\mathcal{A}').$$

Here, the automorphism of $\mathbf{H}_n$ given by $(w, \mathcal{A}) \in ISp$ is

$$(w, \mathcal{A})X = w_0(\mathcal{A}X)w_0^{-1}, \qquad w_0 = (w, 0) \in \mathbf{H}_n.$$

Associated to each $(w, \mathcal{A}) \in ISp$ we have an intertwining operator for the Schrödinger representation, determined up to a phase factor, namely $\rho(w)\mu(\mathcal{A})$ (where $\rho(w) = \rho(w, 0)$ as usual):

$$\rho(w)\mu(\mathcal{A})\rho(X)\mu(\mathcal{A})^{-1}\rho(w)^{-1} = \rho(w)\rho(\mathcal{A}X)\rho(w)^{-1} = \rho\big((w, \mathcal{A})X\big).$$

However,

$$\rho(w)\rho(\mathcal{A})\rho(w')\rho(\mathcal{A}') = \rho(w)\rho(\mathcal{A}w')\mu(\mathcal{A})\mu(\mathcal{A}') = e^{\pi i[w, \mathcal{A}w']}\rho(w + \mathcal{A}w')\mu(\mathcal{A}\mathcal{A}'),$$

where brackets denote the symplectic form. Thus the assignment $(w, \mathcal{A}) \to \rho(w)\mu(\mathcal{A})$ defines only a projective representation of ISp whose ambiguity of phase factors involves the whole circle group. This projective representation does not come from a genuine representation of ISp or a covering group thereof (the Bargmann obstruction in $H^2(\mathrm{isp}, \mathbf{R})$ is nonzero), and the only way to unravel the phase factors is to pass to a suitable central extension of ISp by the circle or its covering group $\mathbf{R}$. The suitable central extension is precisely HSp, and the representation of HSp obtained in this way is ω.

The Groenewold–van Hove Theorems. In quantizing classical observables, we start by declaring the quantizations of the coordinate functions ξ_j and x_j to be the operators D_j and X_j. We wish to extend this correspondence linearly to larger classes of functions on $\mathbf{R}^{2n}$, and one of the desiderata is that Poisson brackets should correspond to commutators (times $2\pi i$). The Weyl correspondence accomplishes this for polynomials of degree ≤ 2, and by Proposition (4.58)—or rather its infinitesimal analogue—it is the only way to do so. Unfortunately it is impossible to extend this scheme to polynomials of higher degree, as the following theorem shows.

(4.59) Theorem. (Groenewold [61]) *Let* $\mathbf{p}_k$ *denote the space of real polynomials of degree* $\leq k$ *on* $\mathbf{R}^{2n}$. *There is no linear map* R *from* $\mathbf{p}_4$ *to the space of Hermitian operators on* $\mathcal{S}(\mathbf{R}^n)$ *such that*

$$R(\xi_j) = D_j, \qquad R(x_j) = X_j,$$
$$R(\{P, Q\}) = 2\pi i\big[R(P), \ R(Q)\big] \text{ for all } P, Q \in \mathbf{p}_3.$$

Proof: We assume $n = 1$ for simplicity, the argument in higher dimensions being the same, and we leave some of the mechanical calculations to the reader. If R existed, by the preceding remarks it would have to coincide with Weyl quantization on p_2, so

$$R(\xi) = D, \quad R(x) = X, \quad R(\xi^2) = D^2, \quad R(x^2) = X^2, \quad R(\xi x) = \tfrac{1}{2}(DX + XD).$$

I claim that $R(x^3) = X^3$ and $R(\xi^3) = D^3$. Indeed, we have

$$\left[X,\, R(x^3)\right] = (2\pi i)^{-1} R(\{x, x^3\}) = 0 = [X, X^3]$$

and

$$\left[D,\, R(x^3)\right] = (2\pi i)^{-1} R(\{\xi, x^3\}) = (2\pi i)^{-1} R(3x^2) = 3(2\pi i)^{-1} X^2 = [D, X^3],$$

so that

$$R(x^3) = X^3 + T \quad \text{where} \quad [X, T] = [D, T] = 0.$$

But then

$$[DX, T] = D[X, T] + [D, T]X = 0$$

and likewise $[XD, T] = 0$, so

$$R(x^3) = \frac{1}{3} R(\{\xi x, x^3\}) = \frac{\pi i}{3}\left[(DX + XD),\, X^3 + T\right]$$

$$= \frac{\pi i}{3}\left[(DX + XD),\, X^3\right] = X^3.$$

Similarly, $R(\xi^3) = D^3$. Next,

$$R(\xi^2 x) = \frac{1}{6} R(\{\xi^3, x^2\}) = \frac{\pi i}{3}[D^3, X^2] = \frac{1}{2}(D^2 X + XD^2)$$

and likewise

$$R(\xi x^2) = \frac{1}{2}(X^2 D + DX^2).$$

Now to the point: we try to calculate $R(\xi^2 x^2)$. On the one hand,

$$R(\xi^2 x^2) = \frac{1}{9} R(\{\xi^3, x^3\}) = \frac{2\pi i}{9}[D^3, X^3],$$

and on the other,

$$R(\xi^2 x^2) = \frac{1}{3} R(\{\xi^2 x, \xi x^2\}) = \frac{\pi i}{6}[D^2 X + XD^2,\, X^2 D + DX^2].$$

If we apply these operators to the constant function 1, we get

$$\frac{2\pi i}{9}[D^3, X^3]1 = \frac{2\pi i}{9} D^3(x^3) = \frac{2}{3(2\pi i)^2}$$

and

$$\frac{\pi i}{6}[D^2 X + XD^2,\, X^2 D + DX^2]1 = \frac{1}{12}(D^2 X + XD^2)(2x) = \frac{1}{3(2\pi i)^2}.$$

Contradiction. ∎

Groenewold's theorem is an important "no-go" result for quantization theory, but it is not the last word on the subject. A deeper analysis, which we now summarize in part, has been given by van Hove [143].

The reason for insisting on the correspondence between Poisson brackets and commutators is that classical and quantum observables define "infinitesimal automorphisms" of their respective systems, and these brackets are the associated Lie algebra structures. However, there are important structural differences between the classical and quantum models. One of these is that, whereas every self-adjoint operator on $L^2(\mathbf{R}^n)$ generates a one-parameter group of unitary operators, not every smooth function on $\mathbf{R}^{2n}$ generates a one-parameter group of canonical transformations. Many of them, including the functions $\xi^2 x$ and ξx^2 that appear in Groenewold's theorem, generate only local flows, i.e., one-parameter pseudogroups of local canonical transformations, and it is not to be expected that these observables will have good quantum counterparts.

Therefore, let us consider the set $\mathcal{G}$ of smooth functions on $\mathbf{R}^{2n}$ that generate global one-parameter groups; if $g \in \mathcal{G}$, we denote the group it generates by Φ_t^g. It comes as an unpleasant surprise that $\mathcal{G}$ is not a vector space, much less a Lie algebra. Nonetheless, one can ask whether there is a map R from $\mathcal{G}$ into the set of self-adjoint operators on $L^2(\mathbf{R}^n)$ with the following properties:

(i) $R(\xi_j) = D_j$ and $R(x_j) = X_j$.

(ii) There is a dense subspace $\mathcal{D} \subset L^2(\mathbf{R}^n)$ that is contained in the domains of all $R(g)$ and is mapped into itself by all $R(g)$ and all $e^{2\pi i t R(g)}$, $g \in \mathcal{G}$, such that on $\mathcal{D}$,

$$R(ag + bh) = aR(g) + bR(h) \text{ when } g, h, ag + bh \in \mathcal{G};$$
$$R(\{g, h\}) = 2\pi i [R(g), R(h)] \text{ when } g, h, \{g, h\} \in \mathcal{G}.$$

(iii) If for some $f, g, h \in \mathcal{G}$ and $s, t \in \mathbf{R}$ we have $\Phi_s^f \Phi_t^g \Phi_{-s}^f = \Phi_t^h$, then

$$e^{2\pi i s R(f)} e^{2\pi i t R(g)} e^{-2\pi i t R(f)} = e^{2\pi i t R(h)}.$$

The answer is negative: van Hove's main result in [143] is that no such mapping R exists. For modern views of van Hove's theorem and related results, see Gotay [59] and the references given there.

Some Applications.   The metaplectic representation intervenes in various ways in the part of mathematics that might be called "symplectic analysis," for example in such things as geometric quantization, geometrical optics and other high-frequency approximations, phenomena connected with the Maslov index, etc. See the books of Guillemin–Sternberg [65], [67] and Leray [96]. See also Guillemin–Sternberg [66] for an elegant application of the metaplectic representation to the spectral analysis of certain pseudodifferential operators, and Gaveau–Greiner–Vauthier [54] for an application to a symbolic calculus on $\mathbf{H}_n$.

The metaplectic representation, or its extension to HSp, solves the Schrödinger equation with a quadratic potential. More precisely, if $P(x)$ is a real quadratic polynomial, the evolution operators

$$U_t : v(x,0) \longrightarrow v(x,t)$$

that give the solutions of the Schrödinger equation

$$\frac{1}{2\pi i} \frac{\partial v}{\partial t} = \left(D^2 + P(X)\right)v$$

form a one-parameter subgroup of the extended metaplectic representation. It is therefore of interest to see how such operators propagate singularities. The answer is rather complicated. For example, it is easy to deduce from (4.24) and (4.25) that

$$\text{if } \mathcal{A} = \begin{pmatrix} A & 0 \\ 0 & A^{*-1} \end{pmatrix} \text{ then } WF(\mu(\mathcal{A})f) = \left\{ (A^{*-1}\xi, Ax) : (\xi, x) \in WF(f) \right\};$$

$$\text{if } \mathcal{A} = \begin{pmatrix} I & 0 \\ C & I \end{pmatrix} \text{ then } WF(\mu(\mathcal{A})f) = WF(f).$$

On the other hand, for a "generic" $\mathcal{A} \in Sp$, namely any $\mathcal{A} = \begin{pmatrix} A & B \\ C & D \end{pmatrix}$ with $\det B \neq 0$, it follows from Theorem (4.53) that $\mu(\mathcal{A})$ transforms all distributions with compact support into C^∞ functions; and by the same token, it can create singularities when applied to smooth functions of non-compact support. This situation has been studied in detail by Zelditch [160] and Weinstein [152], who also prove theorems about propagation of singularities for Schrödinger equations whose potentials are perturbations of quadratic functions.

The metaplectic representation also plays a significant role in the theory of theta functions and related parts of number theory; see Igusa [80] and Weil [151].

5. Gaussians and the Symmetric Space

The symmetric space associated to the group $Sp(1, \mathbf{R}) = SL(2, \mathbf{R})$ is the upper half plane in $\mathbf{C}$, on which $Sp(1, \mathbf{R})$ acts by linear fractional transformations. There is a natural generalization of this for $Sp(n, \mathbf{R})$, which we now explain.

The generalization of the upper half plane is the **Siegel half plane** Σ_n, which despite the name was first described by E. Cartan [29]. Σ_n is the set of all symmetric complex $n \times n$ matrices Z such that $\text{Im } Z > 0$; here "symmetric" means that $Z = Z^t$ (not $Z = Z^*$), and "$\text{Im } Z > 0$" means that $\text{Im } Z$ is positive definite. We shall usually write X and Y for the real and imaginary parts of Z. We observe that if $Z = Z^\dagger$,

$$(4.60) \qquad \mathrm{Re}(\bar{v}Zv) = \bar{v}Xv, \qquad \mathrm{Im}(\bar{v}Zv) = \bar{v}Yv \qquad (v \in \mathbf{C}^n).$$

It follows that if $Z \in \Sigma_n$ then $Zv = 0$ only when $v = 0$, so Z is invertible.

The generalization of linear fractional maps is the action α of $GL(2n, \mathbf{C})$ on $M_n(\mathbf{C})$ defined as follows. If $\mathcal{A} = \begin{pmatrix} A & B \\ C & D \end{pmatrix} \in GL(2n, \mathbf{C})$, the domain of $\alpha(\mathcal{A})$ is the set of all $Z \in M_n(\mathbf{C})$ such that $CZ + D$ is invertible, and for such Z we define

$$\alpha(\mathcal{A})Z = (AZ + B)(CZ + D)^{-1}.$$

We leave it to the reader to verify that α is a homomorphism:

$$\alpha(\mathcal{A})\alpha(\mathcal{B}) = \alpha(\mathcal{A}\mathcal{B}).$$

In this connection we shall frequently encounter the so-called "multiplier"
(4.61)

$$m(\mathcal{A}, Z) = \det{}^{-1/2}(CZ + D) \qquad \left(\mathcal{A} = \begin{pmatrix} A & B \\ C & D \end{pmatrix}, \quad Z \in \mathrm{Dom}\big[\alpha(\mathcal{A})\big] \right).$$

As in the metaplectic representation, there is an ambiguity of sign in $m(\mathcal{A}, Z)$, and hence in most formulas containing $m(\mathcal{A}, Z)$. It will not worry us, but it can be removed by regarding $\mathcal{A}$ as an element of the double cover of $GL(2n, \mathbf{C})$ or by regarding $m(\mathcal{A}, Z)$ as an element of the group of nonzero complex numbers modulo ± 1. The reader may readily verify that m satisfies the "cocycle identity"

$$(4.62) \qquad m(\mathcal{A}\mathcal{B}, Z) = m\big(\mathcal{A}, \alpha(\mathcal{B})Z\big)m(\mathcal{B}, Z).$$

In particular,

$$(4.63) \qquad m\big(\mathcal{A}, \alpha(\mathcal{A}^{-1})Z\big)m(\mathcal{A}^{-1}, Z) = 1.$$

The following theorem shows that Σ_n can be identified with the symmetric space Sp/K where $K = Sp \cap O(2n)$ is the standard maximal compact subgroup of Sp.

(4.64) Theorem. (a) If $Z \in \Sigma_n$ and $\mathcal{A} \in Sp$ then $\alpha(\mathcal{A})Z \in \Sigma_n$.
(b) If $Z_1, Z_2 \in \Sigma_n$, there exists $\mathcal{A} \in Sp$ with $\alpha(\mathcal{A})Z_1 = Z_2$.
(c) $\{\mathcal{A} \in Sp : \alpha(\mathcal{A})(iI) = iI\} = Sp \cap O(2n)$.

Proof: To prove (a) we observe that

$$\alpha\left[\begin{pmatrix} A & 0 \\ 0 & A^{*-1} \end{pmatrix} \right] Z = AZA^*, \qquad \alpha\left[\begin{pmatrix} I & B \\ 0 & I \end{pmatrix} \right] Z = Z + B,$$

$$\alpha\left[\begin{pmatrix} 0 & I \\ -I & 0 \end{pmatrix} \right] Z = -Z^{-1}.$$

If A and B are real, B is symmetric, and $Z \in \Sigma_n$, it is obvious that AZA^* and $Z + B$ are in Σ_n. Also, if $Z \in \Sigma_n$, $-Z^{-1}$ is symmetric, and for any $v = Zw \neq 0 \in \mathbf{C}^n$,

$$\text{Im}(-\bar{v}Z^{-1}v) = \text{Im}(-w\overline{Zw}) = \text{Im}(\bar{w}Zw) > 0,$$

so $-Z^{-1} \in \Sigma_n$ by (4.60). (a) now follows from Proposition (4.10).

To prove (b) it suffices to show that if $Z = X + iY \in \Sigma_n$ then there exists $A \in Sp$ such that $\alpha(A)(iI) = Z$. But since $Y > 0$, there is a positive symmetric matrix A with $Y = A^2$, and then

$$\alpha\left[\begin{pmatrix} I & X \\ 0 & I \end{pmatrix}\begin{pmatrix} A & 0 \\ 0 & A^{-1} \end{pmatrix}\right](iI) = \alpha\left[\begin{pmatrix} I & X \\ 0 & I \end{pmatrix}\right](iY) = X + iY.$$

As for (c), if $A = \begin{pmatrix} A & B \\ C & D \end{pmatrix} \in Sp$, then $\alpha(A)(iI) = iI$ precisely when $iA + B = i(iC + D)$, that is, $A = D$ and $B = -C$. But by Proposition (4.1), this means that

$$A = \begin{pmatrix} A & B \\ C & D \end{pmatrix} = \begin{pmatrix} D & -C \\ -B & A \end{pmatrix} = \mathcal{J}A\mathcal{J}^{-1} = A^{*-1},$$

that is, $A \in O(2n)$. ∎

If Z is any symmetric complex $n \times n$ matrix, we define the function γ_Z on $\mathbf{R}^n$ by

$$\gamma_Z(x) = e^{\pi i x Z x}.$$

It is obvious that

$$\gamma_Z \in \mathcal{S} \iff \gamma_Z \in L^2 \iff Z \in \Sigma_n.$$

Thus, if $Z \in \Sigma_n$, the symplectic group acts on Z by α and on γ_Z by the metaplectic representation, and we are delighted to find that these actions are essentially the same.

(4.65) Theorem. If $A \in Sp$ and $Z \in \Sigma_n$ then

(4.66) $$\mu(A^{*-1})\gamma_Z = m(A, Z)\gamma_{\alpha(A)Z},$$

where $m(A, Z)$ is defined in (4.61).

Proof: If (4.66) is true for A and B then by (4.62),

$$\mu((AB)^{*-1})\gamma_Z = \mu(A^{*-1})m(B, Z)\gamma_{\alpha(B)Z}$$
$$= m(A, \alpha(B)Z)m(B, Z)\gamma_{\alpha(A)\alpha(B)Z} = m(AB, Z)\gamma_{\alpha(AB)Z},$$

so it suffices to prove (4.66) when A is one of the generators of Sp in Proposition (4.10). For this purpose we use formulas (4.24)–(4.26).

If $\mathcal{A} = \begin{pmatrix} A & 0 \\ 0 & A^{*-1} \end{pmatrix}$ then

$$\mu(\mathcal{A}^{*-1})\gamma_Z(x) = (\det^{-1/2} A^{*-1})\gamma_Z(A^*x) = m(\mathcal{A}, Z)\gamma_{\alpha(\mathcal{A})Z}(x),$$

since

$$\gamma_Z(A^*x) = e^{\pi i(A^*x)z(A^*x)} = \gamma_{AZA^*}(x) = \gamma_{\alpha(\mathcal{A})Z}(x).$$

If $\mathcal{A} = \begin{pmatrix} I & B \\ 0 & I \end{pmatrix}$ then

$$\mu(\mathcal{A}^{*-1})\gamma_Z(x) = \mu\left[\begin{pmatrix} I & 0 \\ -B & I \end{pmatrix}\right]\gamma_Z(x) = \pm e^{\pi i x Bx} e^{\pi i x Zx}$$
$$= \pm\gamma_{Z+B}(x) = m(\mathcal{A}, Z)\gamma_{\alpha(\mathcal{A})Z}(x).$$

If $\mathcal{A} = \mathcal{J}$ then

$$\mu(\mathcal{A}^{*-1})\gamma_Z(x) = \mu(\mathcal{J})\gamma_Z(x) = i^{n/2}\mathcal{F}^{-1}\gamma_Z(x) = i^{n/2}\left(\det^{-1/2}(-iZ)\right)\gamma_{-Z^{-1}}(x)$$
$$= \left(\det^{-1/2}(-Z)\right)\gamma_{-Z^{-1}}(x) = m(\mathcal{A}, Z)\gamma_{\alpha(\mathcal{A})Z}(x). \quad \blacksquare$$

Remark. If we had defined the metaplectic representation by using the version $\rho' = \rho \circ \mathcal{J}^{-1}$ of the Schrödinger representation as in (1.27),

$$\mu'(\mathcal{A})\rho'(p,q)\mu'(\mathcal{A})^{-1} = \rho'\left(\mathcal{A}(p,q)\right),$$

we would have $\mu'(\mathcal{A}) = \mu(\mathcal{J}^{-1}\mathcal{A}\mathcal{J}) = \mu(\mathcal{A}^{*-1})$, so formula (4.66) would look a little neater.

As a companion to the Siegel half plane there is also the **Siegel disc** Δ_n, which is the set of $n \times n$ complex symmetric matrices W such that $\|W\| < 1$, where $\|W\|$ is the norm of W as an operator on $\mathbb{C}^n$. Since

$$|z|^2 - |Wz|^2 = \overline{z}(I - W^*W)z,$$

we have

$$W \in \Delta_n \iff I - W^*W > 0.$$

The linear fractional map connecting Σ_n with Δ_n is the **Cayley transform** $\alpha(\mathcal{C})$ where

(4.67)
$$\mathcal{C} = \frac{1}{\sqrt{2}}\begin{pmatrix} iI & I \\ -iI & I \end{pmatrix}.$$

Thus,

$$\alpha(\mathcal{C})Z = (I + iZ)(I - iZ)^{-1} = (I - iZ)^{-1}(I + iZ).$$

(The reason for the factor of $\sqrt{2}$ in $\mathcal{C}$ is to make the multiplier $m(\mathcal{C}, Z)$ come out right.)

(4.67) Proposition. $\alpha(\mathcal{C})$ *is a bijection from* Σ_n *to* Δ_n.

Proof: If $W = \alpha(\mathcal{C})Z$ then

$$I - W^*W = (I - iZ)^{*-1}\Big[(I - iZ)^*(I - iZ) - (I + iZ)^*(I + iZ)\Big](I - iZ)^{-1}$$
$$= (I - iZ)^{*-1}\Big[2i(Z^* - Z)\Big](I - iZ)^{-1}.$$

But $Z^* = \overline{Z}$, so if $A = 2(I - iZ)^{-1}$,

$$I - W^*W = A^*(\operatorname{Im} Z)A.$$

Thus $\operatorname{Im} Z > 0$ if and only if $I - W^*W > 0$. ∎

The group of linear fractional automorphisms of Δ_n is $\mathcal{C}(Sp)\mathcal{C}^{-1}$. This is nothing but the group $Sp_c = W(Sp)W^{-1}$ (cf. (4.11)) that we use in connection with the Fock model. Indeed,

$$\mathcal{C}W^{-1} = \frac{1}{2}\begin{pmatrix} iI & I \\ -iI & I \end{pmatrix}\begin{pmatrix} I & I \\ -iI & iI \end{pmatrix} = i\mathcal{J},$$

and although $i\mathcal{J}$ does not belong to Sp_c, it does belong to the normalizer of Sp_c in $GL(2n, \mathbf{C})$ because it anticommutes with the matrix $\mathcal{K}$ in Proposition (4.17b).

Corresponding to the Gaussians γ_Z attached to Σ_n, there is a family of functions Γ_W attached to Δ_n. Namely, if $W \in M_{2n}(\mathbf{C})$ and $W = W^t$, we define the entire function Γ_W on $\mathbf{C}^n$ by

$$\Gamma_W(\zeta) = e^{(\pi/2)\zeta W \zeta}.$$

(4.69) Proposition. Γ_W *lies in the Fock space* $\mathcal{F}_n$ *if and only if* $W \in \Delta_n$.

Proof: If $\|W\| = c < 1$ then $|\Gamma_W(\zeta)| \le e^{(\pi/2)c|\zeta|^2}$, so $\Gamma_W \in \mathcal{F}_n$. On the other hand, if $\|W\| \ge 1$, there exist $\zeta_0 \in \mathbf{C}^n$ and $\lambda \in \mathbf{C}$ with $W\zeta_0 = \lambda\zeta_0$ and $|\lambda| \ge 1$, and we can choose ζ_0 so that $|\zeta_0| = 1$ and $\lambda\zeta_0^2 = |\lambda|$. Then for any $t \in \mathbf{R}$ and $\zeta' \in \mathbf{C}^n$, if $\zeta = t\zeta_0 + \zeta'$ we have

$$\operatorname{Re}(\zeta W \zeta) = |\lambda|t^2 + t\operatorname{Re}(\lambda\zeta'\zeta_0 + \zeta_0 W\zeta') + \operatorname{Re}(\zeta' W\zeta').$$

It follows that

$$\int_{-\infty}^{\infty} |\Gamma_W(t\zeta_0 + \zeta')|^2 e^{-\pi|t\zeta_0 + \zeta'|^2}\,dt = \int_{-\infty}^{\infty} \exp\Big[\pi\big(|\lambda| - 1\big)t^2 + C_\zeta t + C_\zeta'\Big]\,dt = \infty$$

for any ζ', and hence (by Fubini's theorem) that $\Gamma_W \notin \mathcal{F}_n$. ∎

Now, the Bargmann transform maps $L^2(\mathbf{R}^n)$ to $\mathcal{F}_n$, while the Cayley transform maps Σ_n to Δ_n, and we have:

(4.70) Theorem. $B\gamma_Z = m(\mathcal{C}, Z)\Gamma_{\alpha(\mathcal{C})Z}.$

Proof: By the formula for the Fourier transform of a Gaussian,

$$B\gamma_Z(\zeta) = 2^{n/4} \int e^{\pi i x Z x} e^{2\pi x \zeta - \pi x^2 - (\pi/2)\zeta^2} dx$$

$$= 2^{n/4} e^{-(\pi/2)\zeta^2} \det^{-1/2}(I - iZ) e^{\pi\zeta(I-iZ)^{-1}\zeta}.$$

But

$$2^{n/4} \det^{-1/2}(I - iZ) = m(\mathcal{C}, Z),$$

and

$$-\frac{\pi}{2}\zeta^2 + \pi\zeta(I - iZ)^{-1}\zeta = \frac{\pi}{2}\zeta(I - iZ)^{-1}[2I - (I - iZ)]\zeta = \frac{\pi}{2}\zeta[\alpha(\mathcal{C})Z]\zeta. \quad \blacksquare$$

Combining Theorems (4.65) and (4.70), we obtain the analogue of the former result for the Fock-metaplectic representation ν.

(4.71) Theorem. If $\mathcal{A} \in Sp_c$ and $W \in \Delta_n$ then

$$\nu(\mathcal{A})\Gamma_W = m(\overline{\mathcal{A}}, W)\Gamma_{\alpha(\overline{\mathcal{A}})W}.$$

Proof: Since $\mathcal{B}^{*-1} = \mathcal{J}\mathcal{B}\mathcal{J}^{-1}$ for $\mathcal{B} \in Sp$ and

$$m(\mathcal{C}, \alpha(\mathcal{C}^{-1})W)m(\mathcal{C}^{-1}, W) = 1,$$

with W as in (4.12) we have

$$
\begin{aligned}
\nu(\mathcal{A})\Gamma_W &= B\mu(W^{-1}\mathcal{A}W)B^{-1}\Gamma_W \\
&= m(\mathcal{C}^{-1}, W)B\mu(W^{-1}\mathcal{A}W)\gamma_{\alpha(\mathcal{C}^{-1})W} \\
&= m(\mathcal{C}^{-1}, W)m(\mathcal{J}W^{-1}\mathcal{A}W\mathcal{J}^{-1}, \alpha(\mathcal{C}^{-1})W)B\gamma_{\alpha(\mathcal{J}W^{-1}\mathcal{A}W\mathcal{J}^{-1}\mathcal{C}^{-1})W} \\
&= m(\mathcal{C}\mathcal{J}W^{-1}\mathcal{A}W\mathcal{J}^{-1}\mathcal{C}^{-1}, W)\Gamma_{\alpha(\mathcal{C}\mathcal{J}W^{-1}\mathcal{A}W\mathcal{J}^{-1}\mathcal{C}^{-1})W}.
\end{aligned}
$$

But $\mathcal{C}\mathcal{J}W^{-1} = \begin{pmatrix} 0 & -I \\ -I & 0 \end{pmatrix}$, so if $\mathcal{A} = \begin{pmatrix} P & Q \\ \overline{Q} & \overline{P} \end{pmatrix}$,

$$\mathcal{C}\mathcal{J}W^{-1}\mathcal{A}W\mathcal{J}^{-1}\mathcal{C}^{-1} = \begin{pmatrix} 0 & -I \\ -I & 0 \end{pmatrix} \begin{pmatrix} P & Q \\ \overline{Q} & \overline{P} \end{pmatrix} \begin{pmatrix} 0 & -I \\ -I & 0 \end{pmatrix} = \begin{pmatrix} \overline{P} & \overline{Q} \\ Q & P \end{pmatrix}$$

$$= \overline{\mathcal{A}}. \quad \blacksquare$$

Exercise. Prove Theorem (4.71) directly from the formula (4.38) for $\nu(\mathcal{A})$ by using Theorem 3 of Appendix A and Proposition (4.17).

We may summarize these results in the following commutative diagram:

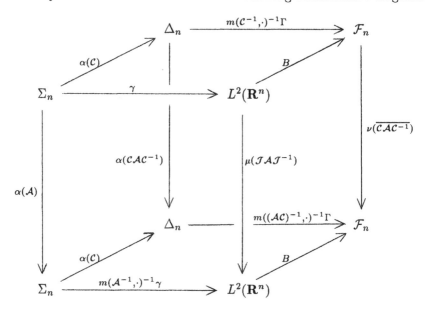

Here $\mathcal{A}$ denotes an arbitrary element of Sp, and, for example, $m(\mathcal{A}^{-1},\cdot)^{-1}\gamma$ denotes the map

$$Z \longrightarrow m(\mathcal{A}^{-1}, Z)^{-1}\gamma_Z = m(\mathcal{A}, \alpha(\mathcal{A}^{-1})Z)\gamma_Z.$$

The (commutativity of the) left face follows from the fact that α is a homomorphism, and the right face expresses the relationship between the representations μ and ν. The front and back faces are embodiments of Theorems (4.65) and (4.71), while the top and bottom faces are both equivalent to Theorem (4.70). (To reconcile the labels on the arrows here with the statements of these results, one uses the cocycle identity (4.62), the fact that $\mathcal{A}^{*-1} = \mathcal{J}\mathcal{A}\mathcal{J}^{-1}$ for $\mathcal{A} \in Sp$, and the fact that $\mathcal{W}\mathcal{J} = \overline{\mathcal{C}}$.)

Characterizations of Gaussians. Gaussian functions have played a conspicuous role in this monograph. We now discuss some of the reasons for their special nature, thereby amplifying some results of earlier sections. To begin with, we need to introduce some precise terminology.

Every quadratic polynomial on $\mathbf{R}^n$ with complex coefficients can be written uniquely in the form

$$(4.72) \quad Q(x) = \pi i x Z x + 2\pi i w x + c \qquad \left(Z \in M_n(\mathbf{C}), \ Z = Z^\dagger, \ w \in \mathbf{C}^n, \ c \in \mathbf{C} \right).$$

For such a Q, it is clear that e^Q belongs to $\mathcal{S}(\mathbf{R}^n)$ if and only if $Z \in \Sigma_n$. We define a **Gaussian** to be a function on $\mathbf{R}^n$ of the form e^Q where Q is of the form (4.72) with $Z \in \Sigma_n$, and a **centered Gaussian** to be a Gaussian for which $w = 0$ in (4.72). The **standard Gaussian** is the function

$$\gamma(x) = \gamma_{iI}(x) = e^{-\pi x^2}.$$

(4.73) Proposition. $f \in \mathcal{S}$ is a Gaussian if and only if $f = C\rho(p,q)\mu(\mathcal{A})\gamma$ for some $C \in \mathbf{C}$, $(p,q) \in \mathbf{R}^{2n}$, and $\mathcal{A} \in Sp$. It is centered if and only if $(p,q) = (0,0)$.

Proof: Since $\mu(\mathcal{A})\gamma = m(\mathcal{A}, iI)\gamma_{\alpha(\mathcal{A})(iI)}$, it is clear that any function of the form $C\rho(p,q)\mu(\mathcal{A})\gamma$ is a Gaussian and is centered if $(p,q) = (0,0)$. Conversely, if $f = e^Q$ where Q is given by (4.72) with $Z = X + iY \in \Sigma_n$ and $w = u + iv \in \mathbf{C}^n$, we can write

$$Q(x) = \pi i(x + Y^{-1}v)Z(x + Y^{-1}v) + 2\pi i(u - XY^{-1}v)x$$
$$+ \pi v Y^{-1}v - \pi i v Y^{-1}XY^{-1}v + c$$

and hence

$$f = C\rho(Y^{-1}v, \, u - XY^{-1}v)\gamma_Z$$

where

$$C = \exp(-\pi i u Y^{-1}v + \pi v Y^{-1}v + c).$$

By Theorems (4.64) and (4.65), for a suitable $\mathcal{A} \in Sp$ we have $Z = \alpha(\mathcal{A})(iI)$ and hence $\gamma_Z = C'\mu(\mathcal{A})\gamma$. Moreover, $Y^{-1}v$ and $u - XY^{-1}v$ vanish precisely when w does, so we are done. ∎

Corollary (4.74). $f \in \mathcal{S}$ is a Gaussian if and only if its Wigner function Wf is nonnegative.

Proof: If $f = C\rho(p,q)\mu(\mathcal{A})\gamma$, by Propositions (1.94) and (4.28) we have

$$Wf(\xi, x) = |C|^2 W\gamma\big(\mathcal{A}^*(\xi - q, \, x + p)\big).$$

But it is easily verified that $W\gamma(\xi, x) = e^{(-\pi/2)(\xi^2 + x^2)}$, so $Wf \geq 0$ (in fact $Wf > 0$). The converse is Hudson's theorem (1.102). ∎

The next result is a generalization of the uncertainty principle (1.38). Given $\mathcal{A} \in Sp$ and $(\alpha, \beta) \in \mathbf{R}^{2n}$, define the operators P_j, Q_j on $\mathcal{S}$ by

$$(P, Q) = \mathcal{A}(D, X) + (\alpha, \beta),$$

that is, if $\mathcal{A} = \begin{pmatrix} A & B \\ C & D \end{pmatrix}$,

$$P_j = \sum(A_{jk}D_k + B_{jk}X_k) + \alpha_j I, \qquad Q_j = \sum(C_{jk}D_k + D_{jk}X_k) + \beta_j I.$$

Then P_j, Q_k satisfy the canonical commutation relations

$$[P_j, P_k] = [Q_j, Q_k] = 0, \qquad [P_j, Q_k] = (2\pi i)^{-1}\delta_{jk}.$$

(4.75) Proposition. *If P_j and Q_j are defined as above, then*

$$\sum_1^n \left(\|P_j f\|_2^2 + \|Q_j f\|_2^2 \right) \geq \frac{n}{2\pi} \|f\|_2^2,$$

with equality if and only if $f = C\rho(\beta, -\alpha)\mu(\mathcal{A}^*)\gamma$.

Proof: By Proposition (2.13) and Theorem (2.15), for any symbol σ we have

$$\begin{aligned}
\sigma\big(\mathcal{A}(D,X) + (\alpha,\beta)\big) &= \rho(\beta,-\alpha)(\sigma \circ \mathcal{A})(D,X)\rho(-\beta,\alpha) \\
&= \rho(\beta,-\alpha)\mu(\mathcal{A}^*)\sigma(D,X)\mu(\mathcal{A}^{*-1})\rho(-\beta,\alpha),
\end{aligned}$$

so taking $\sigma(\xi,x) = \xi_j$ or $\sigma(\xi,x) = x_j$, we obtain

$$\begin{aligned}
P_j &= \rho(\beta,-\alpha)\mu(\mathcal{A}^*)D_j\mu(\mathcal{A}^{*-1})\rho(-\beta,\alpha), \\
Q_j &= \rho(\beta,-\alpha)\mu(\mathcal{A}^*)X_j\mu(\mathcal{A}^{*-1})\rho(-\beta,\alpha).
\end{aligned}$$

The result therefore follows from Proposition (1.37), or rather its generalization to n dimensions (cf. (1.40) et seq.). ∎

For our last group of results, we observe that by Theorem (1.63), the Fock space $\mathcal{F}_n$ is the orthogonal direct sum $\bigoplus_0^\infty \mathcal{P}_k$ where $\mathcal{P}_k$ is the space of homogeneous (holomorphic) polynomials of degree k on $\mathbf{C}^n$. Each $\mathcal{P}_k$ is obviously invariant under the natural action of the unitary group:

$$U \in U(n), \ F \in \mathcal{P}_k \implies F \circ U^{-1} \in \mathcal{P}_k.$$

Moreover, each $\mathcal{P}_k$ is irreducible under the action of $U(n)$. (We omit the proof of this well-known fact. It can be found, for example, in Igusa [80, pp. 36–37].) The representations of $U(n)$ on the spaces $\mathcal{P}_k$ are all inequivalent: if $n > 1$ this is obvious since the spaces $\mathcal{P}_k$ all have different dimensions, while if $n = 1$ it is obvious by inspection. ($U(1)$ acts on $\mathcal{P}_k$ in dimension 1 by the representation $e^{i\theta} \to e^{-ik\theta}$.) It follows that the only one-dimensional $U(n)$-invariant subspaces of $\mathcal{F}_n$ are the spaces $\mathcal{P}_k$ when $n = 1$ and the single space $\mathcal{P}_0$ when $n > 1$. Taking into account Propositions (4.19) and (4.39), we see that we have proved:

(4.76) Proposition. *Let $K = Sp \cap O(2n)$. The only one-dimensional subspaces of $L^2(\mathbf{R}^n)$ that are invariant under the operators $\mu(\mathcal{A})$ for all $\mathcal{A} \in K$ are the spans of the Hermite functions h_j $(0 \leq j \leq \infty)$ when $n = 1$, and the span of $\gamma(x) = e^{-\pi x^2}$ when $n > 1$.*

The following result is due to Klauder [90] in the case $n = 1$.

(4.77) Corollary. *Suppose $f \in L^2(\mathbf{R}^n)$. If $n > 1$, Wf (or equivalently $V(f, f)$) is rotation-invariant precisely when $f = C\gamma$; and if $n = 1$, Wf (or $V(f, f)$) is rotation-invariant precisely when $f = Ch_j$ for some j.*

Proof: By Proposition (4.28), for $\mathcal{A} \in Sp$,

$$W(\mu(\mathcal{A})f) = Wf \circ \mathcal{A}^*, \qquad V(\mu(\mathcal{A})f, \mu(\mathcal{A})f) = V(f, f) \circ \mathcal{A}^{-1}.$$

When $n = 1$ we have $SO(2) \subset Sp(1, \mathbf{R})$, so the assertion follows from Proposition (4.76) together with the fact that Wf or $V(f, f)$ determines f up to a phase factor (Proposition (1.98)). This also proves the necessity that $f = C\gamma$ when $n > 1$, and the sufficiency is verified by inspection: $W\gamma(\xi, x) = e^{-(\pi/2)(\xi^2 + x^2)}$. ∎

We recall from Section 4.4 that the formula

$$\omega(X, \mathcal{A}) = \rho(X)\mu(\mathcal{A})$$

defines a double-valued representation of the extended symplectic group HSp. Since $\mathbf{H}_n$ has no compact subgroups, it is easy to see that every compact subgroup of HSp is of the form

$$\{(X, I)(0, \mathcal{A})(X, I)^{-1} : \mathcal{A} \in K\}$$

for some compact subgroup K of Sp. Moreover, the maximal compact subgroups of Sp are all conjugate to one another. (See Igusa [80, p. 23] for a relatively simple proof of this fact.) We then have the following characterization of Gaussians in dimensions $n \geq 2$, taken from Igusa [80]. (We leave it to the reader to formulate an analogue for $n = 1$.)

(4.78) Theorem. *Suppose $f \in L^2(\mathbf{R}^n)$, $n \geq 2$. The the linear span $\mathbf{C}f$ of f is invariant under $\{\mu(\mathcal{A}) : \mathcal{A} \in K\}$ for some maximal compact subgroup K of Sp if and only if f is a centered Gaussian; and it is invariant under $\{\omega(X, \mathcal{A}) : (X, \mathcal{A}) \in K'\}$ for some maximal compact subgroup K' of HSp if and only if f is a Gaussian.*

Proof: By Proposition (4.5), $K_0 = Sp \cap O(2n)$ is a maximal compact subgroup of Sp, and by Proposition (4.76), $\mathbf{C}\gamma$ is the unique line in L^2 that is invariant under $\mu(K_0)$. Hence, by Theorem (4.65), if $\mathcal{A} \in Sp$ and $Z = \alpha(\mathcal{A})(iI)$, $\mathbf{C}\gamma_Z$ is the unique line that is invariant under $\mu(K)$ where $K = \mathcal{A}K_0\mathcal{A}^{-1}$; moreover, if also $X \in \mathbf{H}_n$, $\mathbf{C}\rho(X)\gamma_Z$ is the unique line that is invariant under $\omega(K')$, where $K' = (X, I)(0, K)(X, I)^{-1}$. Since every maximal compact subgoup of Sp (resp. HSp) is of the form K (resp. K'), the assertions follow from Proposition (4.73). ∎

6. The Disc Model

We now show how the even and odd parts of the metaplectic representation (cf. Theorem (4.56)) can be transferred to spaces of analytic functions on the Siegel disc Δ_n so that the action of Sp essentially reduces to its geometric action on Δ_n. This material is mostly due to Itzykson [81]. We start out with the Fock model, that is, the representation ν of Sp_c on $\mathcal{F}_n$. If $F \in \mathcal{F}_n$, we define a holomorphic function Φ_F on Δ_n by

$$\Phi_F(W) = \langle F, \Gamma_{\overline{W}} \rangle_{\mathcal{F}} = \int F(z) e^{(\pi/2)\overline{z}W\overline{z}} e^{-\pi|z|^2} dz.$$

If $\mathcal{A} \in Sp_c$, then, by Theorem (4.71) we have

$$
\begin{aligned}
\Phi_{\nu(\mathcal{A})F}(W) &= \langle \nu(\mathcal{A})F, \Gamma_{\overline{W}} \rangle_{\mathcal{F}} = \langle F, \nu(\mathcal{A})^{-1}\Gamma_{\overline{W}} \rangle_{\mathcal{F}} \\
&= \langle F, m(\overline{\mathcal{A}^{-1}}, \overline{W})\Gamma_{\alpha(\overline{\mathcal{A}})^{-1}\overline{W}} \rangle_{\mathcal{F}} = m(\mathcal{A}^{-1}, W)\Phi_F\big(\alpha(\mathcal{A})^{-1}W\big).
\end{aligned}
$$

Thus, modulo the multiplier $m(\mathcal{A}^{-1}, W)$, $\Phi_{\nu(\mathcal{A})F}$ is simply the translate of Φ_F by the action of $\mathcal{A}$ on Δ_n. We also observe that Φ_F is essentially a matrix element of ν: any $W \in \Delta_n$ can be written as $W = \alpha(\mathcal{B}) \cdot 0$ for some $\mathcal{B} \in Sp_c$, so

$$\Phi_F(W) = \langle F, \Gamma_{\alpha(\mathcal{B})\cdot 0} \rangle_{\mathcal{F}} = m(\mathcal{B}, 0)^{-1} \langle F, \nu(\mathcal{B})\Gamma_0 \rangle_{\mathcal{F}}.$$

Here Γ_0 is the constant function 1, corresponding to the function $2^{n/4} e^{-\pi x^2}$ in the Schrödinger picture. Thus the above construction is similar to the way we obtained the Fock-Bargmann representation from the Schrödinger representation.

Obviously $\Phi_F = 0$ whenever F is odd, so we have intertwined only the even part of the metaplectic representation with a representation on functions on Δ_n. More precisely, let us denote by $\mathcal{H}^n_{\text{even}}$ the image of $\mathcal{F}_n$ (or its even part) under the map $F \to \Phi_F$, and let us make $\mathcal{H}^n_{\text{even}}$ into a Hilbert space by declaring this map to be an isometry. Then the prescription

$$\pi(\mathcal{A})\Phi(W) = m(\mathcal{A}^{-1}, W)\Phi\big(\alpha(\mathcal{A})^{-1}W\big)$$

defines a double-valued unitary representation of Sp_c on $\mathcal{H}^n_{\text{even}}$, equivalent to the even part of the metaplectic representation.

To describe the space $\mathcal{H}^n_{\text{even}}$ more precisely, we shall find the orthonormal basis corresponding to the basis $\{\zeta_\alpha : |\alpha| \text{ even}\}$ for the even part of $\mathcal{F}_n$. (Recall that $\zeta_\alpha(z) = \sqrt{\pi^{|\alpha|}/\alpha!}\, z^\alpha$.) If we write

$$(4.79) \qquad\qquad e^{(\pi/2)\overline{z}W\overline{z}} = \sum_\alpha P_\alpha(W)\overline{z}^\alpha,$$

it is easy to see that $P_\alpha(W) = 0$ when $|\alpha|$ is odd, and $P_\alpha(W)$ is a polynomial of degree $\frac{1}{2}|\alpha|$ in the entries of the matrix W when $|\alpha|$ is even. Moreover,

$$\Phi_F(W) = \sum P_\alpha(W) \int F(z)\bar{z}^\alpha e^{-\pi|z|^2} dz = \sum \sqrt{\frac{\alpha!}{\pi^{|\alpha|}}} P_\alpha(W)\langle F, \zeta_\alpha\rangle_{\mathcal{F}}.$$

In particular,

$$\Phi_{\zeta_\alpha} = \sqrt{\frac{\alpha!}{\pi^{|\alpha|}}} P_\alpha,$$

so $\left\{ \sqrt{\alpha!/\pi^{|\alpha|}}\, P_\alpha : |\alpha| \text{ even} \right\}$ is an orthonormal basis for $\mathcal{H}^n_{\text{even}}$.

If $n > 1$, a simple dimension count shows that the P_α's do not span all polynomials on $\mathbf{C}^{n(n+1)/2}$ (the space of symmetric $n \times n$ matrices). It can be shown (see Itzykson [81]) that the span of the P_α's is precisely the set of polynomials P such that

$$(4.80) \qquad (L_{jk}L_{lm} - L_{jm}L_{lk})P = 0, \quad \text{where} \quad L_{jk} = L_{kj} = 2^{\delta_{jk}-1}\frac{\partial}{\partial W_{jk}}.$$

Thus $\mathcal{H}^n_{\text{even}}$ is contained in the space of holomorphic functions on Δ_n that satisfy the differential equations (4.80), and we shall not try to pin it down any further here. However, if $n = 1$, we have

$$P_{2k}(w) = \frac{1}{k!}\left(\frac{\pi w}{2}\right)^k,$$

and hence the functions $\{p_{2k} : k = 0, 1, 2, \ldots\}$ defined by

$$(4.81) \qquad p_{2k}(w) = \sqrt{\frac{(2k)!}{\pi^{2k}}}\, P_{2k}(w) = \frac{\sqrt{(2k)!}}{2^k k!}\, w^k = \sqrt{\frac{\Gamma(k+\frac{1}{2})}{\Gamma(k+1)\Gamma(\frac{1}{2})}}\, w^k$$

are an orthonormal basis for $\mathcal{H}^1_{\text{even}}$. We shall use this information later to obtain a closed formula for the norm on $\mathcal{H}^1_{\text{even}}$ and hence a computable description of $\mathcal{H}^1_{\text{even}}$.

A similar but slightly less simple construction works for odd functions. In order to mediate between even and odd functions we employ the creation and annihilation operators

$$A_j^* F(z) = \sqrt{\pi}\, z_j F(z), \qquad A_j F(z) = \frac{1}{\sqrt{\pi}}\frac{\partial F}{\partial z_j}.$$

We write

$$A^* = (A_1^*, \ldots, A_n^*), \qquad A = (A_1, \ldots, A_n),$$

so A^* and A transform $\mathbf{C}$-valued functions into $\mathbf{C}^n$-valued functions. If $W \in \Delta_n$ then clearly $A^*\Gamma_W$ and $A\Gamma_W$ are in $(\mathcal{F}_n)^n$, and

$$(4.82) \qquad A\Gamma_W(z) = \frac{1}{\sqrt{\pi}} \nabla_z e^{(\pi/2)zWz} = \sqrt{\pi}\, Wz e^{(\pi/2)zWz} = WA^*\Gamma_W(z).$$

With this in mind, we define a mapping $F \to \Psi_F$ from $\mathcal{F}_n$ into the space of holomorphic $\mathbf{C}^n$-valued functions on Δ_n by

$$\Psi_F(W) = \langle F, A^*\Gamma_{\overline{W}}\rangle_{\mathcal{F}} = \sqrt{\pi} \int \overline{z} F(z) e^{(\pi/2)\overline{z}W\overline{z}} e^{-\pi|z|^2} dz.$$

Here we obviously have $\Psi_F = 0$ if F is even. We define $\mathcal{H}^n_{\mathrm{odd}}$ to be the image of $\mathcal{F}_n$ (or its odd part) under the map $F \to \Phi_F$ and make $\mathcal{H}^n_{\mathrm{odd}}$ a Hilbert space by declaring this map to be an isomtery.

In order to see what happens to the metaplectic representation, we need to know how A^* transforms.

(4.83) Lemma. If $\mathcal{B} = \left(\begin{smallmatrix} P & Q \\ \overline{Q} & \overline{P} \end{smallmatrix}\right) \in Sp_c$ then

$$\nu(\mathcal{B})^{-1} A^* \nu(\mathcal{B}) = QA + PA^*.$$

Proof: We recall from (1.74) and (1.75) that

$$A = \frac{1}{\sqrt{\pi}} \nabla_w \beta(w, \overline{w})|_{w=0}, \qquad A^* = -\frac{1}{\sqrt{\pi}} \nabla_{\overline{w}} \beta(w, \overline{w})|_{w=0},$$

where we have written $\beta(w, \overline{w})$ rather than $\beta(w)$ for the Fock-Bargmann representation to clarify the calculations. We also recall from Proposition (4.17) that

$$\mathcal{B}^{-1} = \begin{pmatrix} P^* & -Q^\dagger \\ -Q^* & P^\dagger \end{pmatrix}.$$

Hence,

$$\nu(\mathcal{B})^{-1} A^* \nu(\mathcal{B}) = -\frac{1}{\sqrt{\pi}} \nabla_{\overline{w}} \nu(\mathcal{B})^{-1} \beta(w, \overline{w}) \nu(\mathcal{B})|_{w=0}$$

$$= -\frac{1}{\sqrt{\pi}} \nabla_{\overline{w}} \beta \circ \mathcal{B}^{-1}(w, \overline{w})|_{w=0}$$

$$= -\frac{1}{\sqrt{\pi}} \nabla_{\overline{w}} \left[\beta(P^*w - Q^\dagger\overline{w}, -Q^*w + P^\dagger\overline{w}) \right]_{w=0}$$

$$= -\frac{1}{\sqrt{\pi}} \left[-Q\nabla_w\beta + P\nabla_{\overline{w}}\beta \right](0,0) = QA + PA^*. \quad \blacksquare$$

One more bit of notation: if $\mathcal{B} = \left(\begin{smallmatrix} A & B \\ C & D \end{smallmatrix}\right) \in GL(2n, \mathbf{C})$ and $W \in M_n(\mathbf{C})$, we define

$$M(\mathcal{B}, W) = (CW + D)^{-1}$$

whenever $CW + D$ is invertible. $M(\mathcal{B}, W)$ is thus a meromorphic matrix-valued function of W for each $\mathcal{B}$.

(4.84) Proposition. If $F \in \mathcal{F}_n$ and $\mathcal{B} \in Sp_c$,

$$\Psi_{\nu(\mathcal{B})F}(W) = m(\mathcal{B}^{-1}, W)M(\mathcal{B}^{-1}, W)\Psi_F\big(\alpha(\mathcal{B})^{-1}W\big).$$

Proof: Let $\mathcal{B} = \left(\begin{smallmatrix} P & Q \\ \overline{Q} & \overline{P} \end{smallmatrix}\right)$. By Lemma (4.83) and Theorem (4.71),

$$
\begin{aligned}
\Psi_{\nu(\mathcal{B})F}(W) &= \langle \nu(\mathcal{B})F,\, A^*\Gamma_{\overline{W}}\rangle_{\mathcal{F}} = \langle F,\, \nu(\mathcal{B})^{-1}A^*\Gamma_{\overline{W}}\rangle_{\mathcal{F}} \\
&= \langle F,\, (QA + PA^*)\nu(\mathcal{B})^{-1}\Gamma_{\overline{W}}\rangle_{\mathcal{F}} \\
&= \langle F,\, m(\overline{\mathcal{B}^{-1}}, \overline{W})(QA + PA^*)\Gamma_{\alpha(\overline{\mathcal{B}})^{-1}\overline{W}}\rangle_{\mathcal{F}}.
\end{aligned}
$$

But then by formula (4.82),

$$
\begin{aligned}
\Psi_{\nu(\mathcal{B})F}(W) &= \langle F,\, m(\overline{\mathcal{B}^{-1}}, \overline{W})(Q[\alpha(\overline{\mathcal{B}})^{-1}\overline{W}] + P)A^*\Gamma_{\alpha(\overline{\mathcal{B}})^{-1}\overline{W}}\rangle_{\mathcal{F}} \\
&= m(\mathcal{B}^{-1}, W)(\overline{Q}[\alpha(\mathcal{B})^{-1}W] + \overline{P})\Psi_F\big(\alpha(\mathcal{B}^{-1})W\big).
\end{aligned}
$$

Now, by Proposition (4.17b),

$$
\begin{aligned}
\overline{Q}[\alpha(\mathcal{B}^{-1})W] + \overline{P} &= \overline{Q}(P^*W - Q^\dagger)(-Q^*W + P^\dagger)^{-1} + \overline{P} \\
&= (\overline{Q}P^*W - \overline{Q}Q^\dagger - \overline{P}Q^*W + \overline{P}P^\dagger)(-Q^*W + P^\dagger)^{-1},
\end{aligned}
$$

and by Proposition (4.17c), $\overline{Q}P^* = \overline{P}Q^*$ and $\overline{P}P^\dagger - \overline{Q}Q^\dagger = I$. Hence

$$\overline{Q}[\alpha(\mathcal{B})^{-1}W] + \overline{P} = (-Q^*W + P^\dagger)^{-1} = M(\mathcal{B}^{-1}, W),$$

and the proof is complete. ∎

Thus on $\mathcal{H}^n_{\mathrm{odd}}$, the representation of Sp_c is again induced by its geometric action on Δ_n, modified by the matrix-valued multiplier mM.

As with $\mathcal{H}^n_{\mathrm{even}}$, it is easy to identify the image of the orthonormal basis $\{\zeta_\beta : |\beta| \text{ odd}\}$ in $\mathcal{H}^n_{\mathrm{odd}}$. Indeed, if $P_\alpha(W)$ is defined by (4.79), by (1.77) we have

$$
\begin{aligned}
\big(\Psi_{\zeta_\beta}(W)\big)_j &= \sum_\alpha P_\alpha(W)\sqrt{\frac{\alpha!}{\pi^{|\alpha|}}}\,\langle \zeta_\beta,\, A_j^*\zeta_\alpha\rangle_{\mathcal{F}} \\
&= \sum_\alpha P_\alpha(W)\sqrt{\frac{(\alpha + 1_j)!}{\pi^{|\alpha|}}}\,\langle \zeta_\beta,\, \zeta_{\alpha+1_j}\rangle_{\mathcal{F}} = \sqrt{\frac{\beta!}{\pi^{|\beta|-1}}}\,P_{\beta-1_j}(W).
\end{aligned}
$$

Here 1_j is the multi-index with 1 in the jth slot and 0 elsewhere, and $P_{\beta-1_j} = 0$ if $\beta_j = 0$. Thus, if we set

$$Q_\beta(W) = \left(P_{\beta-1_1}(W), \dots, P_{\beta-1_n}(W) \right) \qquad (|\beta| \text{ odd}),$$

Q_β is a $\mathbf{C}^n$-valued polynomial of degree $\frac{1}{2}(|\beta| - 1)$, and $\left\{ \sqrt{\beta!/\pi^{|\beta|-1}}\, Q_\beta : |\beta| \text{ odd} \right\}$ is an orthonormal basis for $\mathcal{H}^n_{\mathrm{odd}}$.

When $n > 1$, the polynomials Q_β still satisfy the differential equations (4.80), and in addition they must satisfy

$$(4.85) \qquad\qquad L_{jk}(Q_\beta)_l = L_{jl}(Q_\beta)_k,$$

where L_{jk} is as in (4.80). Conversely, it can be shown that any $\mathbf{C}^n$-valued polynomial satisfying (4.80) and (4.85) is a linear combination of Q_β's; see Itzykson [81]. Thus $\mathcal{H}^n_{\mathrm{odd}}$ consists of holomorphic $\mathbf{C}^n$-valued functions satisfying (4.80) and (4.85).

When $n = 1$, we have

$$Q_{2k+1}(w) = P_{2k}(w) = \frac{1}{k!} \left(\frac{\pi w}{2} \right)^k,$$

and the corresponding orthonormal basis for $\mathcal{H}^1_{\mathrm{odd}}$ is $\{q_{2k+1}\}_0^\infty$, where

$$q_{2k+1}(w) = \sqrt{\frac{(2k+1)!}{\pi^{2k}}}\, Q_{2k+1}(w) = \sqrt{\frac{\Gamma(k + \frac{3}{2})}{\Gamma(k+1)\Gamma(\frac{3}{2})}}\, w^k.$$

The inner product on $\mathcal{H}^1_{\mathrm{odd}}$ can be given neatly in terms of an integral. Indeed, for $a > 0$ we define the inner product $\langle f, g \rangle_a$ for functions on the disc Δ_1 by

$$(4.86) \qquad\qquad \langle f, g \rangle_a = \frac{1}{\pi} \int_{|w|<1} f(w)\overline{g(w)}(1 - |w|^2)^{a-1}\, dw.$$

Then

$$\langle w^j, w^k \rangle_a = \frac{1}{\pi} \int_0^1 \int_0^{2\pi} e^{i(j-k)\theta} r^{j+k} (1 - r^2)^{a-1} r\, d\theta\, dr$$

$$= \delta_{jk} \int_0^1 s^k (1 - s)^{a-1}\, ds = \delta_{jk} B(k+1, a)$$

$$= \delta_{jk} \frac{\Gamma(k+1)\Gamma(a)}{\Gamma(k+1+a)},$$

so the functions

$$(4.87) \qquad\qquad f_k(w) = \sqrt{\frac{\Gamma(k+a+1)}{\Gamma(k+1)\Gamma(a)}}\, z^k$$

are orthonormal. When $a = \frac{1}{2}$ we have $q_{2k+1} = \sqrt{2}\,f_k$, and we conclude that $\mathcal{H}^1_{\text{odd}}$ is the Hilbert space of holomorphic functions on Δ_1 defined by the norm

$$\|f\|^2_{\text{odd}} = \frac{1}{2\pi} \int_{|w|<1} |f(w)|^2 (1 - |w|^2)^{-1/2}\, dw.$$

Now, what about $\mathcal{H}^1_{\text{even}}$? Comparison of (4.81) with (4.87) suggests that the inner product on $\mathcal{H}^1_{\text{even}}$ should be given by (4.86) with $a = -\frac{1}{2}$, but this makes no sense at all: when $a < 0$ the integral (4.86) diverges for any nonzero polynomials f and g. Rather, let us observe that $q_{2k+1} = \sqrt{2k+1}\,p_{2k}$. This implies that $\mathcal{H}^1_{\text{even}} \subset \mathcal{H}^1_{\text{odd}}$; moreover, if $f \in \mathcal{H}^1_{\text{even}}$,

$$f = \sum \langle f, p_{2k} \rangle_{\text{even}}\, p_{2k} = \sum \langle f, q_{2k+1} \rangle_{\text{odd}}\, q_{2k+1},$$

whence $\langle f, p_{2k} \rangle_{\text{even}} = \sqrt{2k+1}\,\langle f, q_{2k+1} \rangle_{\text{odd}}$, and

$$\|f\|^2_{\text{even}} = \sum |\langle f, p_{2k} \rangle_{\text{even}}|^2 = \sum (2k+1)|\langle f, q_{2k+1} \rangle_{\text{odd}}|^2 = \langle Tf, f \rangle_{\text{odd}}$$

where T is the operator defined by $Tq_{2k+1} = (2k+1)q_{2k+1}$. (Actually, if T is considered as an operator on $\mathcal{H}^1_{\text{odd}}$, its domain does not include all of $\mathcal{H}^1_{\text{even}}$; the latter space is the domain of $T^{1/2}$.) It is easy to express T in a closed form:

$$T = 2w\frac{d}{dw} + 1,$$

from which we conclude that

$$\|f\|^2_{\text{even}} = \frac{1}{2\pi} \int_{|w|<1} \big(2wf'(w) + f(w)\big)\overline{f(w)}(1 - |w|^2)^{-1/2}\, dw.$$

The integral on the right does not look positive definite, but it is. Any doubts about its convergence can be assuaged by considering the integrals over the discs $|w| < 1 - \delta$ and letting $\delta \to 0$. Indeed, the monomials w^k are still orthogonal over these smaller discs, so if $f(w) = \sum a_k w^k$,

$$\frac{1}{2\pi} \int_{|w|<1-\delta} \big(2wf'(w) + f(w)\big)\overline{f(w)}(1 - |w|^2)^{-1/2}\, dw$$

$$= \sum_0^\infty (2k+1)|a_k|^2 \int_0^{1-\delta} r^{2k}(1 - r^2)^{-1/2} r\, dr,$$

which converges as $\delta \to 1$ to

$$\sum_0^\infty (2k+1)|a_k|^2 \frac{\Gamma(k+1)\Gamma(\frac{1}{2})}{2\Gamma(k+\frac{3}{2})} = \sum_0^\infty |a_k|^2 \frac{\Gamma(k+1)\Gamma(\frac{1}{2})}{\Gamma(k+\frac{1}{2})} = \|f\|^2_{\text{even}}.$$

$\|\ \|_{\text{even}}$ is essentially a Sobolev norm of order $\frac{1}{2}$ with respect to the weight function $(1 - |w|^2)^{-1/2}$.

7. Variants and Analogues

This section contains an informal discussion of some representations of other Lie groups that are closely related to the metaplectic representation.

Restrictions of the Metaplectic Representation. Many Lie groups can be realized as subgroups of some symplectic group. When the metaplectic representation is restricted to such a group G, its decomposition into irreducible components often furnishes a wealth of interesting representations of G (or perhaps its double cover).

One important example is the group $U(p, q)$, that is, the subgroup of $GL(p + q, \mathbf{C})$ consisting of transformations that preserve the Hermitian form

$$Q(z, w) = \sum_{1}^{p} z_j \overline{w}_j - \sum_{p+1}^{p+q} z_j \overline{w}_j.$$

The imaginary part of Q is a symplectic form; in fact, it is the standard symplectic form on $\mathbf{R}^{2(p+q)}$ if we identify the latter with $\mathbf{C}^{p+q}$ as follows:

$$(x, y, u, v) \longleftrightarrow (x + iy, u - iv) \qquad (x, y \in \mathbf{R}^p, \ u, v \in \mathbf{R}^q).$$

Consequently, $U(p, q)$ may be considered as a subgroup of $Sp(p + q, \mathbf{R})$. The spaces

$$\mathcal{H}_k = \{ f \in L^2(\mathbf{C}^{p+q}) : f(e^{i\theta} z) = e^{ik\theta} f(z) \}$$

are all invariant under the metaplectic action of $U(p, q)$ on $L^2(\mathbf{C}^{p+q})$, and one can show that they are irreducible. The resulting representations of $U(p, q)$ on the spaces $\mathcal{H}_k$ are called **ladder representations**.

For another example, let us identify $\mathbf{R}^{2nk}$ with the space of $2n \times k$ real matrices. Then $Sp(n, \mathbf{R})$ acts on $\mathbf{R}^{2nk}$ by left matrix multiplication, and thereby becomes a subgroup of $Sp(nk, \mathbf{R})$. The restriction of the metaplectic representation of $Sp(nk, \mathbf{R})$ to $Sp(n, \mathbf{R})$ is the kth tensor power of the metaplectic representation of $Sp(n, \mathbf{R})$, and its irreducible subspaces are related to the subspaces of $L^2(\mathbf{R}^{2nk})$ that transform under the action of $O(k)$ (by right matrix multiplication on $\mathbf{R}^{2nk}$) according to various irreducible representations of $O(k)$.

One can also combine the ingredients of the above examples to obtain representations of $U(p, q)$ by regarding it as a subgroup of $Sp((p + q)k, \mathbf{R})$.

The elucidation of these situations and related ones was the object of a considerable amount of work in the 1970's: see Gross–Kunze [62], [63], Kashiwara–Vergne [89], Sternberg–Wolf [132], and the references given in these papers. More recently, Mantini [101] and Davidson [36] have investigated the representations of $U(p, q)$ in the Fock model, making effective but quite different uses of appropriate variants of the Siegel unit disc.

A general framework in which most of these results can be understood has been provided by Howe's theory of "dual pairs." The main idea is the following. Suppose G and H are Lie subgroups of $Sp(n, \mathbf{R})$ that are each other's centralizers and satisfy some auxiliary algebraic conditions. Then the restriction of the metaplectic representation to GH is a direct sum of irreducibles of the form $\pi_j \otimes \rho_j$ where the π_j's and ρ_j's are distinct irreducible representations of G and H respectively. (In the first example above we have $G = U(p, q)$ and $H = \{e^{i\theta}I\}$, and in the second one we have $G = Sp(n, \mathbf{R})$ and $H = O(k)$.) Howe's original paper [73] on this subject has unfortunately remained in the realm of *samizdat*; a more readily available account has been provided by Sternberg [131].

$U(n, n)$ **as a Complex Symplectic Group.**   There is an imbedding of $U(n, n)$ into $Sp(n, \mathbf{R})$, different from but conjugate to the one described above, which makes $U(n, n)$ appear as the "complex analogue" of $Sp(n, \mathbf{R})$. Namely, let G_n denote the subgroup of $GL(2n, \mathbf{C})$ consisting of those transformations that leave invariant the "Hermitian-symplectic" form

$$A\big((z, w), (z', w')\big) = z\overline{w}' - w\overline{z}' \qquad (z, w, z', w' \in \mathbf{C}^n).$$

On the one hand, if we identify $\mathbf{R}^{4n}$ with $\mathbf{C}^{2n}$ by

$$(x, y, u, v) \longleftrightarrow (x + iy, u + iv) \qquad (x, y, u, v \in \mathbf{R}^n),$$

then Re A is just the standard symplectic form on $\mathbf{R}^{4n}$, so G_n is a subgroup of $Sp(2n, \mathbf{R})$. In fact, it is easily verified that

$$G_n = Sp(2n, \mathbf{R}) \cap GL(2n, \mathbf{C}).$$

On the other hand, G_n is also the group that preserves the Hermitian form iA, which has signature (n, n). (It is positive definite on $\{(z, w) : w = iz\}$ and negative definite on $\{(z, w) : w = -iz\}$.) Thus G_n is isomorphic to $U(n, n)$.

The structure theory of G_n is very much like that of $Sp(n, \mathbf{R})$. In particular, writing $2n \times 2n$ matrices in $n \times n$ blocks, we have

$$\begin{pmatrix} A & B \\ C & D \end{pmatrix} \in G_n \iff A^*C = C^*A, \ B^*D = D^*B, \ \text{and} \ A^*D - C^*B = I.$$

Moreover, G_n is generated by the elements

$$\begin{pmatrix} A & 0 \\ 0 & A^{*-1} \end{pmatrix}, \quad \begin{pmatrix} I & 0 \\ C & I \end{pmatrix}, \quad \begin{pmatrix} 0 & I \\ -I & 0 \end{pmatrix} \qquad (A \in GL(n, \mathbf{C}), \ C = C^*).$$

The proofs are the same as in the real case (Propositions (4.1) and (4.10)). Here, of course, it is important that A^* is the adjoint of A and not the transpose.

The restriction of the metaplectic representation μ of $Sp(2n, \mathbf{R})$ to G_n is a double-valued representation of G_n, but with a better choice of phase factors it can be made single-valued. Namely, let

$$\mu_0(\mathcal{A}) = (\det^{-1/2} \mathcal{A})\mu(\mathcal{A}), \qquad \mathcal{A} \in G_n.$$

Since $|\det \mathcal{A}| = 1$ for $\mathcal{A} \in G_n$, this is still a unitary representation, and it turns out that the sign ambiguities in $\det^{-1/2} \mathcal{A}$ and $\mu(\mathcal{A})$ cancel each other out. Explicitly, for $f \in L^2(\mathbf{C}^n)$ we have

$$(4.88) \quad \mu_0\left[\begin{pmatrix} A & 0 \\ 0 & A^{*-1} \end{pmatrix}\right] f(z) = (\det^{-1} A)f(A^{-1}z) \qquad (A \in GL(n, \mathbf{C})),$$

$$(4.89) \quad \mu_0\left[\begin{pmatrix} I & 0 \\ C & I \end{pmatrix}\right] f(z) = e^{-\pi i \bar{z} C z} f(z) \qquad (C = C^*),$$

$$(4.90) \quad \mu_0\left[\begin{pmatrix} 0 & I \\ -I & 0 \end{pmatrix}\right] f(z) = i^n \mathcal{F}^{-1} f(z),$$

and these formulas completely determine μ_0. (To see that (4.88) is consistent with (4.24), one must recall that if $A \in GL(n, \mathbf{C})$ is regarded as an element of $GL(2n, \mathbf{R})$ then $\det_{\mathbf{R}} A = |\det_{\mathbf{C}} A|^2$.) For more details, see Gross–Kunze [62].

The center Z of G_n is the circle group $\{e^{i\theta}I\}$, and G/Z is simple. The metaplectic action of Z is

$$\mu_0(e^{i\theta}I)f(z) = e^{-in\theta}f(e^{-i\theta}z).$$

$L^2(\mathbf{C}^n)$ has a Fourier decomposition under this circle action:

$$L^2(\mathbf{C}^n) = \bigoplus_{-\infty}^{\infty} \mathcal{H}_k, \qquad \mathcal{H}_k = \{ f : f(e^{i\theta}z) = e^{ik\theta}f(z) \}.$$

The spaces $\mathcal{H}_k$ are clearly all invariant under μ_0. Their irreducibility can be proved by an argument similar to the proof of Theorem (4.56), which we now sketch. For $k \in \mathbf{Z}$, let

$$f_k(z) = \begin{cases} z_1^k e^{-\pi|z|^2} & \text{if } k \geq 0, \\ \bar{z}_1^k e^{-\pi|z|^2} & \text{if } k < 0. \end{cases}$$

Up to scalar multiples, f_k is the unique element of $\mathcal{H}_k$ that satisfies the Hermite equations

$$(4.91) \quad \begin{aligned} \left(\pi|z_1|^2 - \frac{1}{\pi}\frac{\partial^2}{\partial z_1 \partial \bar{z}_1}\right) f &= (|k| + 1)f, \\ \sum_1^n \left(\pi|z_j|^2 - \frac{1}{\pi}\frac{\partial^2}{\partial z_j \partial \bar{z}_j}\right) f &= (|k| + n)f. \end{aligned}$$

(The general L^2 solution of (4.91) is

$$\sum_{j=0}^{|k|} a_j h_j(\operatorname{Re} z_1) h_{|k|-j}(\operatorname{Im} z_1) e^{-\pi|z|^2},$$

which can be written in the form

$$\sum_{l+m\leq|k|} b_{lm} z_1^l \bar{z}_1^m e^{-\pi|z|^2}.$$

The lm-th term of this sum belongs to $\mathcal{H}_{l-m}$.) Since the Hermite operators in (4.91) belong to the infinitesimal representation $d\mu_0$, the argument in the proof of Theorem (4.56) shows that if $\mathcal{M}$ is a closed invariant subspace of $\mathcal{H}_k$ then either $f_k \in \mathcal{M}$ or $f_k \perp \mathcal{M}$. Say $f_k \in \mathcal{M}$: by applying the operators (4.88) we see that for $k \geq 0$ (resp. $k < 0$) we have $P(z)e^{-\pi|z|^2} \in \mathcal{M}$ (resp. $P(\bar{z})e^{-\pi|z|^2} \in \mathcal{M}$) for all homogeneous holomorphic polynomials P of degree $|k|$. Then, by repeatedly applying the multiplication operators $z_i \bar{z}_j$, which are the infinitesimal versions of the operators (4.89), we conclude that $Q(z,\bar{z})e^{-\pi|z|^2} \in \mathcal{M}$ for every polynomial Q of the form

$$Q(z,\bar{z}) = \sum_{|\alpha|-|\beta|=k} c_{\alpha\beta} z^\alpha \bar{z}^\beta.$$

But such functions are dense in $\mathcal{H}_k$, so $\mathcal{M} = \mathcal{H}_k$.

The analogue of the Siegel half plane Σ_n in this setting is the following. If $V \in M_n(\mathbf{C})$, we set $V_+ = (V + V^*)/2$ and $V_- = (V - V^*)/2i$, so that V_+ and V_- are Hermitian and $V = V_+ + iV_-$. We then define the "half plane"

$$\Upsilon_n = \big\{ V \in M_n(\mathbf{C}) : V_- \text{ is positive definite} \big\}.$$

An argument similar to the proof of Theorem (4.64) shows that G_n acts transitively on Υ_n by linear fractional maps, and that the subgroup of G_n that fixes iI is $G_n \cap U(2n)$. To each $V \in \Upsilon_n$ we can associate the Gaussian $\beta_V(z) = e^{\pi i \bar{z} V z}$ on $\mathbf{C}^n$. The analogue of Theorem (4.65) then holds, with the same proof: if $\mathcal{A} = \begin{pmatrix} A & B \\ C & D \end{pmatrix} \in G_n$ and $V \in \Upsilon_n$,

$$\mu_0(\mathcal{A}^{*-1})\beta_V = \det^{-1}(CV + D)\,\beta_{(AV+B)(CV+D)^{-1}}.$$

Υ_n, like Σ_n, is an irreducible Hermitian symmetric space. For a complete list of such spaces, together with realizations of most of them as generalized discs or half planes, see Cartan [29].

The Spin Representation. This last subsection is devoted to an explanation and justification of the following assertion:

The metaplectic representation of $Sp(n, \mathbf{R})$ is to bosons as the spin representation of $SO(2n)$ is to fermions.

The spin representation is a double-valued representation of $SO(2n)$ that first arose in quantum mechanics in the case $n = 2$ and was constructed for general n by Brauer and Weyl [25]. (There is an analogous, but slightly different, representation of $SO(2n + 1)$.) We now sketch a construction of the spin representation, referring for more detailed information to Brauer and Weyl [25] (whose construction, however, differs from ours in appearance). To highlight the analogy with the metaplectic representation, one should consider the latter in the Fock model as a group of intertwining operators, associated not to the representation β of $\mathbf{H}_n$ but to the corresponding representation of its complexified Lie algebra via the creation and annihilation operators A_j^* and A_j.

We recall from Section 1.6 that the Fock space $\mathcal{F}_n$ can be regarded as the state space for collections of identical bosons whose individual state space is $(\mathbf{C}^n)^*$ (or $\mathbf{C}^n$). The analogous state space for collections of fermions is the anti-symmetric tensor algebra or exterior algebra of $\mathbf{C}^n$,

$$\bigwedge \mathbf{C}^n = \bigoplus_{p=0}^{n} \bigwedge^p \mathbf{C}^n.$$

Here $\bigwedge^0 \mathbf{C}^n = \mathbf{C}$, and for $p \geq 1$, $\bigwedge^p \mathbf{C}^n$ is the space of anti-symmetric p-tensors. If $e_1, \ldots, e_n$ is the standard basis for $\mathbf{C}^n$, the tensors

$$e(j_1, \ldots, j_p) = e_{j_1} \wedge \cdots \wedge e_{j_p} \qquad (j_1 < \cdots < j_p)$$

form a basis for $\bigwedge^p(\mathbf{C}^n)$, where $\wedge$ denotes the anti-symmetrized tensor product. We also set $e(\emptyset) = 1$, a basis for $\bigwedge^0(\mathbf{C}^n)$, and we make $\bigwedge \mathbf{C}^n$ into a Hilbert space of dimension 2^n by declaring the basis $\{e(j_1, \ldots, j_p)\}$ to be orthonormal. Physically, if $v_1, \ldots, v_p$ is any orthonormal set in $\mathbf{C}^n$, $v_1 \wedge \cdots \wedge v_p$ represents the state in which there are p particles, one in each of the states v_j; other p-particle states are to be interpreted as superpositions of these. (If the v_j's are not mutually orthogonal, a state containing a particle in each of the states v_j with probability one is not possible. For, if $\langle v_j, v_k \rangle \neq 0$, a particle in state v_j has a positive probability of being in state v_k, so the existence of another particle in state v_k is forbidden by the Pauli exclusion principle.)

We define the creation operators $A_1^*, \ldots, A_n^*$ on $\bigwedge \mathbf{C}^n$ by

$$A_k^* \phi = e_k \wedge \phi,$$

and we define the annihilation operators $A_1, \ldots, A_n$ to be their adjoints. Thus, in terms of the basis $\{e(j_1, \ldots, j_p)\}$ described above, we have

$$A_k^* e(j_1, \ldots, j_p) = \begin{cases} 0 & \text{if } k \in \{j_1, \ldots j_p\}, \\ (-1)^{i-1} e(j_1, \ldots, j_{i-1}, k, j_i, \ldots, j_p) & \text{if } j_{i-1} < k < j_i; \end{cases}$$

$$A_k e(j_1, \ldots, j_p) = \begin{cases} 0 & \text{if } k \notin \{j_1, \ldots, j_p\}, \\ (-1)^{i-1} e(j_1, \ldots, j_{i-1}, j_{i+1}, \ldots, j_p) & \text{if } k = j_i. \end{cases}$$

From this one easily sees that the A_k's and A_k^*'s satisfy the **canonical anti-commutation relations**

$$(4.92) \qquad A_j A_k + A_k A_j = A_j^* A_k^* + A_k^* A_j^* = 0, \qquad A_j A_k^* + A_k^* A_j = \delta_{jk} I.$$

One may also verify without difficulty that the algebra generated by the A_k's and the A_k^*'s is the algebra of all linear transformations on $\bigwedge \mathbf{C}^n$.

Let us now set

$$B_k = A_k + A_k^*, \qquad B_{k+n} = i(A_k - A_k^*) \qquad (1 \le k \le n).$$

Then the relations (4.92) imply that

$$B_j B_k + B_k B_j = 2\delta_{jk} I \qquad (1 \le j, k \le 2n).$$

But these are precisely the relations defining the (complex) Clifford algebra $C(2n)$. More precisely, let $E_1, \ldots, E_{2n}$ be the standard basis for $\mathbf{C}^{2n}$; then $C(2n)$ can be defined as the quotient of the tensor algebra $\bigotimes \mathbf{C}^{2n}$ by the ideal generated by the elements

$$(4.93) \qquad E_j \otimes E_k + E_k \otimes E_j - (2\delta_{jk})1 \qquad (j, k = 1, \ldots, 2n).$$

It is easily verified that the images of

$$1, \quad E_j, \quad E_j \otimes E_k \ (j < k), \quad E_j \otimes E_k \otimes E_l \ (j < k < l), \quad \ldots$$

in $C(2n)$ form a basis for $C(2n)$, which thus has dimension 2^{2n}. The correspondence $E_k \to B_k$ therefore determines a representation ρ of $C(2n)$ as linear operators on $\bigwedge \mathbf{C}^n$. As we have observed above, ρ is surjective, and since $C(2n)$ and $\mathrm{Hom}(\bigwedge \mathbf{C}^n) \cong M_{2^n}(\mathbf{C})$ both have dimension 2^{2n}, ρ is actually an isomorphism.

Now we can construct the spin representation. Suppose $T \in SO(2n)$. As a linear automorphism of $\mathbf{C}^{2n}$, T induces an automorphism of the tensor algebra $\bigotimes \mathbf{C}^{2n}$ which is easily seen to preserve the ideal generated by the elements (4.93); thus T induces an automorphism of $C(2n)$ and hence, via the representation ρ described above, an automorphism α_T of $\mathrm{Hom}(\bigwedge \mathbf{C}^n)$. But the only automorphisms of full matrix algebras are inner ones, so there exists a nonsingular $\sigma(T) \in \mathrm{Hom}(\bigwedge \mathbf{C}^n)$, determined up to scalar multiples, such that $\alpha_T(A) = \sigma(T)A\sigma(T)^{-1}$ for all $A \in \mathrm{Hom}(\bigwedge \mathbf{C}^n)$. σ is a projective representation of $SO(2n)$ in the space $\bigwedge \mathbf{C}^n$, and one can show that when suitably normalized, σ becomes a double-valued representation of $SO(2n)$. This is the spin representation.

Like the metaplectic representation, the spin representation is the direct sum of two inequivalent irreducible subrepresentations, the invariant subspaces being $\bigoplus_k \bigwedge^{2k} \mathbf{C}^n$ and $\bigoplus_k \bigwedge^{2k+1} \mathbf{C}^n$. Indeed, these two spaces are the eigenspaces (with eigenvalues ± 1) of the operator $\sigma(-I)$, which commutes with $\sigma(T)$ for all $T \in SO(2n)$.

One can give a single construction that encompasses both the metaplectic and spin representations, on the Lie algebra level, by using Lie superalgebras: see Sternberg [131].

CHAPTER 5.
THE OSCILLATOR SEMIGROUP

The **oscillator semigroup** Ω, so named by Roger Howe, is the semigroup of integral operators on $\mathbf{R}^n$ whose kernels are centered Gaussians (as defined in Section 4.5), or equivalently, the semigroup of pseudodifferential operators on $\mathbf{R}^n$ whose symbols are centered Gaussians. The name derives from the fact that Ω contains the semigroup generated by the Hermite operator, and the latter is the Hamiltonian for the quantum harmonic oscillator. Ω is an object of considerable aesthetic appeal, and it is closely related to a number of well-known families of operators: not only does it contain the Hermite semigroup, but its closure contains both the heat semigroup and the range of the metaplectic representation μ. More specifically, Ω contains a holomorphic subsemigroup Ω^0 of contraction operators whose distinguished boundary is the range of μ, and which can be regarded as the analytic continuation of μ to a certain open subsemigroup of $Sp(n, \mathbf{C})$.

Despite these features, the oscillator semigroup has received little attention until recently. The one-dimensional case seems to have been first described by de Bruijn [38], and the semigroup Ω^0 was constructed as an analytic continuation of the metaplectic representation on Fock space by Kramer et al. [93]. (More generally, Olshanskii [115] has since constructed similar analytic continuations for a large class of representations of semisimple groups associated to Hermitian symmetric spaces.) However, only in the paper of Howe [76] does a systematic treatment appear.

Section 5.1 contains the basic computations concerning the oscillator semigroup. The formulas there may seem rather forbidding at first sight, but there is an elegant structure underlying them that is revealed by the results—due to Howe [76]—in Section 5.3. The picture is completed by the development of the Fock model in Section 5.4, which is largely new although some of the final results are contained in [93].

1. The Schrödinger Model

In Section 4.5 we introduced the Siegel half-plane Σ_n of symmetric complex $n \times n$ matrices with positive definite imaginary part, and the Gaussian functions

$\gamma_Z(x) = e^{\pi i x Z x}$ on $\mathbf{R}^n$ for $Z \in M_n(\mathbf{C})$, $Z = Z^t$. In this chapter we shall be working with symmetric $2n \times 2n$ matrices and their associated Gaussians on $\mathbf{R}^{2n}$. We shall denote such matrices by $\mathcal{A}$ and $\mathcal{B}$ rather than Z; the reader may be assured that this employment of the letters that we used before for elements of the symplectic group is quite deliberate. We shall now encounter elements of $Sp(2n, \mathbf{R})$ acting on Σ_{2n}; we shall use the letters $\mathbf{S}$ and $\mathbf{T}$ for elements of $Sp(2n, \mathbf{R})$, and we shall denote the action of $\mathbf{S} \in Sp(2n, \mathbf{R})$ on Σ_{2n} (as before) by $\alpha(\mathbf{S})$.

Let T be a continuous linear map from $\mathcal{S}(\mathbf{R}^n)$ to $\mathcal{S}'(\mathbf{R}^n)$. From the discussion in Sections 1.3 and 2.1, we have three ways of describing T:

(i) as an integral operator with kernel $K_T \in \mathcal{S}'(\mathbf{R}^{2n})$:

$$Tf(x) = \int K_T(x, y) f(y) \, dy;$$

(ii) as $\rho(F_T)$ for some $F_T \in \mathcal{S}'(\mathbf{R}^{2n})$, where ρ is the (integrated) Schrödinger representation, in which case we call F_T the **representing function** of T;

(iii) as $\sigma_T(D, X)$ where σ_T is the (Weyl) symbol of T.

We recall that the distributions K_T, F_T, and σ_T are related as follows:

(5.1a) $\qquad\qquad K_T = \mathcal{F}_2^{-1} F_T \circ L = \mathcal{F}_1 \sigma_T \circ L,$

(5.1b) $\qquad\qquad F_T = \mathcal{F}_2(K_T \circ L^{-1}) = \mathcal{F} \sigma_T,$

(5.1c) $\qquad\qquad \sigma_T = \mathcal{F}_1^{-1}(K_T \circ L^{-1}) = \mathcal{F}^{-1} F_T,$

where $\mathcal{F}_1$ and $\mathcal{F}_2$ denote Fourier transformation in the first and second variables, $\mathcal{F} = \mathcal{F}_1 \mathcal{F}_2$, and

(5.2) $\qquad L(x, y) = \left(y - x, \tfrac{1}{2}(y + x)\right), \qquad L^{-1}(x, y) = \left(y - \tfrac{1}{2}x, \, y + \tfrac{1}{2}x\right).$

If $\mathcal{A} \in \Sigma_{2n}$, or more generally if $\mathcal{A}$ is in the closure of Σ_{2n} (i.e., $\mathcal{A} = \mathcal{A}^\dagger$ and $\operatorname{Im} \mathcal{A} \geq 0$) so that $\gamma_\mathcal{A}$ is tempered, we denote by $T_\mathcal{A}$ the operator whose kernel is $\gamma_\mathcal{A}$:

(5.3) $\qquad\qquad T_\mathcal{A} f(x) = \int \gamma_\mathcal{A}(x, y) f(y) \, dy.$

From the properties of Gaussian integrals it is easily verified (and we shall perform the calculations below) that the six transformations of (5.1) take functions of the form $\gamma_\mathcal{A}$, $\mathcal{A} \in \Sigma_{2n}$, into constant multiples of other such functions. It follows that

(5.4) $\qquad \{\, c T_\mathcal{A} : \mathcal{A} \in \Sigma_{2n}, \ c \in \mathbf{C} \backslash \{0\} \,\} = \{\, c \rho(\gamma_\mathcal{A}) : \mathcal{A} \in \Sigma_{2n}, \ c \in \mathbf{C} \backslash \{0\} \,\}$

$$= \{\, c \gamma_\mathcal{A}(D, X) : \mathcal{A} \in \Sigma_{2n}, \ c \in \mathbf{C} \backslash \{0\} \,\}.$$

We denote this collection of operators by Ω_n—or just Ω for short—and call it the **oscillator semigroup**. It is a subset of the algebra of Hilbert-Schmidt operators on $L^2(\mathbf{R}^n)$. We shall verify below that it is indeed a semigroup, and in fact a holomorphic semigroup: that is, if Ω is parametrized by the complex manifold $M = \Sigma_{2n} \times (\mathbf{C}\backslash\{0\})$ in any of the above three ways, the composition law in Ω is given by a holomorphic mapping from $M \times M$ to M.

At this point there are two calculations facing us: to make explicit the relations between the three presentations of Ω in (5.4), and to compute the semigroup law in each of these presentations. These calculations can be performed in either order, but we shall tackle the former one first. There are two possible ways to proceed. One is simply to use the formulas (5.1) directly. The other is to observe that the partial Fourier transforms and the compositions with L and L^{-1} in (5.1) are all of the form $c\mu(\mathbf{S})$ for some $c \in \mathbf{C}$ and $\mathbf{S} \in Sp(2n, \mathbf{R})$ where μ is the metaplectic representation, and then to use Theorem (4.65). We shall utilize both methods; the latter one has the advantage of showing clearly that the correspondences among kernels, symbols, and representing functions for operators in Ω are essentially implemented by linear fractional maps of Σ_{2n}, and that the normalization constants are given by the multiplier m defined by (4.61). The following theorem gives a complete list of the results.

(5.5) Theorem. *Suppose* $\mathcal{A} = \begin{pmatrix} A & B \\ B^\dagger & D \end{pmatrix} \in \Sigma_{2n}$ *(so* $A = A^\dagger$ *and* $D = D^\dagger$*).* *Then:*

(a) *The symbol of* $T_\mathcal{A}$ *is* $c_1\gamma_{\mathcal{B}_1}$ *where, with* $E = \frac{1}{4}(A - B - B^\dagger + D)$, $c_1 = \det^{-1/2}(-iE)$ *and*

$$\mathcal{B}_1 = \frac{1}{2}\begin{pmatrix} -2E^{-1} & E^{-1}(A+B-B^\dagger-D) \\ (A-B+B^\dagger-D)E^{-1} & (A+B^\dagger)E^{-1}(D-B)+(D+B)E^{-1}(A-B^\dagger) \end{pmatrix}.$$

(b) *The kernel of* $\gamma_\mathcal{A}(D, X)$ *is* $c_2\gamma_{\mathcal{B}_2}$ *where* $c_2 = \det^{-1/2}(-iA)$ *and*

$$\mathcal{B}_2 = \frac{1}{4}\begin{pmatrix} D - (B^\dagger+2I)A^{-1}(B+2I) & D - (B^\dagger+2I)A^{-1}(B-2I) \\ D - (B^\dagger-2I)A^{-1}(B+2I) & D - (B^\dagger-2I)A^{-1}(B-2I) \end{pmatrix}.$$

(c) *The representing function of* $T_\mathcal{A}$ *is* $c_3\gamma_{\mathcal{B}_3}$ *where, with* $F = A + B + B^\dagger + D$, $c_3 = \det^{-1/2}(-iF)$ *and*

$$\mathcal{B}_3 = \frac{1}{2}\begin{pmatrix} (A-B^\dagger)F^{-1}(B+D)+(D-B)F^{-1}(A+B^\dagger) & (-A+B^\dagger-B+D)F^{-1} \\ F^{-1}(-A+B-B^\dagger+D) & -2F^{-1} \end{pmatrix}.$$

(d) *The kernel of* $\rho(\gamma_\mathcal{A})$ *is* $c_4\gamma(\mathcal{B}_4)$ *where* $c_4 = \det^{-1/2}(-iD)$ *and*

$$\mathcal{B}_4 = \frac{1}{4}\begin{pmatrix} 4A - (I-2B)D^{-1}(I-2B^\dagger) & -4A - (I-2B)D^{-1}(I+2B^\dagger) \\ -4A - (I+2B)D^{-1}(I-2B^\dagger) & 4A - (I+2B)D^{-1}(I+2B^\dagger) \end{pmatrix}.$$

(e) *The representing function of* $\gamma_A(D, X)$ *and the symbol of* $\rho(\gamma_A)$ *are both* $c_5\gamma_{B_5}$ *where*

$$c_5 = \det^{-1/2}(-iA), \qquad B_5 = -A^{-1}.$$

(f) *If m and α are defined as in Section 4.5, in the above cases we also have*

$$c_j = \epsilon_j m(S_j, A), \qquad B_j = \alpha(S_j)A \qquad (j = 1, \ldots, 5)$$

where

$$\epsilon_1 = \epsilon_4 = i^{n/2}, \qquad \epsilon_2 = \epsilon_3 = i^{-n/2}, \qquad \epsilon_5 = i^n,$$

and the S_j's are the followng $4n \times 4n$ matrices, written in $n \times n$ blocks:

$$S_1 = \begin{pmatrix} 0 & 0 & -I & I \\ I & I & 0 & 0 \\ \frac{1}{2}I & -\frac{1}{2}I & 0 & 0 \\ 0 & 0 & \frac{1}{2}I & \frac{1}{2}I \end{pmatrix}, \qquad S_2 = \begin{pmatrix} 0 & \frac{1}{2}I & I & 0 \\ 0 & \frac{1}{2}I & -I & 0 \\ -\frac{1}{2}I & 0 & 0 & I \\ \frac{1}{2}I & 0 & 0 & I \end{pmatrix},$$

$$S_3 = \begin{pmatrix} -\frac{1}{2}I & \frac{1}{2}I & 0 & 0 \\ 0 & 0 & -\frac{1}{2}I & -\frac{1}{2}I \\ 0 & 0 & -I & I \\ I & I & 0 & 0 \end{pmatrix}, \qquad S_4 = \begin{pmatrix} -I & 0 & 0 & \frac{1}{2}I \\ I & 0 & 0 & \frac{1}{2}I \\ 0 & -I & -\frac{1}{2}I & 0 \\ 0 & -I & \frac{1}{2}I & 0 \end{pmatrix},$$

$$S_5 = \begin{pmatrix} 0 & 0 & -I & 0 \\ 0 & 0 & 0 & -I \\ I & 0 & 0 & 0 \\ 0 & I & 0 & 0 \end{pmatrix}.$$

Remark. The matrices A, D, E, and F that appear as arguments of $\det^{-1/2}(-i\,\cdot)$ in the above formulas all belong to Σ_n, as may be verified by applying the inequality $w(\operatorname{Im} A)\overline{w} > 0$ (valid for all nonzero $w \in \mathbf{C}^{2n}$) to vectors of the form $(z, 0)$, $(0, z)$, $(z, -z)$, and (z, z) with $z \in \mathbf{C}^n$. In all cases, the sign of the square root is determined by the requirement that $\det^{-1/2}(-i\,\cdot)$ should be continuous on the simply connected set Σ_n and positive when its argument is pure imaginary.

Proof: We begin with (a). By (5.1c), the symbol of T_A is

$$\sigma(\xi, x) = \mathcal{F}_1^{-1}(\gamma_A \circ L^{-1})(\xi, x) = \int e^{2\pi i \xi t} \gamma_A(x - \tfrac{1}{2}t,\, x + \tfrac{1}{2}t)\, dt$$

$$= \int \exp \pi i \Big[2\xi t + (x - \tfrac{1}{2}t)A(x - \tfrac{1}{2}t) + (x - \tfrac{1}{2}t)B(x + \tfrac{1}{2}t)$$

$$+ (x + \tfrac{1}{2}t)B^{\dagger}(x - \tfrac{1}{2}t) + (x + \tfrac{1}{2}t)D(x + \tfrac{1}{2}t) \Big]\, dt$$

$$= e^{\pi i x (A + B + B^{\dagger} + D)x} \int e^{(\pi i/4)t(A - B - B^{\dagger} + D)t} e^{2\pi i t[\xi - (1/2)(A + B - B^{\dagger} - D)x]}\, dt.$$

With $E = \frac{1}{4}(A - B - B^\dagger + D)$, this equals

$$\det{}^{-1/2}(-iE) \cdot \exp \pi i \Big[x(A+B+B^\dagger+D)x$$
$$+(\xi - \tfrac{1}{2}(A+B-B^\dagger-D)x)(-E^{-1})(\xi - \tfrac{1}{2}(A+B-B^\dagger-D)x)\Big]$$
$$= \det{}^{-1/2}(-iE)\,\gamma_{\mathcal{B}_1}(\xi, x).$$

Here

$$\mathcal{B}_1 = \frac{1}{2}\begin{pmatrix} 2E^{-1} & E^{-1}(A+B-B^\dagger-D) \\ (A+B^\dagger-B-D)E^{-1} & \widetilde{D} \end{pmatrix},$$

where the lower right entry $\widetilde{D}$ equals

$$2(A+B+B^\dagger+D) - 2(A+B^\dagger-B-D)(A-B-^\dagger+D)^{-1}(A+B-B^\dagger-D)$$
$$=2(A + B^\dagger)(A - B - B^\dagger + D)^{-1}[(A - B - B^\dagger + D) - (A + B - B^\dagger - D)]$$
$$+ 2(B + D)(A - B - B^\dagger + D)^{-1}[(A - B - B^\dagger + D) + (A + B - B^\dagger - D)]$$
$$=\tfrac{1}{2}(A + B^\dagger)E^{-1}(D - B) + \tfrac{1}{2}(B + D)E^{-1}(A - B^\dagger).$$

This proves (a); now let us turn to (b). By (5.1a), the kernel of $\gamma_A(D, X)$ is

$$K(x,y) = (\mathcal{F}_1\gamma_A)(y - x, \tfrac{1}{2}(y + x)) = \int e^{2\pi i(x-y)t}\gamma_A(t, \tfrac{1}{2}(x + y))\, dt$$

$$= \int \exp(\pi i)\Big[2(x-y)t + tAt + \tfrac{1}{2}tB(x+y) + \tfrac{1}{2}(x+y)B^\dagger t + \tfrac{1}{4}(x+y)D(x+y)\Big]\, dt$$

$$= e^{(\pi i/4)(x+y)D(x+y)} \int e^{\pi i t A t}e^{2\pi i t[(x-y)+(1/2)B(x+y)]}\, dt$$

$$= \det{}^{-1/2}(-iA)\, e^{(\pi i/4)\{(x+y)D(x+y)-[(B+2I)x+(B-2I)y]A^{-1}[(B+2I)x+(B-2I)y]\}},$$

which is $\det{}^{-1/2}(-iA)\gamma_{\mathcal{B}_2}(x, y)$ as claimed.

The proofs of (c) and (d) are similar to the proofs of (a) and (b); we leave them as exercises for the reader. (e) is obvious: since γ_A is even, the symbol of $\rho(\gamma_A)$ and the representing function of $\gamma_A(D, X)$ are both $\widehat{\gamma}_A$.

Now on to (f). With reference to formula (5.2), by (4.24) and (4.26) we have, for any $F \in \mathcal{S}(\mathbf{R}^{2n})$,

$$F \circ L^{-1} = (\det{}^{-1/2} L)\mu(\mathbf{T}_1)F = (-1)^{n/2}\mu(\mathbf{T}_1)F, \qquad \mathcal{F}_1 F = i^{-n/2}\mu(\mathbf{T}_2)F,$$

where

$$\mathbf{T}_1 = \begin{pmatrix} -I & I & 0 & 0 \\ \tfrac{1}{2}I & \tfrac{1}{2}I & 0 & 0 \\ 0 & 0 & -\tfrac{1}{2}I & \tfrac{1}{2}I \\ 0 & 0 & I & I \end{pmatrix}, \qquad \mathbf{T}_2 = \begin{pmatrix} 0 & 0 & I & 0 \\ 0 & I & 0 & 0 \\ -I & 0 & 0 & 0 \\ 0 & 0 & 0 & I \end{pmatrix}.$$

(Here the entries are $n \times n$ blocks; $\mathbf{T}_1 = \left(\begin{smallmatrix} L & 0 \\ 0 & L^{-1} \end{smallmatrix} \right)$; and if $\mathbf{R}^{4n} = \prod_1^4 \mathbf{R}^n_{(j)}$, $\mathbf{T}_2$ is the direct sum of $\mathcal{J}$ on $\mathbf{R}^n_{(1)} \times \mathbf{R}^n_{(3)}$ and the identity on $\mathbf{R}^n_{(2)} \times \mathbf{R}^n_{(4)}$.) Hence, by Theorem (4.65),

$$\mathcal{F}_1^{-1}(\gamma_{\mathcal{A}} \circ L^{-1}) = i^{n/2} \mu(\mathbf{T}_2)\mu(\mathbf{T}_1)\gamma_{\mathcal{A}} = i^{n/2} m(\mathbf{S}_1, \mathcal{A})\gamma_{\alpha(\mathbf{S}_1)\mathcal{A}},$$
$$(\mathcal{F}_1 \gamma_{\mathcal{A}}) \circ L = i^{-n/2} \mu(\mathbf{T}_1^{-1})\mu(\mathbf{T}_2^{-1})\gamma_{\mathcal{A}} = i^{-n/2} m(\mathbf{S}_2, \mathcal{A})\gamma_{\alpha(\mathbf{S}_2)\mathcal{A}},$$

where

$$\mathbf{S}_1 = \mathbf{T}_2^{*-1}\mathbf{T}_1^{*-1} \qquad \mathbf{S}_2 = \mathbf{T}_1^{*}\mathbf{T}_2^{*}.$$

Also, if

$$\mathbf{T}_3 = \begin{pmatrix} I & 0 & 0 & 0 \\ 0 & 0 & 0 & I \\ 0 & 0 & I & 0 \\ 0 & -I & 0 & 0 \end{pmatrix}$$

so that $\mu(\mathbf{T}_3) = i^{n/2}\mathcal{F}_2^{-1}$, we have

$$\mathcal{F}_2(\gamma_{\mathcal{A}} \circ L^{-1}) = i^{-n/2} \mu(\mathbf{T}_3^{-1})\mu(\mathbf{T}_1)\gamma_{\mathcal{A}} = i^{-n/2} m(\mathbf{S}_3, \mathcal{A})\gamma_{\alpha(\mathbf{S}_3)\mathcal{A}},$$
$$(\mathcal{F}_2^{-1}\gamma_{\mathcal{A}}) \circ L = i^{n/2} \mu(\mathbf{T}_1^{-1})\mu(\mathbf{T}_3)\gamma_{\mathcal{A}} = i^{n/2} m(\mathbf{S}_4, \mathcal{A})\gamma_{\alpha(\mathbf{S}_4)\mathcal{A}},$$

where

$$\mathbf{S}_3 = \mathbf{T}_3^{*}\mathbf{T}_1^{*-1}, \qquad \mathbf{S}_4 = \mathbf{T}_1^{*}\mathbf{T}_3^{*-1} = \mathbf{T}_1^{*}\mathbf{T}_3.$$

Finally,

$$\mathcal{F}\gamma_{\mathcal{A}} = \mathcal{F}^{-1}\gamma_{\mathcal{A}} = i^{2n/2}\mu(\mathbf{S}_5)\gamma_{\mathcal{A}} = i^n m(\mathbf{S}_5, \mathcal{A})\gamma_{\alpha(\mathbf{S}_5)\mathcal{A}}$$

where $\mathbf{S}_5$ is $\mathcal{J}$ in $2n$ dimensions.

The reader may verify that these formulas for $\mathbf{S}_1, \ldots, \mathbf{S}_5$ agree with those in the statement of the theorem. Granted this, (f) follows from the above calculations and (5.1). ∎

In order to complete the picture, it is instructive to verify directly that the formulas of (a)–(d) in Theorem (5.5) agree with those of (f). Consider (a): if $\mathcal{A} = \left(\begin{smallmatrix} A & B \\ B^\dagger & D \end{smallmatrix} \right)$ then $\alpha(\mathbf{S}_1)\mathcal{A}$ equals

$$\left[\begin{pmatrix} 0 & 0 \\ I & I \end{pmatrix} \begin{pmatrix} A & B \\ B^\dagger & D \end{pmatrix} + \begin{pmatrix} -I & I \\ 0 & 0 \end{pmatrix} \right] \left[\begin{pmatrix} \frac{1}{2}I & -\frac{1}{2}I \\ 0 & 0 \end{pmatrix} \begin{pmatrix} A & B \\ B^\dagger & D \end{pmatrix} + \begin{pmatrix} 0 & 0 \\ \frac{1}{2}I & \frac{1}{2}I \end{pmatrix} \right]^{-1}$$

$$= \begin{pmatrix} -I & I \\ A+B^\dagger & B+D \end{pmatrix} \begin{pmatrix} \frac{1}{2}(A-B^\dagger) & \frac{1}{2}(B-D) \\ \frac{1}{2}I & \frac{1}{2}I \end{pmatrix}^{-1}.$$

A routine calculation yields

$$\begin{pmatrix} \frac{1}{2}(A-B^\dagger) & \frac{1}{2}(B-D) \\ \frac{1}{2}I & \frac{1}{2}I \end{pmatrix}^{-1} = \frac{1}{2} \begin{pmatrix} E^{-1} & E^{-1}(D-B) \\ -E^{-1} & E^{-1}(A-B^\dagger) \end{pmatrix},$$

where $E = \frac{1}{4}(A - B - B^\dagger + D)$, which gives the formula of (a) for $\alpha(\mathbf{S}_1)\mathcal{A}$. Moreover, by Lemma 4 of Appendix A,

$$i^{n/2}m(\mathbf{S}_1, \mathcal{A}) = i^{n/2}\det^{-1/2}\begin{pmatrix} \frac{1}{2}(A - B^\dagger) & \frac{1}{2}(B - D) \\ \frac{1}{2}I & \frac{1}{2}I \end{pmatrix}$$

$$= i^{n/2}\det^{-1/2}E = \det^{-1/2}(-iE).$$

Likewise, for (b),

$$\alpha(\mathbf{S}_2)\mathcal{A} = \begin{pmatrix} \frac{1}{2}B^\dagger + I & \frac{1}{2}D \\ \frac{1}{2}B^\dagger - I & \frac{1}{2}D \end{pmatrix}\begin{pmatrix} -\frac{1}{2}A & -\frac{1}{2}B + I \\ \frac{1}{2}A & \frac{1}{2}B + I \end{pmatrix}^{-1},$$

and we have

$$\begin{pmatrix} -\frac{1}{2}A & -\frac{1}{2}B + I \\ \frac{1}{2}A & \frac{1}{2}B + I \end{pmatrix}^{-1} = \begin{pmatrix} -A^{-1}(\frac{1}{2}B + I) & -A^{-1}(\frac{1}{2}B - I) \\ \frac{1}{2}I & \frac{1}{2}I \end{pmatrix}$$

and

$$i^{-n/2}m(\mathbf{S}_2, \mathcal{A}) = i^{-n/2}\det^{-1/2}\begin{pmatrix} -\frac{1}{2}A & -\frac{1}{2}B + I \\ \frac{1}{2}A & \frac{1}{2}B + I \end{pmatrix} = \det^{-1/2}(-iA).$$

The cases (c) and (d) are similar.

We now turn to the task of computing the composition law for the oscillator semigroup. In terms of symbols or representing functions, of course, composition is given by twisted multiplication or twisted convolution. The formulas in these cases can be written in several different ways; we shall give one version of them in the following theorem and massage them algebraically afterwards. A further insight into the composition law will emerge in Section 5.3.

(5.6) Theorem. *Suppose*

$$A_j = \begin{pmatrix} A_j & B_j \\ B_j^\dagger & D_j \end{pmatrix} \in \Sigma_{2n}, \qquad j = 1, 2.$$

Then

$$T_{\mathcal{A}_1}T_{\mathcal{A}_2} = \det^{-1/2}[(-i)(D_1 + A_2)]T_{\mathcal{A}_3},$$

$$\gamma_{\mathcal{A}_1} \natural \gamma_{\mathcal{A}_2} = \det^{-1/2}(I + \tfrac{1}{4}A_1\mathcal{J}A_2\mathcal{J})\gamma_{\mathcal{A}_3'},$$

$$\gamma_{\mathcal{A}_1} \natural \gamma_{\mathcal{A}_2} = \det^{-1/2}[(-i)(A_1 + A_2)]\gamma_{\mathcal{A}_3''},$$

where $\mathcal{J} = \begin{pmatrix} 0 & I \\ -I & 0 \end{pmatrix}$ *and*

(5.7) $A_3 = \begin{pmatrix} A_1 - B_1(D_1 + A_2)^{-1}B_1^\dagger & -B_1(D_1 + A_2)^{-1}B_2 \\ -B_2^\dagger(D_1 + A_2)^{-1}B_1^\dagger & D_2 - B_2^\dagger(D_1 + A_2)^{-1}B_2 \end{pmatrix},$

(5.8) $A_3' = \left[(\tfrac{1}{2}A_2\mathcal{J} - I)(I + \tfrac{1}{4}A_1\mathcal{J}A_2\mathcal{J})^{-1}\right.$

$$\left. + (\tfrac{1}{2}A_1\mathcal{J} + I)(I + \tfrac{1}{4}A_2\mathcal{J}A_1\mathcal{J})^{-1}\right](-2\mathcal{J}),$$

(5.9) $A_3'' = A_1 - (A_1 - \tfrac{1}{2}\mathcal{J})(A_1 + A_2)^{-1}(A_1 + \tfrac{1}{2}\mathcal{J}).$

Proof: The kernel of $T_{\mathcal{A}_1} T_{\mathcal{A}_2}$ is

$$\int \gamma_{\mathcal{A}_1}(x,t)\gamma_{\mathcal{A}_2}(t,y)\,dt$$

$$= \int e^{\pi i(x A_1 x + 2x B_1 t + t D_1 t + t A_2 t + 2t B_2 y + y D_2 y)}\,dt$$

$$= e^{\pi i(x A_1 x + y D_2 y)} \int e^{\pi i t(D_1 + A_2)t} e^{2\pi i t(B_2 y + B_1^{\dagger} x)}\,dt$$

$$= \det^{-1/2}\left[(-i)(D_1 + A_2)\right] e^{\pi i[x A_1 x + y D_2 y - (B_1^{\dagger} x + B_2 y)(D_1 + A_2)^{-1}(B_1^{\dagger} x + B_2 y)]}$$

$$= \det^{-1/2}\left[(-i)(D_1 + A_2)\right] \gamma_{\mathcal{A}_3}(x,y),$$

where $\mathcal{A}_3$ is as in (5.7).

Next let us consider twisted convolution, which is a bit easier than twisted multiplication. We denote points in $\mathbf{R}^{2n}$ by w, w', so the symplectic form is $[w, w'] = wJw'$. We then have

$$\gamma_{\mathcal{A}_1} \natural \gamma_{\mathcal{A}_2}(w) = \int e^{\pi i(w-w')A_1(w-w')} e^{\pi i w' A_2 w'} e^{\pi i w J w'}\,dw'$$

$$= e^{\pi i w A_1 w} \int e^{\pi i w'(A_1 + A_2)w' - 2\pi i w'[A_1 + (J/2)]w}\,dw'$$

$$= \det^{-1/2}\left[\frac{A_1 + A_2}{i}\right] \exp \pi i\left[w A_1 w - (A_1 + \tfrac{1}{2}J)w \cdot (A_1 + A_2)^{-1}(A_1 + \tfrac{1}{2}J)w\right]$$

$$= \det^{-1/2}\left[(-i)(A_1 + A_2)\right] \gamma_{\mathcal{A}_3''}(w),$$

where, since $A_1 = A_1^{\dagger}$ and $J = -J^{\dagger}$,

$$\mathcal{A}_3'' = A_1 - (A_1 - \tfrac{1}{2}J)(A_1 + A_2)^{-1}(A_1 + \tfrac{1}{2}J).$$

(If we had used the formula

$$f \natural g(w) = \int f(w')g(w - w')e^{-\pi i w J w'}\,dw'$$

for twisted convolution, we would have got

$$\mathcal{A}_3'' = A_2 - (A_1 + \tfrac{1}{2}J)(A_1 + A_2)^{-1}(A_2 - \tfrac{1}{2}J).$$

The reader is invited to check directly that these formulas are equivalent.)

As for twisted multiplication, we have

$$f \natural\!\!\!\natural\, g(w) = 4^n \iint f(w')g(w'')e^{4\pi i(w - w'')J(w - w')}\,dw'\,dw'',$$

from which we obtain

$$\gamma_{\mathcal{A}_1} \natural \gamma_{\mathcal{A}_2}(w) = 4^n \int e^{\pi i Y \mathbf{A} Y + 2\pi i Y \mathbf{B} X} dY$$

where

$$X = (w, w), \quad Y = (w', w''), \quad \mathbf{A} = \begin{pmatrix} \mathcal{A}_1 & -2\mathcal{J} \\ 2\mathcal{J} & \mathcal{A}_2 \end{pmatrix}, \quad \mathbf{B} = \begin{pmatrix} 2\mathcal{J} & 0 \\ 0 & -2\mathcal{J} \end{pmatrix}.$$

Hence

$$\gamma_{\mathcal{A}_1} \natural \gamma_{\mathcal{A}_2}(w) = 4^n \det{}^{-1/2}(-i\mathbf{A}) e^{-\pi i X \mathbf{B}^\dagger \mathbf{A}^{-1} \mathbf{B} X}.$$

But by Lemma 4 of Appendix A,

$$\det(-i\mathbf{A}) = \det \left[\begin{pmatrix} \frac{1}{2}\mathcal{A}_1\mathcal{J} & I \\ -I & \frac{1}{2}\mathcal{A}_2\mathcal{J} \end{pmatrix} \begin{pmatrix} 2i\mathcal{J} & 0 \\ 0 & 2i\mathcal{J} \end{pmatrix} \right]$$

$$= 2^{4n} \det \begin{pmatrix} \frac{1}{2}\mathcal{A}_1\mathcal{J} & I \\ -I & \frac{1}{2}\mathcal{A}_1\mathcal{J} \end{pmatrix} = 2^{4n} \det(I + \tfrac{1}{4}\mathcal{A}_1\mathcal{J}\mathcal{A}_2\mathcal{J}),$$

and hence

$$4^n \det{}^{-1/2}(-i\mathbf{A}) = \det{}^{-1/2}(I + \tfrac{1}{4}\mathcal{A}_1\mathcal{J}\mathcal{A}_2\mathcal{J}).$$

Moreover,

$$\mathbf{B}^\dagger \mathbf{A}^{-1} = 4\mathbf{B}^{-1}\mathbf{A}^{-1} = (\tfrac{1}{4}\mathbf{A}\mathbf{B})^{-1} = \begin{pmatrix} \frac{1}{2}\mathcal{A}_1\mathcal{J} & -I \\ -I & -\frac{1}{2}\mathcal{A}_2\mathcal{J} \end{pmatrix}^{-1},$$

which, by a routine calculation, equals

$$\begin{pmatrix} \frac{1}{2}\mathcal{A}_2\mathcal{J}(I + \frac{1}{4}\mathcal{A}_1\mathcal{J}\mathcal{A}_2\mathcal{J})^{-1} & -(I + \frac{1}{4}\mathcal{A}_2\mathcal{J}\mathcal{A}_1\mathcal{J})^{-1} \\ -(I + \frac{1}{4}\mathcal{A}_1\mathcal{J}\mathcal{A}_2\mathcal{J})^{-1} & -\frac{1}{2}\mathcal{A}_1\mathcal{J}(I + \frac{1}{4}\mathcal{A}_2\mathcal{J}\mathcal{A}_1\mathcal{J})^{-1} \end{pmatrix}.$$

It then follows easily that

$$X\mathbf{B}^\dagger \mathbf{A}^{-1}\mathbf{B}X = -w\mathcal{A}_3'w \qquad (X = (w, w)),$$

where $\mathcal{A}_3'$ is given by (5.8). ∎

The formula (5.8) for $\mathcal{A}_3'$ can be made to look more pleasant by making the substitution $\mathcal{B} = \frac{1}{2}\mathcal{A}\mathcal{J}$. That is, if

$$(5.10) \qquad \mathcal{B}_1 = \tfrac{1}{2}\mathcal{A}_1\mathcal{J}, \qquad \mathcal{B}_2 = \tfrac{1}{2}\mathcal{A}_2\mathcal{J}, \qquad \mathcal{B}_3 = \tfrac{1}{2}\mathcal{A}_3'\mathcal{J},$$

we clearly have

$$(5.11) \qquad \mathcal{B}_3 = (\mathcal{B}_2 - I)(I + \mathcal{B}_1\mathcal{B}_2)^{-1} + (\mathcal{B}_1 + I)(I + \mathcal{B}_2\mathcal{B}_1)^{-1}.$$

The following variants of (5.11) will be useful later.

(5.12) Proposition. If $\mathcal{B}_1, \mathcal{B}_2, \mathcal{B}_3$ are given by (5.10) and (5.8) with $\mathcal{A}_j \in \Sigma_{2n}$, then

$$(5.13) \qquad I - \mathcal{B}_3 = (I - \mathcal{B}_2)(I + \mathcal{B}_1 \mathcal{B}_2)^{-1}(I - \mathcal{B}_1).$$

If also $I - \mathcal{B}_1$ is invertible, then

$$(5.14) \qquad \mathcal{B}_3 = (I - \mathcal{B}_1)^{-1}(\mathcal{B}_1 + \mathcal{B}_2)(I + \mathcal{B}_1 \mathcal{B}_2)^{-1}(I - \mathcal{B}_1).$$

Proof: From (5.11), we have

$$
\begin{aligned}
I - \mathcal{B}_3 &= -(\mathcal{B}_2 - I)(I + \mathcal{B}_1 \mathcal{B}_2)^{-1} + \left[(I + \mathcal{B}_2 \mathcal{B}_1) - (\mathcal{B}_1 + I)\right](I + \mathcal{B}_2 \mathcal{B}_1)^{-1} \\
&= (I - \mathcal{B}_2)\left[(I + \mathcal{B}_1 \mathcal{B}_2)^{-1} - \mathcal{B}_1(I + \mathcal{B}_2 \mathcal{B}_1)^{-1}\right] \\
&= (I - \mathcal{B}_2)(I + \mathcal{B}_1 \mathcal{B}_2)^{-1}(I - \mathcal{B}_1),
\end{aligned}
$$

since $\mathcal{B}_1 = \frac{1}{2}\mathcal{A}_1 \mathcal{J}$ is invertible and

$$\mathcal{B}_1(I + \mathcal{B}_2 \mathcal{B}_1)^{-1} = (\mathcal{B}_1^{-1} + \mathcal{B}_2)^{-1} = (I + \mathcal{B}_1 \mathcal{B}_2)^{-1}\mathcal{B}_1.$$

From this formula, if $I - \mathcal{B}_1$ is invertible we have

$$
\begin{aligned}
\mathcal{B}_3 &= I - (I - \mathcal{B}_2)(I + \mathcal{B}_1 \mathcal{B}_2)^{-1}(I - \mathcal{B}_1) \\
&= (I - \mathcal{B}_1)^{-1}\left[(I + \mathcal{B}_1 \mathcal{B}_2) - (I - \mathcal{B}_1)(I - \mathcal{B}_2)\right](I + \mathcal{B}_1 \mathcal{B}_2)^{-1}(I - \mathcal{B}_1) \\
&= (I - \mathcal{B}_1)^{-1}(\mathcal{B}_1 + \mathcal{B}_2)(I + \mathcal{B}_1 \mathcal{B}_2)^{-1}(I - \mathcal{B}_1). \quad \blacksquare
\end{aligned}
$$

The formula for $\mathcal{A}_3''$ can also be advantageously reformulated by making the substitution $\mathcal{C} = 2\mathcal{J}\mathcal{A}$. Namely, if

$$(5.15) \qquad \mathcal{C}_1 = 2\mathcal{J}\mathcal{A}_1, \qquad \mathcal{C}_2 = 2\mathcal{J}\mathcal{A}_2, \qquad \mathcal{C}_3 = 2\mathcal{J}\mathcal{A}_3'',$$

then (5.9) becomes

$$(5.16) \qquad \mathcal{C}_3 = \mathcal{C}_1 - (\mathcal{C}_1 + I)(\mathcal{C}_1 + \mathcal{C}_2)^{-1}(\mathcal{C}_1 - I).$$

There is an analogue of Proposition (5.12):

(5.17) Proposition. If $\mathcal{C}_1, \mathcal{C}_2, \mathcal{C}_3$ are given by (5.15) and (5.9) with $\mathcal{A}_j \in \Sigma_{2n}$, then

$$(5.18) \qquad \mathcal{C}_3 - I = (\mathcal{C}_2 - I)(\mathcal{C}_1 + \mathcal{C}_2)^{-1}(\mathcal{C}_1 - I).$$

If also $\mathcal{C}_1 - I$ is invertible, then

$$(5.19) \qquad \mathcal{C}_3 = (\mathcal{C}_1 - I)^{-1}(\mathcal{C}_1 \mathcal{C}_2 + I)(\mathcal{C}_1 + \mathcal{C}_2)^{-1}(\mathcal{C}_1 - I).$$

Proof: From (5.16) we have

$$C_3 - I = [(C_1 + C_2) - (C_1 + I)](C_1 + C_2)^{-1}(C_1 - I)$$
$$= (C_2 - I)(C_1 + C_2)^{-1}(C_1 - I),$$

whence, if $C_1 - I$ is invertible,

$$C_3 = I + (C_2 - I)(C_1 + C_2)^{-1}(C_1 - I)$$
$$= (C_1 - I)^{-1}[(C_1 + C_2) + (C_1 - I)(C_2 - I)](C_1 + C_2)^{-1}(C_1 - I)$$
$$= (C_1 - I)^{-1}(I + C_1 C_2)(C_1 + C_2)^{-1}(C_1 - I). \quad \blacksquare$$

Twisted multiplication and twisted convolution are intertwined by the Fourier transform, so the formulas (5.8) and its variants (5.11), (5.13), and (5.14) for A_3' must be related to the formulas (5.9), (5.16), (5.18), and (5.19) for A_3''. For the sake of completeness, we now verify directly that this is the case.

Suppose $A_1, A_2 \in \Sigma_{2n}$. It suffices to prove that the formulas match up when $I - \frac{1}{2}A_1 J$, and hence also $-2J A_1^{-1} - I = -2J A_1^{-1}(I - \frac{1}{2}A_1 J)$, is invertible, since the set of A_1 satisfying this condition is dense in Σ_{2n}. On the one hand,

$$(\gamma_{A_1} \natural \gamma_{A_2})\widehat{} = \widehat{\gamma}_{A_1} \natural \widehat{\gamma}_{A_2} = \det^{-1/2}(-iA_1)\det^{-1/2}(-iA_2)\widehat{\gamma}_{-A_1^{-1}} \natural \widehat{\gamma}_{-A_2^{-1}}$$
$$= \det^{-1/2}(-iA_1)\det^{-1/2}(-iA_2)\det^{-1/2}[(-i)(-A_1^{-1} - A_2^{-1})]\gamma_D$$

where, according to formula (5.19), if $C_j = 2J(-A_j^{-1})$,

$$2J D = (C_1 - I)^{-1}(I + C_1 C_2)(C_1 + C_2)^{-1}(C_1 - I).$$

On the other hand,

$$(\gamma_{A_1} \natural \gamma_{A_2})\widehat{} = \det^{-1/2}(I + \tfrac{1}{4}A_1 J A_2 J)\widehat{\gamma}_{A_3'}$$
$$= \det^{-1/2}(I + \tfrac{1}{4}A_1 J A_2 J)\det^{-1/2}(-iA_3')\gamma_{-(A_3')^{-1}},$$

where, according to formula (5.14), if $B_j = \frac{1}{2}A_j J$,

(5.20) $$\tfrac{1}{2}A_3' J = (I - B_1)^{-1}(B_1 + B_2)(I + B_1 B_2)^{-1}(I - B_1).$$

So we must have $D = -(A_3')^{-1}$, and the determinant factors must also match up. The former assertion is true because $C_j = B_j^{-1}$ and (as is easily seen) the expression on the right of (5.20) is unchanged when B_j is replaced by B_j^{-1}:

$$-2J(A_3')^{-1} = (\tfrac{1}{2}A_3' J)^{-1} = [(I - C_1)^{-1}(C_1 + C_2)(I + C_1 C_2)^{-1}(I - C_1)]^{-1} = 2J D.$$

The latter one is true because, from (5.20),

$$\det A_3' = \det(A_1 + A_2)\det^{-1}(I + \tfrac{1}{4}A_1 J A_2 J)$$
$$= \det(A_1)\det(A_1^{-1} + A_2^{-1})\det(A_2)\det^{-1}(I + \tfrac{1}{4}A_1 J A_2 J).$$

Lastly, we remark that the semigroup Ω is closed under adjoints. More precisely, we have the following result, whose proof is an easy exercise.

(5.21) Proposition. *If* $\mathcal{A} \in \Sigma_{2n}$ *then* $T_{\mathcal{A}}^* = T_{\mathcal{B}}$ *where*

$$\mathcal{A} = \begin{pmatrix} A & B \\ B^\dagger & D \end{pmatrix}, \qquad \mathcal{B} = \begin{pmatrix} \overline{D} & \overline{B}^\dagger \\ \overline{B} & \overline{A} \end{pmatrix}.$$

Moreover,

$$\left[\gamma_{\mathcal{A}}(D,X)\right]^* = \gamma_{-\overline{\mathcal{A}}}(D,X), \qquad \left[\rho(\gamma_{\mathcal{A}})\right]^* = \rho(\gamma_{-\overline{\mathcal{A}}}).$$

The Extended Oscillator Semigroup. The oscillator semigroup consists of integral operators whose kernels are of the form $e^{Q(x,y)}$ where Q is a quadratic polynomial with no linear term. It is natural to generalize by allowing a linear term in the exponent, thus arriving at integral kernels of the form
(5.22)
$$c\gamma_{\mathcal{A}}^{\lambda\mu}(x,y) = c\gamma_{\mathcal{A}}(x,y)e^{2\pi i(\lambda x + \mu y)} \qquad (c \in \mathbf{C}\backslash\{0\}, \ \mathcal{A} \in \Sigma_{2n}, \ \lambda,\mu \in \mathbf{C}^n).$$

As in Section 4.5, we shall use the term "Gaussian" to refer to functions of precisely this form. We denote the set of integral operators whose kernels are Gaussians by $E\Omega$ and call it the **extended oscillator semigroup**. $E\Omega$ is indeed a semigroup; in fact, if $T_{\mathcal{A}}^{\lambda\mu}$ denotes the integral operator whose kernel $\gamma_{\mathcal{A}}^{\lambda\mu}$ is defined by (5.22), we have:

(5.23) Proposition. *Suppose*

$$\mathcal{A}_1 = \begin{pmatrix} A_1 & B_1 \\ B_1^\dagger & D_1 \end{pmatrix}, \qquad \mathcal{A}_2 = \begin{pmatrix} A_2 & B_2 \\ B_2^\dagger & D_2 \end{pmatrix}.$$

Then

$$T_{\mathcal{A}_1}^{\lambda_1\mu_1} T_{\mathcal{A}_2}^{\lambda_2\mu_2} = \det{}^{-1/2}\left[(-i)(D_1 + A_2)\right] T_{\mathcal{A}_3}^{\lambda_3\mu_3}$$

where $\mathcal{A}_3$ *is given by (5.7) and*

$$\lambda_3 = \lambda_1 - B_1(D_1 + A_2)^{-1}(\mu_1 + \lambda_2), \qquad \mu_3 = \mu_2 - B_2^\dagger(D_1 + A_2)^{-1}(\mu_1 + \lambda_2).$$

The proof is essentially the same as the proof of the first part of Theorem (5.6); details are left to the reader.

The structure of $E\Omega$ is illuminated by the following consideration. According to Proposition (4.73), every Gaussian on $\mathbf{R}^{2n}$ can be written uniquely in the form

$$K(x,y) = ce^{\pi i(a\alpha + b\beta) + 2\pi i(\alpha x + \beta y)}\gamma_{\mathcal{A}}(x + a, \ y + b)$$

where $c \in \mathbf{C}\backslash\{0\}$, $\mathcal{A} \in \Sigma_{2n}$, and $a,b,\alpha,\beta \in \mathbf{R}^n$. Let T be the operator whose kernel is K. The substitution $y \to y - b$ in the integral defining T quickly reveals that

$$T = \rho(a,\alpha)T_{\mathcal{A}}\rho(-b,\beta).$$

In other words, we have

(5.24) $E\Omega = \{ \rho(w_1)T\rho(w_2) : T \in \Omega \text{ and } w_1, w_2 \in \mathbf{R}^{2n} \}.$

It is now easy to obtain the description of $E\Omega$ in terms of symbols or representing functions. The symbol and representing function of the operator $\rho(w_0)$ ($w_0 \in \mathbf{R}^{2n}$) are e_{w_0} and δ_{w_0} respectively, where $e_{w_0}(w) = e^{2\pi i w_0 w}$ and δ_{w_0} is the delta function with pole at w_0. Thus $E\Omega$ can be identified with the semigroup of functions of the form $ce_{w_1} \natural \gamma_A \natural e_{w_2}$ under twisted multiplication, and to the semigroup of functions of the form $c\delta_{w_1} \natural \gamma_A \natural \delta_{w_2}$ under twisted convolution. Explicitly, simple calculations show that

$$e_{w_1} \natural \gamma_A \natural e_{w_2}(w) = e^{-\pi i w_2 \mathcal{J} w_1} e^{2\pi i w(w_1 + w_2)} \gamma_A(w + \tfrac{1}{2}\mathcal{J}(w_2 - w_1)),$$
$$\delta_{w_1} \natural \gamma_A \natural \delta_{w_2}(w) = e^{-\pi i w_2 \mathcal{J} w_1} e^{\pi i w \mathcal{J}(w_2 - w_1)} \gamma_A(w - w_1 - w_2).$$

It follows easily from these formulas and Proposition (4.73) that the functions of the form $ce_{w_1} \natural \gamma_A \natural e_{w_2}$, or of the form $c\delta_{w_1} \natural \gamma_A \natural \delta_{w_2}$, are precisely the Gaussians. Thus:

(5.25) $E\Omega = \{ G(D, X) : G \text{ is a Gaussian} \} = \{ \rho(G) : G \text{ is a Gaussian} \}$
$$= \{ c\rho(w_1)\gamma_A(D, X)\rho(w_2) : w_1, w_2 \in \mathbf{R}^{2n}, \ A \in \Sigma_{2n} \}$$
$$= \{ c\rho(w_1)\rho(\gamma_A)\rho(w_2) : w_1, w_2 \in \mathbf{R}^{2n}, \ A \in \Sigma_{2n} \}.$$

We leave to the reader the task of computing the semigroup law on $E\Omega$ in terms of the above descriptions. It is rather messy.

We have met $E\Omega$—or a small piece of it—once before, in our second proof of Lemma (2.75) (the main step in the proof of the Calderón-Vaillancourt (0,0) estimate). There the point was to consider operators of the form

$$\int f(w)\rho(-w)\rho(\gamma_A)\rho(w)\, dw$$

where $A = \tfrac{1}{2}iI$. One can also consider more general superpositions of elements of $E\Omega$,

(5.26) $$\int \rho(w_1)\rho(\gamma_A)\rho(w_2)d\mu(w_1, w_2, A)$$

where μ is a measure on $\mathbf{R}^{2n} \times \mathbf{R}^{2n} \times \Sigma_{2n}$. In Section 5.3 we shall derive the estimate
$$\|\rho(w_1)\rho(\gamma_A)\rho(w_2)\| = \|\rho(\gamma_A)\| < \left|\det{}^{-1/2}(A + \tfrac{1}{2}\mathcal{J})\right|$$

for the operator norm of $\rho(w_1)\rho(\gamma_A)\rho(w_2)$ on L^2. From this one can obtain L^2 estimates for operators of the form (5.26) from Cotlar's lemma once one has a good grip on the semigroup law in $E\Omega$. Indeed, we have

$$\rho(w_1)\rho(\gamma_A)\rho(w_2)\rho(w_1')\rho(\gamma_{A'})\rho(w_2') = C\rho(w_1'')\rho(\gamma_{A''})\rho(w_2'')$$

where C, w_1'', w_2'', and A'' depend on the w's and A's on the left, and what one needs in order to apply Cotlar's lemma is to calculate $|\det^{-1/2}(A'' + \frac{1}{2}J)|$ and $|C|$ effectively. As Howe [76] has shown, this technique can be used to produce L^2 estimates for a wide variety of operators, including various sorts of pseudodifferential and Fourier integral operators. However, here we shall content ourselves with the illustration of this method provided by the proof of Lemma (2.75) and refer the reader to Howe [76] for a fuller discussion.

Another approach to the synthesis of large classes of operators from operators with Gaussian kernels can be found in Unterberger [140], [141].

2. The Hermite Semigroup

At this point we pause to calculate the semigroup of operators

$$\left\{ e^{-2\pi s(D^2+X^2)} : \operatorname{Re} s > 0 \right\}$$

generated by the Hermite operator $2\pi(D^2 + X^2)$. We have already done this in the limiting case $\operatorname{Re} s = 0$ in Corollary (4.55), in which $s = \frac{1}{2}i\theta$. After looking at the formula there it is easy to conjecture the following result, in which we employ the notation T_A defined in (5.3).

(5.27) Theorem. *We have* $e^{-2\pi s(D^2+X^2)} = (\operatorname{csch} 2s)^{n/2} T_A$ *where*

$$A = \begin{pmatrix} (i \coth 2s)I & -(i \operatorname{csch} 2s)I \\ -(i \operatorname{csch} 2s)I & (i \coth 2s)I \end{pmatrix}.$$

Proof: We recall Mehler's formula (1.87): if h_α is the αth Hermite function, the series

$$k(x, y, w) = \sum w^{|\alpha|} h_\alpha(x) h_\alpha(y) \qquad (|w| < 1)$$

converges uniformly on compact sets, and in $L^2(y)$ for each x and w, to

$$\left(\frac{2}{1 - w^2} \right)^{n/2} \exp\left[\frac{-\pi(1 + w^2)(x^2 + y^2) + 4\pi wxy}{1 - w^2} \right].$$

Now, $2\pi(D^2+X^2)h_\alpha = (2|\alpha|+n)h_\alpha$, so $e^{-2\pi s(D^2+X^2)}h_\alpha = e^{-s(2|\alpha|+n)}h_\alpha$. Thus, if $f \in L^2$ and $f = \sum c_\alpha h_\alpha$, then

$$e^{-2\pi s(D^2+X^2)}f(x) = \sum c_\alpha e^{-s(2|\alpha|+n)}h_\alpha(x)$$

$$= \sum e^{-s(2|\alpha|+n)}h_\alpha(x) \int f(y)h_\alpha(y)\,dy = e^{-ns}\int f(y)k(x,y,e^{-2s})\,dy,$$

in view of the nice convergence properties of the series k. Thus, by Mehler's formula, the kernel of $e^{-2\pi s(D^2+X^2)}$ is

$$e^{-ns}k(x,y,e^{-2s}) = \left(\frac{2e^{-2s}}{1-e^{-4s}}\right)^{n/2} \exp\left[\frac{-(1+e^{-4s})\pi(x^2+y^2)+4\pi e^{-2s}xy}{1-e^{-4s}}\right]$$

$$= \left(\frac{2}{e^{2s}-e^{-2s}}\right)^{n/2} \exp\left[\frac{-(e^{2s}+e^{-2s})\pi(x^2+y^2)+4\pi xy}{e^{2s}-e^{-2s}}\right]$$

$$= (\operatorname{csch} 2s)^{n/2}\exp(-\pi)[(\coth 2s)(x^2+y^2)-2(\operatorname{csch} 2s)xy],$$

and this is $(\operatorname{csch} 2s)^{n/2}\gamma_{\mathcal{A}}(x,y)$ with $\mathcal{A}$ defined as above. ∎

Thus the Hermite semigroup is contained in the oscillator semigroup. Having calculated the kernel of $e^{-2\pi s(D^2+X^2)}$, we can now find its symbol and representing function.

(5.28) Theorem. *The symbol of $e^{-2\pi s(D^2+X^2)}$ is*

$$(\operatorname{sech} s)^n \gamma_{2i(\tanh s)I}(\xi, x) = (\operatorname{sech} s)^n e^{-2\pi(\tanh s)(\xi^2+x^2)},$$

and its representing function is

$$(\tfrac{1}{2}\operatorname{csch} s)^n \gamma_{(i/2)(\coth s)I}(p,q) = (\tfrac{1}{2}\operatorname{csch} s)^n e^{-(\pi/2)(\coth s)(p^2+q^2)}.$$

Proof: We prove the first assertion by means of Theorem (5.5a) and the hyperbolic identities

$$\coth 2s + \operatorname{csch} 2s = \coth s, \qquad \coth 2s - \operatorname{csch} 2s = \tanh s.$$

In the notation of Theorem (5.5a), if $A = D = (i\coth 2s)I$ and $B = B^\dagger = -(i\operatorname{csch} 2s)I$ then

$$E = \tfrac{1}{4}(A-B-B^\dagger+D) = \tfrac{1}{2}(i\coth s)I, \qquad A+B-B^\dagger-D = 0,$$

$$(A+B^\dagger)E^{-1}(D-B)+(D+B)E^{-1}(A-B^\dagger) = (2i\tanh s)I,$$

so that

$$\mathcal{B}_1 = \begin{pmatrix} (2i\tanh s)I & 0 \\ 0 & (2i\tanh s)I \end{pmatrix}, \qquad c_1 = \det^{-1/2}(-iE) = (2\tanh s)^{n/2}.$$

Thus the symbol of $e^{-2\pi s(D^2+X^2)}$ is $c'\gamma_{(2i\tanh s)I}$ where

$$c' = (2\operatorname{csch} 2s \tanh s)^{n/2} = \left(\frac{1}{\sinh s \cosh s}\frac{\sinh s}{\cosh s}\right)^{n/2} = (\operatorname{sech} s)^n.$$

The representing function is the Fourier transform of the symbol, so the second assertion follows immediately. ∎

Viewed from a different angle, Theorem (5.28) expresses all the operators $\sigma_\zeta(D, X)$, where

$$\sigma_\zeta(\xi, x) = e^{-2\pi\zeta(\xi^2 + x^2)}, \qquad \operatorname{Re}\zeta > 0,$$

in terms of the Hermite semigroup, except for $\zeta = 1$. Indeed, if $\zeta = \tanh s$ we have

$$\zeta = \frac{1 - e^{-2s}}{1 + e^{-2s}}, \qquad e^{-2s} = \frac{1 - \zeta}{1 + \zeta}, \qquad \operatorname{sech}^2 s = 1 - \zeta^2.$$

The function $w = e^{-2s}$ maps the half plane $\operatorname{Re} s > 0$ onto the punctured disc $0 < |w| < 1$, and the function $\zeta = (1 - w)/(1 + w)$ maps the punctured disc onto the punctured half plane $\operatorname{Re}\zeta > 0$, $\zeta \neq 1$. Thus:

(5.29) Corollary. *If* $\operatorname{Re}\zeta > 0$, $\zeta \neq 1$, *and* $\sigma_\zeta(\xi, x) = e^{-2\pi\zeta(\xi^2 + x^2)}$, *then*

$$\sigma_\zeta(D, X) = (1 - \zeta^2)^{-n/2} \left(\frac{1 - \zeta}{1 + \zeta}\right)^{\pi(D^2 + X^2)} = \frac{1}{(1 + \zeta)^n} \left(\frac{1 - \zeta}{1 + z}\right)^{\pi(D^2 + X^2) - (n/2)}$$

The second version of this formula makes it clear than no ambiguity concerning the branch of $\log[(1 - \zeta)/(1 + \zeta)]$ is involved, for the spectrum of $\pi(D^2 + X^2) - (n/2)$ consists of integers. Indeed, we have

$$(5.30) \qquad\qquad \sigma_\zeta(D, X)h_\alpha = \frac{1}{(1 + \zeta)^n} \left(\frac{1 - \zeta}{1 + \zeta}\right)^{|\alpha|} h_\alpha.$$

As for the exceptional point $\zeta = 1$, we already know from (2.29) that $2^n \sigma_1$ is the symbol of the orthogonal projection onto Ch_0. Ch_0 is the nullspace of $\pi(D^2 + X^2) - (n/2)$, so the orthogonal projection onto it is the limit as $w \to 0$ of $w^{-\pi(D^2 + X^2) - (n/2)}$; thus Corollary (5.29) also gives the right result in the limit as $\zeta \to 1$.

Incidentally, if we take the scalar product of both sides of (5.30) with h_α and use Proposition (2.5), we obtain a formula of de Bruijn [37]:

$$\int \sigma_\zeta(\xi, x) W h_\alpha(\xi, x) \, d\xi \, dx = \frac{1}{(1 + \zeta)^n} \left(\frac{1 - \zeta}{1 + \zeta}\right)^{|\alpha|}.$$

One other interesting point also follows easily from (5.30): when $0 < \zeta < 1$ the operator $\sigma_\zeta(D, X)$ is positive definite; when $\zeta = 1$ it is positive semidefinite; when $\zeta > 1$ it is positive and negative on the spans of $\{h_\alpha : |\alpha| \text{ even}\}$ and $\{h_\alpha : |\alpha| \text{ odd}\}$ respectively.

Generalizing Corollary (5.29), Geller [57, Theorem 2.3] has computed the operators $\sigma_{\zeta, P}(D, X)$ in terms of the Hermite semigroup, where $\sigma_{\zeta, P}(\xi, x) =$

$e^{-2\pi\zeta(\xi^2+x^2)}P(\xi,x)$ and P is a homogeneous harmonic polynomial. (Geller computes $\rho(\sigma_{\zeta,P})$, but this makes little difference because $\sigma_{\zeta,P}$ is essentially its own Fourier transform, by the Hecke-Bochner formula. In fact, Geller's result is a sort of operator-theoretic analogue of this formula.)

The Hermite semigroup has been used by de Bruijn [38] to construct a theory of generalized functions that are more general than tempered distributions but still are suitable for doing Fourier analysis. Roughly speaking, for de Bruijn a generalized function is the boundary value as $t \to 0$ of a solution $u(x,t)$ of the Hermite heat equation

$$\frac{\partial u}{\partial t} = -2\pi(D^2 + X^2)u, \qquad t > 0.$$

The Hermite semigroup also plays an important role in the study of L^2 estimates for (generalized) pseudodifferential operators in Unterberger [140], [141].

3. Normalization and the Cayley Transform

In Theorem (5.6) and Proposition (5.12) we showed that

$$(5.31) \qquad \gamma_{\mathcal{A}_1} \natural \gamma_{\mathcal{A}_2} = \det^{-1/2}(I + \mathcal{B}_1\mathcal{B}_2)\gamma_{\mathcal{A}_3}$$

where the matrices $\mathcal{B}_j = \frac{1}{2}\mathcal{A}_j\mathcal{J}$ $(j = 1, 2, 3)$ are related by

$$(5.32) \qquad I - \mathcal{B}_3 = (I - \mathcal{B}_2)(I + \mathcal{B}_1\mathcal{B}_2)^{-1}(I - \mathcal{B}_1).$$

This suggests a way of getting rid of the determinant factor in the composition formula. Namely, let

$$\Sigma_{2n}^0 = \big\{ \mathcal{A} \in \Sigma_{2n} : \det(I - \tfrac{1}{2}\mathcal{A}\mathcal{J}) \neq 0 \big\},$$

and for $\mathcal{A} \in \Sigma_{2n}^0$, define

$$\gamma_{\mathcal{A}}^0 = \det^{1/2}(I - \tfrac{1}{2}\mathcal{A}\mathcal{J})\gamma_{\mathcal{A}}.$$

Here, because of the square root, $\gamma_{\mathcal{A}}^0$ actually denotes a pair of functions differing from each other by a factor of -1; as in the discussion of the metaplectic representation in Chapter 4, we shall refuse to let this ambiguity bother us. From formulas (5.31) and (5.32) it is clear that

$$\gamma_{\mathcal{A}_1}^0 \natural \gamma_{\mathcal{A}_2}^0 = (\pm)\gamma_{\mathcal{A}_3}^0.$$

Thus, if we define

$$\Omega^0 = \big\{ (\pm)\gamma_{\mathcal{A}}^0(D, X) : \mathcal{A} \in \Sigma_{2n}^0 \big\},$$

Ω^0 is a subsemigroup of Ω that is parametrized by a double cover of Σ_{2n}^0. We call Ω^0 the **normalized oscillator semigroup**.

Ω^0 is closed under adjoints; more precisely, we have:

(5.33) Proposition. $\left[\gamma_{\mathcal{A}}^0(D,X)\right]^* = \gamma_{-\overline{\mathcal{A}}}^0(D,X).$

Proof: We have already observed in Proposition (5.21) that

$$\left[\gamma_{\mathcal{A}}(D,X)\right]^* = \gamma_{-\overline{\mathcal{A}}}(D,X),$$

so we must check that

$$\overline{\det(I - \tfrac{1}{2}\mathcal{A}\mathcal{J})} = \det(I + \tfrac{1}{2}\overline{\mathcal{A}}\mathcal{J}).$$

What is obvious is that

$$\overline{\det(I - \tfrac{1}{2}\mathcal{A}\mathcal{J})} = \det(I - \tfrac{1}{2}\overline{\mathcal{A}}\mathcal{J}),$$

since $\mathcal{J}$ is real; but then by Lemma 4 of Appendix A,

$$\det(I - \tfrac{1}{2}\overline{\mathcal{A}}\mathcal{J}) = \det\begin{pmatrix} -\tfrac{1}{2}\overline{\mathcal{A}} & -I \\ I & \mathcal{J} \end{pmatrix} = \det\begin{pmatrix} -\tfrac{1}{2}\overline{\mathcal{A}} & -I \\ I & \mathcal{J} \end{pmatrix}^\dagger$$

$$= \det\begin{pmatrix} -\tfrac{1}{2}\overline{\mathcal{A}} & I \\ -I & -\mathcal{J} \end{pmatrix} = \det(I + \tfrac{1}{2}\overline{\mathcal{A}}\mathcal{J}). \quad \blacksquare$$

The Hermite semigroup is contained in the normalized oscillator semigroup. Indeed, if $\mathcal{A} = (2i\tanh s)I$ (Re $s > 0$) then

$$\det(I - \tfrac{1}{2}\mathcal{A}\mathcal{J}) = \det\begin{pmatrix} I & (-i\tanh s)I \\ (i\tanh s)I & I \end{pmatrix} = (1 - \tanh^2 s)^n = (\operatorname{sech} s)^{2n},$$

and hence by Theorem (5.28),

$$(5.34) \qquad\qquad e^{-2\pi s(D^2 + X^2)} = \gamma_{(2i\tanh s)I}^0(D,X).$$

From this we can now derive the crucial fact that the normalization factor $\det^{1/2}(I - \tfrac{1}{2}\mathcal{A}\mathcal{J})$ not only simplifies the semigroup law but also controls the operator norm of $\gamma_{\mathcal{A}}^0(D,X)$ on L^2.

(5.35) Theorem. If $T \in \Omega^0$ then T is a strict contraction on $L^2(\mathbf{R}^n)$. In other words, $\|\gamma_{\mathcal{A}}^0(D,X)\| < 1$ for all $\mathcal{A} \in \Sigma_{2n}^0$.

Proof: First, suppose that

$$\mathcal{A} = \begin{pmatrix} A & 0 \\ 0 & A \end{pmatrix} \quad \text{with} \quad A = \operatorname{diag}(2i\lambda_1, \ldots, 2i\lambda_n), \quad \lambda_j > 0.$$

Then

$$\gamma_\mathcal{A}^0(\xi, x) = \prod_1^n (1 - 4\lambda_j^2)^{1/2} e^{-2\pi\lambda_j(\xi_j^2 + x_j^2)} = \prod_1^n \gamma_{2i\lambda_j I}^0(\xi_j, x_j),$$

where $2i\lambda_j I$ is a 2×2 matrix. Hence, by (5.34),

$$\gamma_\mathcal{A}^0(D, X) = \prod_1^n \gamma_{2i\lambda_j I}^0(D_j, X_j) = \prod_1^n e^{-2\pi s_j(D_j^2 + X_j^2)},$$

where $\mathrm{Re}\, s_j > 0$ and $\lambda_j = \tanh s_j$. The Hermite functions are an orthonormal eigenbasis for this operator, and its eigenvalue on h_α is $\prod_1^n e^{-s_j(2\alpha_j + 1)}$, so its operator norm (the modulus of its largest eigenvalue) is $\prod_1^n e^{-\mathrm{Re}\, s_j} < 1$.

Next, suppose that $\mathcal{A} \in \Sigma_{2n}^0$ and $\mathrm{Re}\, \mathcal{A} = 0$. Then $-i\mathcal{A}$ is positive definite, so by Proposition (4.22), there exists $S \in Sp(n, \mathbf{R})$ such that $S^* \mathcal{A} S$ is of the diagonal form above. But $\gamma_{S^* \mathcal{A} S} = \gamma_\mathcal{A} \circ S$, and since S is symplectic,

$$\det(I - \tfrac{1}{2} S^* \mathcal{A} S \mathcal{J}) = \det(I - \tfrac{1}{2} S^* \mathcal{A} \mathcal{J} S^{*-1})$$
$$= \det[S^*(I - \tfrac{1}{2}\mathcal{A}\mathcal{J})S^{*-1}] = \det(I - \tfrac{1}{2}\mathcal{A}\mathcal{J}).$$

Hence, by Theorem (2.15),

$$\gamma_\mathcal{A}^0(D, X) = \gamma_{S^* \mathcal{A} S} \circ S^{-1}(D, X) = \mu(S^{*-1})\gamma_{S^* \mathcal{A} S}(D, X)\mu(S^*),$$

so by the preceding calculation and the unitarity of $\mu(S^*)$,

$$\|\gamma_\mathcal{A}^0(D, X)\| = \|\gamma_{S^* \mathcal{A} S}^0(D, X)\| < 1.$$

(Note that we need only the existence of $\mu(S^*)$ as a unitary intertwining operator and not any specific formula for it.)

Finally, let $\mathcal{A} \in \Sigma_{2n}^0$ be arbitrary, and let $T = \gamma_\mathcal{A}^0(D, X)$. Then by Proposition (5.33), $T^* = \gamma_{-\overline{\mathcal{A}}}^0(D, X)$ and hence $T^* T = \gamma_\mathcal{B}^0(D, X)$ where $\mathrm{Re}\, \mathcal{B} = 0$ since $T^* T$ is self-adjoint. But then $\|T^* T\| < 1$ by the result of the preceding paragraph, and

$$\|Tf\|_2^2 = \langle T^* Tf, f \rangle \leq \|T^* T\| \, \|f\|_2^2,$$

so $\|T\|^2 \leq \|T^* T\| < 1$ and the proof is complete. ∎

We now develop a remarkable connection between the normalized oscillator semigroup Ω^0 and the complexified symplectic group

$$Sp(n, \mathbf{C}) = \{ \mathcal{A} \in GL(2n, \mathbf{C}) : \mathcal{A}^\dagger \mathcal{J} \mathcal{A} = \mathcal{J} \}$$

that is mediated by the complexified symplectic Lie algebra

$$\mathsf{sp}(n, \mathbf{C}) = \{ \mathcal{B} \in M_{2n}(\mathbf{C}) : \mathcal{J}\mathcal{B} + \mathcal{B}^\dagger \mathcal{J} = 0 \}.$$

We remark to begin with that the algebraic structures of $Sp(n, \mathbf{C})$ and $\mathsf{sp}(n, \mathbf{C})$ are very similar to those of $Sp(n, \mathbf{R})$ and $\mathsf{sp}(n, \mathbf{R})$; in particular, the obvious analogues of Propositions (4.1) and (4.2) hold, in which the matrices are taken to be complex and adjoints are replaced by transposes.

The connection between Ω^0 and $Sp(n, \mathbf{C})$ is effected in two steps. First, the map $\mathcal{A} \to \frac{1}{2}\mathcal{A}\mathcal{J}$, which has already forced itself on our attention, is a bijection from the set of all symmetric matrices in $M_{2n}(\mathbf{C})$ to $\mathsf{sp}(n, \mathbf{C})$, for if $\mathcal{A} \in M_{2n}(\mathbf{C})$ then

$$\mathcal{J}(\tfrac{1}{2}\mathcal{A}\mathcal{J}) + (\tfrac{1}{2}\mathcal{A}\mathcal{J})^{\dagger}\mathcal{J} = \tfrac{1}{2}\mathcal{J}(\mathcal{A} - \mathcal{A}^{\dagger})\mathcal{J}.$$

Second, there is a linear fractional map, a variant of the Cayley transform that we used in Section 4.5, that gives a bijection between open dense subsets of $\mathsf{sp}(n, \mathbf{C})$ and $Sp(n, \mathbf{C})$. Namely, if $\mathcal{B} \in M_{2n}(\mathbf{C})$ and $\det(I - \mathcal{B}) \neq 0$ we define

(5.36) $$\mathrm{c}(\mathcal{B}) = (I + \mathcal{B})(I - \mathcal{B})^{-1}.$$

Proposition (5.37). c *is a bijection from*

$$\left\{ \mathcal{B} \in \mathsf{sp}(n, \mathbf{C}) : \det(I - \mathcal{B}) \neq 0 \right\}$$

to

$$\left\{ \mathcal{A} \in Sp(n, \mathbf{C}) : \det(I + \mathcal{A}) \neq 0 \right\}.$$

Its inverse is

$$\mathrm{c}^{-1}(\mathcal{A}) = -\mathrm{c}(\mathcal{A})^{-1} = (\mathcal{A} - I)(\mathcal{A} + I)^{-1}.$$

Proof: We have

$$I + \mathrm{c}(\mathcal{B}) = \big[(I - \mathcal{B}) + (I + \mathcal{B})\big](I - \mathcal{B})^{-1} = 2(I - \mathcal{B})^{-1},$$

so the range of c consists of matrices $\mathcal{A}$ with $\det(\mathcal{A} + I) \neq 0$; the asserted formula for c^{-1} follows by an easy calculation. Moreover,

$$\mathrm{c}(\mathcal{B}) \in Sp(n, \mathbf{C}) \iff (I + \mathcal{B})^{\dagger}\mathcal{J}(I + \mathcal{B}) = (I - \mathcal{B})^{\dagger}\mathcal{J}(I - \mathcal{B})$$
$$\iff \mathcal{B}^{\dagger}\mathcal{J} + \mathcal{J}\mathcal{B} = -\mathcal{B}^{\dagger}\mathcal{J} - \mathcal{J}\mathcal{B},$$

which happens precisely when $\mathcal{B}^{\dagger}\mathcal{J} + \mathcal{J}\mathcal{B} = 0$, i.e., $\mathcal{B} \in \mathsf{sp}(n, \mathbf{C})$. ∎

The final link is forged by calculating what the Cayley transform c does to matrix multiplication:

$$\mathrm{c}^{-1}\big[\mathrm{c}(\mathcal{B}_1)\mathrm{c}(\mathcal{B}_2)\big]$$
$$= \mathrm{c}^{-1}\big[(I + \mathcal{B}_1)(I - \mathcal{B}_1)^{-1}(I + \mathcal{B}_2)(I - \mathcal{B}_2)^{-1}\big]$$
$$= \big[(I + \mathcal{B}_1)(I - \mathcal{B}_1)^{-1}(I + \mathcal{B}_2)(I - \mathcal{B}_2)^{-1} - I\big] \times$$
$$\big[(I + \mathcal{B}_1)(I - \mathcal{B}_1)^{-1}(I + \mathcal{B}_2)(I - \mathcal{B}_2)^{-1} + I\big]^{-1}$$
$$= (I - \mathcal{B}_1)^{-1}\big[(I + \mathcal{B}_1)(I + \mathcal{B}_2) - (I - \mathcal{B}_1)(I - \mathcal{B}_2)\big](I - \mathcal{B}_2)^{-1} \times$$
$$(I - \mathcal{B}_2)\big[(I + \mathcal{B}_1)(I + \mathcal{B}_2) + (I - \mathcal{B}_1)(I - \mathcal{B}_2)\big]^{-1}(I - \mathcal{B}_1)$$
$$= (I - \mathcal{B}_1)^{-1}(2\mathcal{B}_1 + 2\mathcal{B}_2)(2I + 2\mathcal{B}_1\mathcal{B}_2)^{-1}(I - \mathcal{B}_1)$$
$$= (I - \mathcal{B}_1)^{-1}(\mathcal{B}_1 + \mathcal{B}_2)(I + \mathcal{B}_1\mathcal{B}_2)^{-1}(I - \mathcal{B}_1).$$

Behold, this last expression is identical to the right side of (5.14)! Moreover, if $\mathcal{B} = \frac{1}{2}\mathcal{AJ}$, the condition for $\gamma_{\mathcal{A}}(D, X)$ to be normalizable is the same as the condition for $\mathcal{B}$ to be in the domain of c, namely $\det(I - \mathcal{B}) \neq 0$. We have therefore proved:

(5.38) Theorem. *The map*

$$(\pm)\gamma_{\mathcal{A}}^0(D, X) \longrightarrow c(\tfrac{1}{2}\mathcal{AJ}) = (I + \tfrac{1}{2}\mathcal{AJ})(I - \tfrac{1}{2}\mathcal{AJ})^{-1}$$

is a 2-to-1 homomorphism from the normalized oscillator semigroup Ω^0 onto an open subsemigroup of $Sp(n, \mathbf{C})$.

For future reference, we note that

$$(5.39) \quad \begin{aligned} c(\tfrac{1}{2}\mathcal{AJ}) &= (I + \tfrac{1}{2}\mathcal{AJ})\mathcal{JJ}^{-1}(I - \tfrac{1}{2}\mathcal{AJ})^{-1} = (-\tfrac{1}{2}\mathcal{A} + \mathcal{J})(\tfrac{1}{2}\mathcal{A} + \mathcal{J})^{-1} \\ &= \alpha(\mathbf{T})\mathcal{A} \end{aligned}$$

where

$$(5.40) \qquad\qquad \mathbf{T} = \begin{pmatrix} -\tfrac{1}{2}I & \mathcal{J} \\ \tfrac{1}{2}I & \mathcal{J} \end{pmatrix}.$$

Moreover,

$$\begin{aligned} m(\mathbf{T}, \mathcal{A}) &= \det^{-1/2}(\tfrac{1}{2}\mathcal{A} + \mathcal{J}) \\ &= \det^{-1/2}(\tfrac{1}{2}\mathcal{A} + \mathcal{J})\det^{-1/2}(-\mathcal{J}) = \det^{-1/2}(I - \tfrac{1}{2}\mathcal{AJ}), \end{aligned}$$

so the normalization factor for $\gamma_{\mathcal{A}}^0$ is

$$(5.41) \qquad \det^{1/2}(I - \tfrac{1}{2}\mathcal{AJ}) = m(\mathbf{T}, \mathcal{A})^{-1} = m(\mathbf{T}^{-1}, \alpha(\mathbf{T})\mathcal{A}).$$

We have described the normalized oscillator semigroup in terms of symbols, but we could equally well have used representing functions. Let us sketch briefly how this goes. We have

$$\begin{aligned} \gamma_{\mathcal{A}}^0(D, X) &= \det^{1/2}(I - \tfrac{1}{2}\mathcal{JA})\gamma_{\mathcal{A}}(D, X) \\ &= \det^{1/2}(I - \tfrac{1}{2}\mathcal{JA})\det^{-1/2}(-i\mathcal{A})\rho(\gamma_{-\mathcal{A}^{-1}}) \\ &= \det^{-1/2}\big[(-i)(\tfrac{1}{2}\mathcal{J} - \mathcal{A}^{-1})\big]\rho(\gamma_{-\mathcal{A}^{-1}}). \end{aligned}$$

Thus if we make the substitution $\mathcal{A} \to -\mathcal{A}^{-1}$ and set

$$\begin{aligned} \Sigma_{2n}^1 &= \big\{ \mathcal{A} \in \Sigma_{2n} : \det(\tfrac{1}{2}\mathcal{J} + \mathcal{A}) \neq 0 \big\}, \\ \gamma_{\mathcal{A}}^1 &= \det^{1/2}\big[(-i)(\tfrac{1}{2}\mathcal{J} + \mathcal{A})\big]\gamma_{\mathcal{A}} \quad \text{for} \quad \mathcal{A} \in \Sigma_{2n}^1, \end{aligned}$$

we have

(5.42) $\Omega^0 = \{\, \rho(\gamma_A^1) : A \in \Sigma_{2n}^1 \,\}$

and

(5.43) $\gamma_{A_1}^1 \, \natural \, \gamma_{A_2}^1 = \gamma_{A_3}^1,$

where the matrices $C_j = 2\mathcal{J}A_j$ $(j = 1, 2, 3)$ are related by (5.18):

$$C_3 - I = (C_2 - I)(C_1 + C_2)^{-1}(C_1 - I).$$

The determinant factors all match up, for

$$\tfrac{1}{2}\mathcal{J} + A = -\tfrac{1}{2}\mathcal{J}(2\mathcal{J}A - I),$$

whence

$$\det(C_j - I) = 2^{2n} \det(\tfrac{1}{2}\mathcal{J} + A_j) \quad \text{and} \quad \det(C_1 + C_2) = 2^{2n} \det(A_1 + A_2),$$

so that (5.43) is in accordance with Theorem (5.6). The Cayley transform from Σ_{2n}^1 to $Sp(n, \mathbf{C})$ is

$$A \longrightarrow \mathrm{c}\big[\tfrac{1}{2}(-A^{-1})\mathcal{J}\big] = \mathrm{c}\big[(2\mathcal{J}A)^{-1}\big] = -\mathrm{c}(2\mathcal{J}A)$$
$$= (2\mathcal{J}A + I)(2\mathcal{J}A - I)^{-1},$$

and the homomorphism of Theorem (5.38) is given by

$$(\pm)\rho(\gamma_A^1) \longrightarrow (2\mathcal{J}A + I)(2\mathcal{J}A - I)^{-1}.$$

We can also compute the normalized oscillator semigroup in terms of integral kernels, and this will be important for what follows. If $\mathbf{T}$ is given by (5.40), $\mathbf{S}_1$ is as in Theorem (5.5), and $A \in \Sigma_{2n}$, by Theorem (5.5f) and (5.41) we have

$$T_A = i^{n/2} m(\mathbf{S}_1, A)\gamma_{\alpha(\mathbf{S}_1)A}(D, X)$$
$$= i^{n/2} m(\mathbf{S}_1, A)m\big(\mathbf{T}, \alpha(\mathbf{S}_1)A\big)\gamma_{\alpha(\mathbf{S}_1)A}^0(D, X) = i^{n/2} m(\mathbf{TS}_1, A)\gamma_{\alpha(\mathbf{S}_1)A}^0(D, X),$$

so the normalization factor for T_A is $i^{-n/2} m(\mathbf{TS}_1, A)^{-1}$. But one readily computes that

(5.44) $\mathbf{TS}_1 = \begin{pmatrix} 0 & 0 & I & 0 \\ -I & 0 & 0 & 0 \\ 0 & 0 & 0 & I \\ 0 & I & 0 & 0 \end{pmatrix},$

so that

$$m(\mathbf{TS}_1, \mathcal{A}) = \det^{-1/2} \left[\begin{pmatrix} 0 & 0 \\ 0 & I \end{pmatrix} \begin{pmatrix} A & B \\ B^\dagger & D \end{pmatrix} + \begin{pmatrix} 0 & I \\ 0 & 0 \end{pmatrix} \right]$$

$$= \det^{-1/2} \begin{pmatrix} 0 & I \\ B^\dagger & D \end{pmatrix} = \det^{-1/2}(-B).$$

Thus, if we define

$$T_\mathcal{A}^0 = \det^{1/2}(iB)T_\mathcal{A} \quad \text{for} \quad \mathcal{A} = \begin{pmatrix} A & B \\ B^\dagger & D \end{pmatrix},$$

we have

$$T_{\mathcal{A}_1}^0 T_{\mathcal{A}_2}^0 = (\pm) T_{\mathcal{A}_3}^0$$

where $\mathcal{A}_3$ is as in Theorem (5.6). The correctness of the normalization is clear from Theorem (5.6), since formula (5.7) gives

$$iB_3 = (iB_1)\left[(-i)(D_1 + A_2)\right]^{-1}(iB_2).$$

In other words,

$$(5.45) \qquad \Omega^0 = \left\{ T_\mathcal{A}^0 : \mathcal{A} = \begin{pmatrix} A & B \\ B^\dagger & D \end{pmatrix} \in \Sigma_{2n}, \ \det B \neq 0 \right\}.$$

The homomorphism $\Omega^0 \to Sp(n, \mathbf{C})$ is given by

$$T_\mathcal{A}^0 \longrightarrow \alpha(\mathbf{TS}_1)\mathcal{A} = \begin{pmatrix} -B^{\dagger-1}D & B^{\dagger-1} \\ AB^{\dagger-1}D & -AB^{\dagger-1} \end{pmatrix}.$$

(The second equation follows from a calculation which we leave to the reader.)
It is actually more interesting to go the other way: from (5.44) we have $(\mathbf{TS}_1)^{-1} = (\mathbf{TS}_1)^\dagger$ and hence, if $\mathcal{A} \in Sp(n, \mathbf{C})$,

$$\alpha(\mathbf{TS}_1)^{-1}\mathcal{A}$$

$$= \left[\begin{pmatrix} 0 & -I \\ 0 & 0 \end{pmatrix} \begin{pmatrix} A & B \\ B^\dagger & D \end{pmatrix} + \begin{pmatrix} 0 & 0 \\ 0 & I \end{pmatrix} \right] \left[\begin{pmatrix} I & 0 \\ 0 & 0 \end{pmatrix} \begin{pmatrix} A & B \\ B^\dagger & D \end{pmatrix} + \begin{pmatrix} 0 & 0 \\ I & 0 \end{pmatrix} \right]^{-1}$$

$$= \begin{pmatrix} -B^\dagger & -D \\ 0 & I \end{pmatrix} \begin{pmatrix} A & B \\ I & 0 \end{pmatrix}^{-1} = \begin{pmatrix} -B^\dagger & -D \\ 0 & I \end{pmatrix} \begin{pmatrix} 0 & I \\ B^{-1} & -B^{-1}A \end{pmatrix}$$

$$= \begin{pmatrix} -DB^{-1} & -B^\dagger + DB^{-1}A \\ B^{-1} & -B^{-1}A \end{pmatrix}.$$

(The condition $\mathcal{A} \in Sp(n, \mathbf{C})$ guarantees that $\alpha(\mathbf{TS}_1)^{-1}\mathcal{A}$ is symmetric, although it does not appear so at first glance; this follows from the complex analogue of Proposition (4.1e,f). In particular, $-B^\dagger + DB^{-1}A = B^{\dagger-1}$.) Thus the integral kernel of the operator in Ω^0 corresponding to $\mathcal{A} \in Sp(n, \mathbf{C})$ is

$$(5.46) \quad K(x,y) = \det{}^{1/2}(iB^{-1}) \exp\left[(\pi i)(-xDB^{-1}x + 2yB^{-1}x - yB^{-1}Ay)\right].$$

At this point the reader will find it rewarding to compare formula (5.46) with Theorem (4.53).

Clearly some connection with the metaplectic representation is lurking here. In fact, the situation is as follows. The boundary $\partial\Sigma_{2n}$ of the Siegel half space consists of symmetric matrices $\mathcal{B}$ such that $\operatorname{Im}\mathcal{B} \geq 0$ but $\det(\operatorname{Im}\mathcal{B}) = 0$; the so-called *distinguished boundary* (essentially the Shilov boundary) is the set of symmetric $\mathcal{B}$ such that $\operatorname{Im}\mathcal{B} = 0$. Now, the closure of the semigroup Ω^0 in the strong operator topology will include operators of the form $T^0_{\mathcal{B}}$ with $\mathcal{B} \in \partial\Sigma_{2n}$, and by Theorem (4.53) these operators belong to the metaplectic representation when $\operatorname{Im}\mathcal{B} = 0$. On the other hand, the transformation $\alpha(\mathbf{TS}_1)$ maps real symmetric matrices to the real symplectic group, and the upshot is that the inverse map $\mathcal{A} \to T^0_{\alpha(\mathcal{TS}_1)^{-1}\mathcal{A}}$, $\mathcal{A} \in Sp(n, \mathbf{R})$, is the metaplectic representation—or at least part of it. The trouble is that the map $\alpha(\mathbf{TS}_1)^{-1}$ is only defined on those $\mathcal{A} = \begin{pmatrix} A & B \\ B^\dagger & D \end{pmatrix}$ for which $\det B \neq 0$, and the remaining metaplectic operators come from "points at infinity" in $\partial\Sigma_{2n}$. In order to get a clear picture it is best to move over to Fock space, where Σ_{2n} is replaced by the Siegel disc Δ_{2n} and there are no points at infinity. This we now do.

4. The Fock Model

We define the Fock oscillator semigroup $\Omega_{\mathcal{F}}$ by

$$\Omega_{\mathcal{F}} = \left\{ BTB^{-1} : T \in \Omega \right\},$$

where B is the Bargmann transform. The explicit description of $\Omega_{\mathcal{F}}$ follows easily from results we have already proved. First, if $\mathcal{A} \in \Sigma_{2n}$, by Proposition (1.81) the integral kernel of $BT_{\mathcal{A}}B^{-1}$ is $B\gamma_{\mathcal{A}}(z, \overline{w})$. Second, by Theorem (4.70), we have

$$(5.47) \quad B\gamma_{\mathcal{A}} = 2^{n/2} \det{}^{-1/2}(I - i\mathcal{A})\Gamma_{(I+i\mathcal{A})(I-i\mathcal{A})^{-1}} = m(\mathbf{U}^{-1}, \mathcal{A})\Gamma_{\alpha(\mathbf{U}^{-1})\mathcal{A}},$$

where $\Gamma_{\mathcal{B}}(\zeta) = e^{(\pi/2)\zeta\mathcal{B}\zeta}$ and $\mathbf{U}^{-1}$ is the matrix which we called $\mathcal{C}$ in Section 4.5, but in dimension $4n$ rather than $2n$, that is,

$$(5.48) \qquad \mathbf{U} = \frac{1}{\sqrt{2}}\begin{pmatrix} -iI & iI \\ I & I \end{pmatrix} \qquad (2n \times 2n \text{ blocks}).$$

Third, by Proposition (4.67), the map $\alpha(\mathbf{U}^{-1})$ is a bijection from Σ_{2n} to the Seigel disc Δ_{2n}. Therefore, we have

(5.49) $$\Omega_{\mathcal{F}} = \{cS_{\mathcal{B}} : \mathcal{B} \in \Delta_{2n}, \ c \in \mathbf{C}\setminus\{0\}\},$$

where $S_{\mathcal{B}}$ is the operator on $\mathcal{F}_n$ defined by

$$S_{\mathcal{B}}F(z) = \int \Gamma_{\mathcal{B}}(z,\overline{w})F(w)e^{-\pi|w|^2}\,dw.$$

We shall use the notation $S_{\mathcal{B}}$ in this way throughout this section.

Let us compute the composition law in $\Omega_{\mathcal{F}}$.

(5.50) Proposition. *Suppose*

$$\mathcal{A}_j = \begin{pmatrix} A_j & B_j \\ B_j^\dagger & D_j \end{pmatrix} \in \Delta_{2n} \text{ for } j = 1, 2.$$

Then

$$S_{\mathcal{A}_1}S_{\mathcal{A}_2} = \det{}^{-1/2}(I - A_2 D_1)S_{\mathcal{A}_3},$$

where

$$\mathcal{A}_3 = \begin{pmatrix} A_1 + B_1 A_2(I - D_1 A_2)^{-1}B_1^\dagger & B_1(I - A_2 D_1)^{-1}B_2 \\ B_2^\dagger(I - D_1 A_2)^{-1}B_1^\dagger & D_2 + B_2^\dagger D_1(I - A_2 D_1)^{-1}B_2 \end{pmatrix}.$$

Proof: The integral kernel of $S_{\mathcal{A}_1}S_{\mathcal{A}_2}$ is

$$\int \Gamma_{\mathcal{A}_1}(z,\overline{u})\Gamma_{\mathcal{A}_2}(u,\overline{w})e^{-\pi|u|^2}\,du$$

$$= \exp\frac{\pi}{2}\left[zA_1 z + \overline{w}D_2\overline{w}\right]\int \exp\frac{\pi}{2}\left[uA_2 u + \overline{u}D_1\overline{u} + 2\overline{u}B_1^\dagger z + 2uB_2\overline{w}\right]e^{-\pi|u|^2}\,du.$$

By Theorem 3 of Appendix A, this last integral equals

$$\det{}^{-1/2}(I - A_2 D_1)\exp\frac{\pi}{2}\Big[B_2\overline{w}\cdot D_1(I - A_2 D_1)^{-1}B_2\overline{w}$$

$$+ 2B_1^\dagger z\cdot(I - A_2 D_1)^{-1}B_2\overline{w} + B_1^\dagger z\cdot A_2(I - D_1 A_2)^{-1}B_1^\dagger z\Big],$$

which yields the desired formula for $\mathcal{A}_3$. ∎

So far we have been getting along with writing $A, B, B^\dagger, D$ for the $n \times n$ blocks of a $2n \times 2n$ symmetric matrix, but in what follows it will be well to have a more canonical notation at hand. Henceforth we shall write

$$\mathcal{A} = \begin{pmatrix} \mathcal{A}_{11} & \mathcal{A}_{12} \\ \mathcal{A}_{21} & \mathcal{A}_{22} \end{pmatrix}.$$

We now identify the normalized oscillator semigroup in Fock space. Examination of Proposition (5.50) will suggest that the correct normalization factor should be $\det^{1/2} \mathcal{A}_{12}$. Thus, we set

$$\Delta_{2n}^0 = \{\, \mathcal{A} \in \Delta_{2n} : \det \mathcal{A}_{12} \neq 0 \,\},$$
$$\Gamma_{\mathcal{A}}^0 = (\det^{1/2} \mathcal{A}_{12})\Gamma_{\mathcal{A}}, \qquad S_{\mathcal{A}}^0 = (\det^{1/2} \mathcal{A}_{12})S_{\mathcal{A}},$$
$$\Omega_{\mathcal{F}}^0 = \{\, (\pm)S_{\mathcal{A}}^0 : \mathcal{A} \in \Delta_{2n}^0 \,\}.$$

If $\mathcal{A}_1$, $\mathcal{A}_2$, and $\mathcal{A}_3$ are as in Proposition (5.50), it follows that

$$S_{\mathcal{A}_1}^0 S_{\mathcal{A}_2}^0 = \pm S_{\mathcal{A}_3}^0.$$

(5.51) Theorem. $\Omega_{\mathcal{F}}^0 = B\Omega^0 B^{-1}$; more precisely, if $\mathbf{U}$ is defined by (5.48) and $\mathcal{A} \in \Delta_{2n}^0$, $B^{-1}S_{\mathcal{A}}^0 B = T_{\alpha(\mathbf{U}^{-1})\mathcal{A}}^0$.

Proof: By the discussion at the beginning of this section, we know that the kernel of $B^{-1}S_{\mathcal{A}}^0 B$ is

$$(\det^{1/2} \mathcal{A}_{12})B^{-1}\Gamma_{\mathcal{A}} = (\det^{1/2} \mathcal{A}_{12})m(\mathbf{U}, \mathcal{A})\gamma_{\alpha(\mathbf{U})\mathcal{A}}.$$

On the other hand, in the previous section we showed that the kernel of $T_{\alpha(\mathbf{U})\mathcal{A}}^0$ is

$$i^{-n/2}m\big(\mathbf{TS}_1, \alpha(\mathbf{U})\mathcal{A}\big)^{-1}\gamma_{\alpha(\mathbf{U})\mathcal{A}}$$

where $\mathbf{TS}_1$ is given by (5.44), so our assertion is that

$$i^{n/2}m(\mathbf{U}, \mathcal{A})m\big(\mathbf{TS}_1, \alpha(\mathbf{U})\mathcal{A}\big) = \det^{-1/2} \mathcal{A}_{12}.$$

But the expression on the left equals $i^{n/2}m(\mathbf{TS}_1\mathbf{U}, \mathcal{A})$, and

(5.52)
$$\mathbf{TS}_1\mathbf{U} = \frac{1}{\sqrt{2}}\begin{pmatrix} 0 & 0 & I & 0 \\ -I & 0 & 0 & 0 \\ 0 & 0 & 0 & I \\ 0 & I & 0 & 0 \end{pmatrix}\begin{pmatrix} -iI & 0 & iI & 0 \\ 0 & -iI & 0 & iI \\ I & 0 & I & 0 \\ 0 & I & 0 & I \end{pmatrix}$$
$$= \frac{1}{\sqrt{2}}\begin{pmatrix} I & 0 & I & 0 \\ iI & 0 & -iI & 0 \\ 0 & I & 0 & I \\ 0 & -iI & 0 & iI \end{pmatrix},$$

so by Lemma 4 of Appendix A,

$$m(\mathbf{TS}_1\mathbf{U}, \mathcal{A}) = \det^{-1/2}\frac{1}{\sqrt{2}}\left[\begin{pmatrix} 0 & I \\ 0 & -iI \end{pmatrix}\begin{pmatrix} \mathcal{A}_{11} & \mathcal{A}_{12} \\ \mathcal{A}_{21} & \mathcal{A}_{22} \end{pmatrix} + \begin{pmatrix} 0 & I \\ 0 & iI \end{pmatrix}\right]$$
$$= \det^{-1/2}\frac{1}{\sqrt{2}}\begin{pmatrix} \mathcal{A}_{21} & \mathcal{A}_{22} + I \\ -i\mathcal{A}_{21} & -i\mathcal{A}_{22} + iI \end{pmatrix} = \det^{-1/2}(i\mathcal{A}_{21}),$$

which is what we want since $\mathcal{A}_{21} = \mathcal{A}_{12}^{\dagger}$. ∎

As an example, let us compute the Hermite semigroup in Fock space. We recall from Theorem (5.27) that

$$e^{-2\pi s(D^2+X^2)} = T^0_{\mathcal{A}(s)} \quad \text{where} \quad \mathcal{A}(s) = i \begin{pmatrix} \coth 2s & -\operatorname{csch} 2s \\ -\operatorname{csch} 2s & \coth 2s \end{pmatrix}.$$

Since

$$\frac{\operatorname{csch} 2s}{1 + \coth 2s} = \frac{1}{\sinh 2s + \cosh 2s} = e^{-2s},$$

an easy computation yields

$$\bigl(I + i\mathcal{A}(s)\bigr)\bigl(I - i\mathcal{A}(s)\bigr)^{-1} = \begin{pmatrix} 0 & e^{-2s} \\ e^{-2s} & 0 \end{pmatrix},$$

so that

$$\Gamma^0_{\alpha(\mathbf{U}^{-1})\mathcal{A}(s)}(z, \overline{w}) = e^{-ns} \exp(\pi e^{-2s} z \overline{w}).$$

Therefore,

$$B e^{-2\pi s(D^2+X^2)} B^{-1} F(z) = e^{-ns} \int e^{\pi e^{-2s} z \overline{w}} F(w) e^{-\pi|w|^2} \, dw = e^{-ns} F(e^{-2s} z).$$

Let us denote by $\overline{\Omega}^0_{\mathcal{F}}$ the closure of $\Omega^0_{\mathcal{F}}$ in the strong operator topology; by Theorem (5.35), this is a subset of the unit ball in the algebra of bounded operators on $\mathcal{F}_n$. We also denote by $\overline{\Delta}_{2n}$ the closure of Δ_{2n}, namely,

$$\overline{\Delta}_{2n} = \bigl\{\, \mathcal{A} \in M_{2n}(\mathbf{C}) : \mathcal{A} = \mathcal{A}^\dagger, \ \|\mathcal{A}\| \le 1 \,\bigr\},$$

and set

$$\overline{\Delta}^0_{2n} = \bigl\{\, \mathcal{A} \in \overline{\Delta}_{2n} : \det \mathcal{A}_{12} \ne 0 \,\bigr\}.$$

The operators $S_{\mathcal{A}}$ and $S^0_{\mathcal{A}}$ are defined for $\mathcal{A} \in \overline{\Delta}^0_{2n}$ just as before. It is not immediately obvious that these operators are bounded on $\mathcal{F}_n$ when $\|\mathcal{A}\| = 1$. However, if $\|\mathcal{A}\| = 1$ and $z \in \mathbf{C}^n$,

$$|\mathcal{A}_{12} z|^2 + |\mathcal{A}_{22} z|^2 = |\mathcal{A}(0, z)|^2 \le |z|^2,$$

so the condition $\det \mathcal{A}_{12} \ne 0$ implies that $\|\mathcal{A}_{22}\| < 1$. Thus $\Gamma_{\mathcal{A}}(z, \cdot) \in \mathcal{F}_n$ for each z, so the integral defining $S^0_{\mathcal{A}} F(z)$ is convergent for all $F \in \mathcal{F}_n$ and $z \in \mathbf{C}^n$. It turns out that $S_{\mathcal{A}}$ is bounded on $\mathcal{F}_n$ for all $\mathcal{A} \in \overline{\Delta}^0_{2n}$, and in fact we have the following result.

(5.53) Theorem. $\overline{\Omega}^0_{\mathcal{F}} = \{ S^0_{\mathcal{A}} : \mathcal{A} \in \overline{\Delta}^0_{2n} \} \cup \{0\}.$

Proof: If $\mathcal{A} \in \overline{\Delta}^0_{2n}$ and $\|\mathcal{A}\| = 1$, let $\mathcal{A}_k = (1 - k^{-1})\mathcal{A}$. Then $\mathcal{A} \in \Delta^0_{2n}$ and

$$e^{-3\pi(|z|^2+|w|^2)}(\Gamma^0_{\mathcal{A}_k} - \Gamma^0_{\mathcal{A}})(z,\overline{w}) \longrightarrow 0 \text{ uniformly on } \mathbf{C}^{2n}.$$

It follows easily that $S^0_{\mathcal{A}_k} F \to S^0_{\mathcal{A}} F$ whenever F is a polynomial, and therefore—since $\|S^0_{\mathcal{A}_k}\| < 1$ for all k by Theorems (5.35) and (5.51)—that $S^0_{\mathcal{A}_k} F \to S^0_{\mathcal{A}} F$ for all $F \in \mathcal{F}_n$. Hence $S^0_{\mathcal{A}} \in \overline{\Omega}^0_{\mathcal{F}}$.

To see that $0 \in \overline{\Omega}^0_{\mathcal{F}}$, let $\mathcal{A}_k = 2^{-k}\left(\begin{smallmatrix} 0 & I \\ I & 0 \end{smallmatrix}\right)$. Then $\mathcal{A}_k \in \Delta^0_{2n}$ for $k \geq 1$ and $\{\Gamma_{\mathcal{A}_k}\}_{k \geq 1}$ is bounded in $\mathcal{F}_{2n}$. It follows that the operator norms (in fact, the Hilbert-Schmidt norms) of the operators $S_{\mathcal{A}_k}$ are bounded, and hence $S^0_{\mathcal{A}_k} = 2^{-kn/2}S_{\mathcal{A}_k} \to 0$ in norm.

On the other hand, suppose $\mathcal{A}_k \in \Delta^0_{2n}$ and $S^0_{\mathcal{A}_k} \to T$ strongly. Let K be the integral kernel of T, according to Proposition (1.68). Then for each $z, w \in \mathbf{C}^n$,

$$K(z,\overline{w}) = \langle TE_w, E_z \rangle_{\mathcal{F}} = \lim \langle S^0_{\mathcal{A}_k} E_w, E_z \rangle = \lim \Gamma^0_{\mathcal{A}_k}(z,\overline{w}),$$

and $|K(z,\overline{w})|$ and $|\Gamma^0_{\mathcal{A}_k}(z,\overline{w})|$ are all bounded by $e^{(\pi/2)(|z|^2+|w|^2)}$ [Proposition (1.68b)], so $\Gamma^0_{\mathcal{A}_k} \to K$ in the topology of analytic functions. Since $\Gamma^0_{\mathcal{A}_k}$ is nonvanishing for all k, by applying Hurwitz's theorem or Rouché's theorem on every complex line through the origin we see that K is either nonvanishing or identically zero. In the latter case, $T = 0$. In the former case, by passing to a subsequence we may assume that $\mathcal{A}_k$ converges to some $\mathcal{A} \in \overline{\Delta}_{2n}$. We have

$$\det \mathcal{A}_{12} = \lim \det [(\mathcal{A}_k)_{12}] = \lim \Gamma^0_{\mathcal{A}_k}(0,0) = K(0,0) \neq 0,$$

so $\mathcal{A} \in \overline{\Delta}^0_{2n}$, and $\Gamma^0_{\mathcal{A}_k} \to \Gamma^0_{\mathcal{A}}$ in the topology of analytic functions. Thus $K = \Gamma^0_{\mathcal{A}}$ and $T = S^0_{\mathcal{A}}$. ∎

Now we can see how the metaplectic representation fits in. We recall from Sections 4.1 and 4.2 that

$$\mathcal{W} = \frac{1}{\sqrt{2}}\begin{pmatrix} I & iI \\ I & -iI \end{pmatrix}, \qquad Sp_c = \mathcal{W}(Sp)\mathcal{W}^{-1},$$

and the metaplectic representation ν of Sp_c is given by

$$\nu\left[\begin{pmatrix} P & Q \\ \overline{Q} & \overline{P} \end{pmatrix}\right] F(z)$$

$$= (\det^{-1/2} P) \int e^{(\pi/2)(z\overline{Q}P^{-1}z + 2\overline{w}P^{-1}z - \overline{w}P^{-1}Q\overline{w})} F(w) e^{-\pi|w|^2} dw.$$

Thus, in our current notation,

$$(5.54) \qquad \nu\left[\begin{pmatrix} P & Q \\ \overline{Q} & \overline{P} \end{pmatrix}\right] = S_T^0, \qquad T = \begin{pmatrix} \overline{Q}P^{-1} & P^{\dagger-1} \\ P^{-1} & -P^{-1}Q \end{pmatrix}.$$

(5.55) Proposition. *The map*

$$V\left[\begin{pmatrix} P & Q \\ \overline{Q} & \overline{P} \end{pmatrix}\right] = \begin{pmatrix} \overline{Q}P^{-1} & P^{\dagger-1} \\ P^{-1} & -P^{-1}Q \end{pmatrix}$$

is a bijection from Sp_c *to* $\overline{\Delta}_{2n}^0 \cap U(2n)$.

Proof: This argument relies entirely on Proposition (4.17). If $\begin{pmatrix} P & Q \\ \overline{Q} & \overline{P} \end{pmatrix} \in$ Sp_c and $T = V\left[\begin{pmatrix} P & Q \\ \overline{Q} & \overline{P} \end{pmatrix}\right]$, then T is symmetric, $\det T_{12} \neq 0$, and

$$TT^* = \begin{pmatrix} \overline{Q}P^{-1}P^{*-1}Q^{\dagger} + P^{\dagger-1}P^{-1} & \overline{Q}P^{-1}P^{*-1} - P^{\dagger-1}Q^*P^{*-1} \\ P^{-1}P^{*-1}Q^{\dagger} - P^{-1}Q\overline{P}^{-1} & P^{-1}P^{*-1} + P^{-1}QQ^*P^{*-1} \end{pmatrix}.$$

Now, $\overline{Q}P^{-1} = P^{\dagger-1}Q^*$ and $Q\overline{P}^{-1} = P^{*-1}Q^{\dagger}$, so the off-diagonal entries are zero. The lower right entry is

$$P^{-1}(I + QQ^*)P^{*-1} = P^{-1}(PP^*)P^{*-1} = I,$$

and the upper left entry is

$$\overline{Q}(I + Q^{\dagger}\overline{Q})Q^{\dagger} + (I + \overline{Q}Q^{\dagger})^{-1} = ((QQ^{\dagger})^{-1} + I)^{-1} + (I + \overline{Q}Q^{\dagger})^{-1}$$
$$= (I + \overline{Q}Q^{\dagger})^{-1}\overline{Q}Q^{\dagger} + (I + \overline{Q}Q^{\dagger})^{-1} = I.$$

(This calculation is valid if $\det Q \neq 0$; the result is then true in general by continuity.)

On the other hand, suppose $T = \begin{pmatrix} A & B \\ B^{\dagger} & D \end{pmatrix}$ is unitary and symmetric, and $\det B \neq 0$. Then

$$TT^* = T\overline{T} = \begin{pmatrix} A\overline{A} + BB^* & A\overline{B} + B\overline{D} \\ B^*A + DB^* & B^{\dagger}\overline{B} + D\overline{D} \end{pmatrix} = \begin{pmatrix} I & 0 \\ 0 & I \end{pmatrix}.$$

Define $P = B^{\dagger-1}$ and $Q = -B^{\dagger-1}D$. The equation $B^{\dagger}\overline{A} + DB^* = 0$ implies that $Q = \overline{A}B^{*-1}$, so $T = V\left[\begin{pmatrix} P & Q \\ \overline{Q} & \overline{P} \end{pmatrix}\right]$. Moreover, $P^{\dagger}\overline{Q} = Q^*P$ since $\overline{Q}P^{-1} = A$ is symmetric, and

$$P^*P - Q^{\dagger}\overline{Q} = \overline{B}^{-1}(I - \overline{A}A)B^{\dagger-1} = \overline{B}^{-1}\overline{B}B^{\dagger}B^{\dagger-1} = I,$$

so $\begin{pmatrix} P & Q \\ \overline{Q} & \overline{P} \end{pmatrix} \in Sp_c$. ∎

(5.56) Proposition. *Suppose $\mathcal{A} \in \overline{\Delta}_{2n}^0$. Then $S_{\mathcal{A}}^0$ is unitary (on $\mathcal{F}_n$) if and only if $\mathcal{A}$ is unitary (on $\mathbf{C}^n$).*

Proof: Let $\mathcal{A} = \begin{pmatrix} A & B \\ B^\dagger & D \end{pmatrix}$. It is easily verified that

$$(S_{\mathcal{A}}^0)^* = S_{\mathcal{A}'}^0, \quad \text{where } \mathcal{A}' = \begin{pmatrix} \overline{D} & B^* \\ \overline{B} & \overline{A} \end{pmatrix},$$

so by Proposition (5.50), $S_{\mathcal{A}}^0 (S_{\mathcal{A}}^0)^* = S_{\mathcal{B}}^0$ where

$$\mathcal{B} = \begin{pmatrix} A + B\overline{D}(I - D\overline{D})^{-1}B^\dagger & B(I - \overline{D}D)^{-1}B^* \\ \overline{B}(I - D\overline{D})^{-1}B^\dagger & \overline{A} + \overline{B}D(I - \overline{D}D)^{-1}B^* \end{pmatrix}.$$

On the other hand, the identity operator on $\mathcal{F}_n$ is $S_{\mathcal{C}}^0$ where $\mathcal{C} = \begin{pmatrix} 0 & I \\ I & 0 \end{pmatrix}$, so $S_{\mathcal{A}}^0$ is unitary if and only if

$$\overline{B}(I - D\overline{D})^{-1}B^\dagger = I, \qquad A + B\overline{D}(I - D\overline{D})^{-1}B^\dagger = 0.$$

The first equation is equivalent to

$$(I - D\overline{D})^{-1} = \overline{B}^{-1}B^{\dagger -1}, \quad \text{or} \quad I - D\overline{D} = B^\dagger \overline{B}.$$

When this holds, the second equation becomes

$$A + B\overline{D}\,\overline{B}^{-1} = 0, \quad \text{or} \quad A\overline{B} + B\overline{D} = 0,$$

and we have

$$A\overline{A} = (-B\overline{D}\,\overline{B}^{-1})(-\overline{B}DB^{-1}) = B\overline{D}DB^{-1} = B(I - B^*B)B^{-1} = I - BB^*.$$

Since $\mathcal{A}$ is symmetric, these equations are equivalent to the unitarity of $\mathcal{A}$. ∎

Combining Propositions (5.55) and (5.56) with (5.54), we have proved:

(5.57) Theorem. *The set of unitary operators in $\overline{\Omega}_{\mathcal{F}}^0$ is the range of the metaplectic representation ν.*

We can now transfer these results back to the Schrödinger model. In summary, the strong closure of Ω^0 contains all operators $T_{\mathcal{A}}^0$ with $\mathcal{A} \in M_{2n}(\mathbf{C})$, $\mathcal{A} = \mathcal{A}^\dagger$, $\operatorname{Im}\mathcal{A} \geq 0$, and $\det \mathcal{A}_{12} \neq 0$, plus some others coming from "points at infinity" in the closure of Σ_{2n}^0; equivalently, it contains all operators $\gamma_{\mathcal{A}}^0(D, X)$ with $\mathcal{A} \in M_{2n}(\mathbf{C})$, $\mathcal{A} = \mathcal{A}^\dagger$, $\operatorname{Im}\mathcal{A} \geq 0$, and $\det(I - \frac{1}{2}\mathcal{A}J) \neq 0$, plus others coming from points at infinity. The unitary elements of the closure are the metaplectic operators $\mu(\mathcal{A})$, $\mathcal{A} \in Sp$.

It is worth remarking that the closure of Ω^0 contains the classical heat-diffusion semigroup $\{H_t : \operatorname{Re} t > 0\}$ defined by

$$H_t f(x) = t^{-n/2} \int e^{-\pi(x-y)^2/t} f(y) \, dy$$

that solves the initial value problem $\partial u/\partial t = -\pi D_x^2 u$, $u(x,0) = f(x)$. Indeed,

$$H_t = T^0_{\mathcal{A}(t)} = \gamma^0_{\mathcal{B}(t)}(D,X)$$

where

$$\mathcal{A}(t) = \frac{i}{t}\begin{pmatrix} I & -I \\ -I & I \end{pmatrix}, \qquad \mathcal{B}(t) = it \begin{pmatrix} I & 0 \\ 0 & 0 \end{pmatrix}.$$

One more ingredient is necessary to complete the picture. The map V defined in Proposition (5.55) is not linear fractional, but it agrees with a linear fractional map on Sp_c. Namely, let

$$\mathbf{V} = \begin{pmatrix} 0 & I & 0 & 0 \\ 0 & 0 & I & 0 \\ I & 0 & 0 & 0 \\ 0 & 0 & 0 & I \end{pmatrix} \in M_{4n}(\mathbf{C}).$$

Then for any $\mathcal{A} = \begin{pmatrix} A & B \\ C & D \end{pmatrix} \in M_{2n}(\mathbf{C})$ with $\det A \neq 0$,

$$\alpha(\mathbf{V})\mathcal{A} = \left[\begin{pmatrix} 0 & I \\ 0 & 0 \end{pmatrix}\begin{pmatrix} A & B \\ C & D \end{pmatrix} + \begin{pmatrix} 0 & 0 \\ I & 0 \end{pmatrix} \right]\left[\begin{pmatrix} I & 0 \\ 0 & 0 \end{pmatrix}\begin{pmatrix} A & B \\ C & D \end{pmatrix} + \begin{pmatrix} 0 & 0 \\ 0 & I \end{pmatrix} \right]^{-1}$$

$$= \begin{pmatrix} C & D \\ I & 0 \end{pmatrix}\begin{pmatrix} A & B \\ 0 & I \end{pmatrix}^{-1} = \begin{pmatrix} C & D \\ I & 0 \end{pmatrix}\begin{pmatrix} A^{-1} & -A^{-1}B \\ 0 & I \end{pmatrix}$$

$$= \begin{pmatrix} CA^{-1} & -CA^{-1}B + D \\ A^{-1} & -A^{-1}B \end{pmatrix}.$$

In particular, if $\mathcal{A} = \begin{pmatrix} P & Q \\ \overline{Q} & \overline{P} \end{pmatrix} \in Sp_c$,

$$\alpha(\mathbf{V})\mathcal{A} = \begin{pmatrix} \overline{Q}P^{-1} & -\overline{Q}P^{-1}Q + \overline{P} \\ P^{-1} & -P^{-1}Q \end{pmatrix},$$

and by Proposition (4.17),

$$-\overline{Q}P^{-1}Q + \overline{P} = -P^{\dagger-1}Q^*Q + \overline{P} = -P^{\dagger-1}(P^{\dagger}\overline{P} - I) + \overline{P} = P^{\dagger-1}.$$

Thus $\alpha(\mathbf{V})\mathcal{A} = V(\mathcal{A})$. Moreover, if we combine $\mathbf{V}$ with the matrices $\mathbf{T}$, $\mathbf{S}_1$, and $\mathbf{U}$ that we have used in our previous calculations, by (5.52) we have

$$\mathbf{TS}_1\mathbf{UV} = \frac{1}{\sqrt{2}}\begin{pmatrix} I & I & 0 & 0 \\ -iI & iI & 0 & 0 \\ 0 & 0 & I & I \\ 0 & 0 & -iI & iI \end{pmatrix} = \begin{pmatrix} \mathcal{W}^{-1} & 0 \\ 0 & \mathcal{W}^{-1} \end{pmatrix},$$

so that

$$(5.58) \qquad\qquad \alpha(\mathbf{TS}_1\mathbf{UV})\mathcal{A} = \mathcal{W}^{-1}\mathcal{A}\mathcal{W}.$$

In other words, $\alpha(\mathbf{TS}_1\mathbf{UV})$ is the canonical isomorphism from Sp_c to Sp.

Therefore, we have the following situation. Let Osc denote the image of Σ_{2n}^0 in $Sp(n,\mathbf{C})$ under the Cayley transform

$$\mathcal{A} \longrightarrow \mathsf{c}(\tfrac{1}{2}\mathcal{A}\mathcal{J}) = (I + \tfrac{1}{2}\mathcal{A}\mathcal{J})(I - \tfrac{1}{2}\mathcal{A}\mathcal{J})^{-1}$$

as in Theorem (5.38), and let $\mathsf{Osc}_c = \mathcal{W}(\mathsf{Osc})\mathcal{W}^{-1}$ denote the corresponding subsemigroup of the complexification of Sp_c in $GL(2n,\mathbf{C})$. The equation (5.58) shows that $\alpha(\mathbf{V})$ maps Osc_c onto Δ_{2n}^0, since $\alpha(\mathbf{TS}_1\mathbf{U})$ maps Δ_{2n}^0 onto Osc. Thus, if we set

$$\Sigma_{2n}' = \{\, \mathcal{A} \in \Sigma_{2n} : \det \mathcal{A}_{12} \neq 0 \,\},$$

we have a commutative diagram:

$$\begin{array}{ccccccccc}
\mathsf{Osc}_c & \xrightarrow{\alpha(\mathbf{V})} & \Delta_{2n}^0 & \xrightarrow{\alpha(\mathbf{U})} & \Sigma_{2n}' & \xrightarrow{\alpha(\mathbf{S}_1)} & \Sigma_{2n}^0 & \xrightarrow{\alpha(\mathbf{T})} & \mathsf{Osc} \\
\downarrow{\scriptstyle I} & & \downarrow{\scriptstyle F_1} & & \downarrow{\scriptstyle F_2} & & \downarrow{\scriptstyle F_3} & & \downarrow{\scriptstyle I} \\
\mathsf{Osc}_c & \xrightarrow{G_1} & \Omega_{\mathcal{F}}^0 & \xrightarrow{G_2} & \Omega^0 & \xrightarrow{G_3} & \Omega^0 & \xrightarrow{G_4} & \mathsf{Osc}
\end{array}$$

Here, the first and last vertical arrows are the identity map, and the middle three are the correspondences between matrices and operators:

$$F_1(\mathcal{A}) = S_{\mathcal{A}}^0, \qquad F_2(\mathcal{A}) = T_{\mathcal{A}}^0, \qquad F_3(\mathcal{A}) = \gamma_{\mathcal{A}}^0(D, X).$$

On the bottom line, $G_1(\mathcal{A}) = S_{\alpha(\mathbf{V})\mathcal{A}}^0$, $G_2(T) = B^{-1}TB$ where B is the Bargmann transform, G_3 is the identity, and G_4 is the map of Theorem (5.38). The maps F_1, F_2, F_3, and G_1 are double-valued, and G_4 is 2-to-1; the others are bijections. Moreover, the composite of all the horizontal maps on either top or bottom is the isomorphism $\mathcal{A} \to \mathcal{W}^{-1}\mathcal{A}\mathcal{W}$ from Osc_c to Osc.

The maps G_j on the bottom row are all semigroup homomorphisms. (We know this is true of G_2, G_3, G_4, and the composite $G_1 G_2 G_3 G_4$; hence it is also true of G_1.) They extend continuously to the closures of these semigroups:

$$\overline{\mathrm{Osc}_c} \xrightarrow{G_1} \overline{\Omega}^0_{\mathcal{F}} \xrightarrow{G_2} \overline{\Omega}^0 \xrightarrow{G_3} \overline{\Omega}^0 \xrightarrow{G_4} \overline{\mathrm{Osc}}.$$

Finally, the "distinguished boundaries" of Osc_c and Osc are Sp_c and Sp respectively, and we have

$$G_1 | Sp_c = \nu, \qquad G_4^{-1} | Sp = \mu$$

where ν and μ are the Fock and Schrödinger models of the metaplectic representation.

APPENDIX A.
GAUSSIAN INTEGRALS
AND A LEMMA ON DETERMINANTS

The main result of this appendix is the following theorem.

Theorem 1. *Let A be an $n \times n$ complex matrix such that $A = A^\dagger$ and $\operatorname{Re} A$ is positive definite. Then for any $z \in \mathbf{C}^n$,*

$$(1) \qquad \int e^{-\pi x A x - 2\pi i z x} dx = (\det^{-1/2} A) e^{-\pi z A^{-1} z},$$

where the branch of the square root is determined by the requirement that $\det^{-1/2} A > 0$ when A is real and positive definite.

Proof: We proceed in four steps.

Step 1. Suppose $n = 1$ and A (which is now a number) is real. Let

$$I(z) = \int e^{-\pi A x^2 - 2\pi i z x} dx \qquad (A > 0, \; z \in \mathbf{C}).$$

Then

$$I'(z) = \int (-2\pi i x) e^{-\pi A x^2 - 2\pi i z x} dx = \frac{i}{A} \int \frac{d}{dx}(e^{-\pi A x^2}) e^{-2\pi i z x} dx$$

$$= -\frac{i}{A} \int e^{-\pi A x^2} \frac{d}{dx}(e^{-2\pi i z x}) dx = -\frac{2\pi z}{A} I(z).$$

It follows that

$$\frac{d}{dz}\left[I(z) e^{\pi z^2 / A} \right] = 0,$$

so that

$$I(z) = C e^{-\pi z^2 / A}$$

where

$$C = I(0) = \int e^{-\pi A x^2} dx = A^{-1/2}.$$

Step 2. Back to the case of general n, suppose A is real and diagonal. Then the integral in (1) is a product of one-dimensional integrals, each of which can be evaluated by Step 1. The result follows.

Step 3. Now suppose A is real but not diagonal. Since $A = A^\dagger$ there is a rotation R such that $R^\dagger A R = D$ is diagonal. Setting $x = Ry$ and using the fact that $R^\dagger = R^{-1}$, by Step 2 we obtain

$$\int e^{-\pi x A x - 2\pi i z x} dx = \int e^{-\pi y D y - 2\pi i z R y} dy = (\det^{-1/2} D) e^{-\pi (R^{-1}z) \cdot D^{-1}(R^{-1}z)}$$

$$= (\det^{-1/2} A) e^{-\pi z A^{-1} z}.$$

Step 4. We observe that the the integral on the left of (1) is an analytic function of the entries $A_{ij} = A_{ji}$ of A in the region $\mathrm{Re}\, A > 0$, as is the expression on the right since this region is simply connected. The general case therefore follows from Step 3 by analytic continuation. ∎

The branch of the square root in (1) can be described a bit more explicitly as follows. Since $A = A^\dagger$, one easily sees that $\mathrm{Re}(\overline{w}Aw) = \overline{w}(\mathrm{Re}\, A)w$ for all $w \in \mathbf{C}^n$, and from this it follows that the eigenvalues $\lambda_1, \ldots, \lambda_n$ of A all have positive real part. We then have

$$\det^{-1/2} A = \lambda_1^{-1/2} \lambda_2^{-1/2} \cdots \lambda_n^{-1/2},$$

where $\lambda_j^{-1/2}$ is the square root of λ_j^{-1} with positive real part.

Although the integral in (1) converges only when $\mathrm{Re}\, A > 0$, the formula (1) remains valid for $\mathrm{Re}\, A$ positive semidefinite provided it is interpreted in the sense of distributions. The most important case is the following.

Theorem 2. *Let B be a real, invertible, symmetric $n \times n$ matrix, and let $F_B(x) = e^{\pi i x B x}$. Then the distribution Fourier transform of F_B is the function $\widehat{F}_B$ defined by*

$$(2) \qquad \widehat{F}_B(\xi) = \det^{-1/2}(-iB) e^{-\pi i \xi B^{-1} \xi} = e^{\pi i \sharp(B)/4} |\det B| e^{-\pi i \xi B^{-1} \xi},$$

where $\sharp(B)$ is the number of positive eigenvalues of B minus the number of negative eigenvalues.

Proof: By Theorem 1, for $\epsilon > 0$ the Fourier transform of

$$F_{B+i\epsilon I}(x) = e^{\pi i x B x - \pi \epsilon x^2}$$

is

$$\widehat{F}_{B+i\epsilon I}(\xi) = \det^{-1/2}(\epsilon I - iB) e^{-\pi \xi(\epsilon I - iB)^{-1} \xi}.$$

As $\epsilon \to 0$, $F_{B+i\epsilon I}$ and $\widehat{F}_{B+i\epsilon I}$ converge pointwise and boundedly, and hence in $\mathcal{S}'$, to F_B and the function on the right of (2), respectively. The specification of the square root follows from the remarks preceding the theorem. ∎

We shall also need the following formula for Gaussian integrals in complex coordinates. In this theorem, dw denotes the volume element in $\mathbf{C}^n$.

Theorem 3. *Let A and D be complex $n \times n$ matrices such that*

$$A = A^\dagger, \qquad D = D^\dagger, \qquad \|A\| \le 1, \qquad \|D\| \le 1, \qquad \|A\|\,\|D\| < 1.$$

Then for any $u, v \in \mathbf{C}^n$,

$$\int \exp \frac{\pi}{2}\left[wAw + \overline{w}D\overline{w} + 2uw + 2v\overline{w}\right] e^{-\pi|w|^2}\, dw$$

$$= \det{}^{-1/2}(I-AD)\exp\frac{\pi}{2}\left[uD(I-AD)^{-1}u + 2v(I-AD)^{-1}u + vA(I-DA)^{-1}v\right].$$

Remark. This formula can be restated in several ways. Since $A = A^\dagger$ and $D = D^\dagger$ we have $(AD)^\dagger = DA$, so

$$\det{}^{-1/2}(I - AD) = \det{}^{-1/2}(I - DA) \quad \text{and} \quad v(I - AD)^{-1}u = u(I - DA)^{-1}v.$$

Moreover,

$$D(I - AD)^{-1} = (I - DA)^{-1}D \quad \text{and} \quad A(I - DA)^{-1} = (I - AD)^{-1}A,$$

since, for example,

$$(I - DA)D = D - DAD = D(I - AD).$$

Proof: Let

$$\xi = (\operatorname{Re} w, \operatorname{Im} w) \in \mathbf{R}^{2n}, \qquad \eta = \tfrac{1}{2}\left(u + v, i(u - v)\right) \in \mathbf{C}^{2n},$$
$$\omega = (w, \overline{w}) \in \mathbf{C}^{2n}, \qquad \zeta = (u, v) \in \mathbf{C}^{2n}.$$

Then $\xi = \mathcal{K}\omega$ and $\eta = i\mathcal{L}\zeta$ where

$$\mathcal{K} = \frac{1}{2}\begin{pmatrix} I & I \\ -iI & iI \end{pmatrix}, \qquad \mathcal{L} = \frac{1}{2}\begin{pmatrix} -iI & -iI \\ I & -I \end{pmatrix} \qquad (n \times n \text{ blocks}).$$

Moreover,

$$uw + v\overline{w} = 2\xi\eta = 2i\xi\mathcal{L}\zeta,$$

and

$$wAw + \overline{w}D\overline{w} - 2|w|^2 = -2\omega T\omega = -2\xi\mathcal{K}^{\dagger -1}T\mathcal{K}^{-1}\xi$$

where

$$T = \frac{1}{2}\begin{pmatrix} -A & I \\ I & -D \end{pmatrix}.$$

Thus,

$$\int_{\mathbf{C}^n} e^{(\pi/2)(wAw+\overline{w}D\overline{w}+2uw+2v\overline{w})} e^{-\pi|w|^2}\, dw = \int_{\mathbf{R}^{2n}} e^{-\pi\xi\mathcal{K}^{\dagger-1}\mathcal{T}\mathcal{K}^{-1}\xi+2\pi i\xi\mathcal{L}\zeta}\, d\xi$$

$$= \det^{-1/2}(\mathcal{K}^{\dagger-1}\mathcal{T}\mathcal{K}^{-1}) e^{-\pi\zeta\mathcal{L}^{\dagger}\mathcal{K}\mathcal{T}^{-1}\mathcal{K}^{\dagger}\mathcal{L}\zeta}.$$

Here we have used Theorem 1, which applies since the hypotheses on A and D guarantee that for some $\epsilon > 0$,

$$\operatorname{Re}\xi\mathcal{K}^{\dagger-1}\mathcal{T}\mathcal{K}^{-1}\xi = |w|^2 - \tfrac{1}{2}\operatorname{Re}(wAw + \overline{w}D\overline{w}) > \epsilon|w|^2.$$

Routine calculations show that

$$\mathcal{L}^T\mathcal{K} = \mathcal{K}^{\dagger}\mathcal{L} = -\tfrac{1}{2}iI, \qquad \mathcal{T}^{-1} = 2\begin{pmatrix} D(I-AD)^{-1} & (I-DA)^{-1} \\ (I-AD)^{-1} & A(I-DA)^{-1} \end{pmatrix},$$

so that

$$\zeta\mathcal{L}^{\dagger}\mathcal{K}\mathcal{T}^{-1}\mathcal{K}^{\dagger}\mathcal{L}\zeta = -\tfrac{1}{4}\zeta\mathcal{T}^{-1}\zeta$$

$$= -\tfrac{1}{2}\left[uD(I-AD)^{-1}u + 2v(I-AD)^{-1}u + vA(I-DA)^{-1}v\right].$$

Also, clearly $\det\mathcal{K} = (i/2)^n$, so

$$\det(\mathcal{K}^{\dagger-1}\mathcal{T}\mathcal{K}^{-1}) = (-1)^n 2^{2n} \det\frac{1}{2}\begin{pmatrix} -A & I \\ I & -D \end{pmatrix} = \det\begin{pmatrix} A & -I \\ I & -D \end{pmatrix}.$$

The proof is therefore completed by invoking the following lemma, which is also useful in other situations.

Lemma 4. *Let* A, B, C, D *be complex* $n \times n$ *matrices such that* $AC = CA$. *Then*

$$\det\begin{pmatrix} A & B \\ C & D \end{pmatrix} = \det(AD - CB).$$

Proof: First suppose that A is invertible: then

$$\begin{pmatrix} A & B \\ C & D \end{pmatrix} = \begin{pmatrix} A & 0 \\ C & I \end{pmatrix}\begin{pmatrix} I & A^{-1}B \\ 0 & D-CA^{-1}B \end{pmatrix}.$$

The matrices on the right are of a simple block-triangular form, so it follows that

$$\det\begin{pmatrix} A & B \\ C & D \end{pmatrix} = (\det A)(\det(D - CA^{-1}B)) = \det(AD - ACA^{-1}B).$$

But $ACA^{-1} = C$, so we are done. If A is not invertible, then $A + \epsilon I$ is invertible for sufficiently small $\epsilon > 0$, and it still commutes with C, so we can apply this argument with A replaced by $A + \epsilon I$ and then let $\epsilon \to 0$. ∎

Remark. The same argument shows that if we assume that $AB = BA$, $BD = DB$, or $CD = DC$ instead of $AC = CA$, we get the same result with $AD - CB$ replaced by $DA - CB$, $DA - BC$, or $AD - BC$ respectively.

APPENDIX B.
SOME HILBERT SPACE RESULTS

The main object of this appendix is to present two useful theorems about operators on Hilbert space, commonly known as Schur's lemma and Cotlar's lemma although they are substantial generalizations of the results originally proved by Schur and Cotlar. Along the way, we shall review some standard facts about spectral theory. (Reed–Simon [122] and Nelson [114] are good references for the latter material.)

The spectral theorem can be stated in several forms, of which the most generally useful is the following: if A is a self-adjoint operator on a Hilbert space $\mathcal{H}$, there exists a measure space (X, μ), a measurable function $\phi : X \to \mathbf{R}$, and a unitary map $U : \mathcal{H} \to L^2(X, \mu)$ such that

$$UAU^{-1}\psi(x) = \phi(x)\psi(x), \qquad \psi \in L^2(X, \mu).$$

If $f : \mathbf{R} \to \mathbf{R}$ is a Borel function, the operator $f(A)$ is defined by

$$Uf(A)U^{-1}\psi(x) = f(\phi(x))\psi(x),$$

the domain of $f(A)$ being the set of all $U^{-1}\psi$ such that ψ and $(f \circ \phi)\psi$ are in $L^2(X, \mu)$. Although (X, μ), ϕ, and U are not canonically determined by A, the operators $f(A)$ are.

If we take $f = \chi_E$, the characteristic function of $E \subset \mathbf{R}$, we obtain the spectral projections for A:

$$P(E) = \chi_E(A).$$

These are orthogonal projections with the property that $P(\mathbf{R}) = I$ and if $E_1, E_2, \dots$ are disjoint then

$$P(E_j)P(E_k) = 0 \quad (j \neq k), \qquad P\left(\bigcup E_j\right) = \sum P(E_j).$$

A can be recovered from the projections $P(E)$ by the formula

$$A = \int \lambda \, dP(\lambda).$$

In more detail, if $\psi_1, \psi_2 \in L^2(X, \mu)$ and $\psi_1 \in \text{Dom}(UAU^{-1})$,

$$
\begin{aligned}
\langle UAU^{-1}\psi_1, \psi_2 \rangle &= \int \phi(x)\psi_1(x)\overline{\psi_2(x)}\, d\mu(x) \\
&= \int_{\mathbf{R}} \int_{\phi^{-1}[\lambda,\, \lambda+d\lambda)} \phi(x)\psi_1(x)\overline{\psi_2(x)}\, d\mu(x)\, d\lambda \\
&= \int_{\mathbf{R}} \int_X \lambda \chi_{[\lambda,\, \lambda+d\lambda)}(\phi(x))\psi_1(x)\overline{\psi_2(x)}\, d\mu(x)\, d\lambda \\
&= \int_{\mathbf{R}} \lambda \langle UdP(\lambda)U^{-1}\psi_1, \psi_2 \rangle = \langle U\int \lambda dP(\lambda)\, U^{-1}\psi_1, \psi_2 \rangle.
\end{aligned}
$$

Proposition 1. *Let $\{f_j\}$ be a uniformly bounded sequence of Borel functions on* **R**. *If $f_j \to f$ uniformly then $f_j(A) \to f(A)$ in the norm topology, and if $f_j \to f$ pointwise then $f_j(A) \to f(A)$ in the strong topology.*

Proof: The first assertion follows from the obvious fact that $\|g(A)\| \le \|g\|_\infty$, while the second one is a corollary of the Lebesgue dominated convergence theorem. ∎

Proposition 2. *If A is bounded, the norm closure of the algebra generated by A contains $f(A)$ for all continuous functions f.*

Proof: If A is bounded, $f(A)$ is determined by the restriction of f to the compact set $[-\|A\|, \|A\|]$. Thus, if f is continuous, $f(A)$ is the uniform limit of polynomials in A by Proposition 1 and the Weierstrass approximation theorem. ∎

In the next two propositions, the projections $P(E)$ could be replaced by $f(A)$ for any bounded Borel function f, but the proof would be more involved.

Proposition 3. *If A is bounded and E is an interval, the projection $P(E) = \chi_E(A)$ is in the strong closure of the algebra generated by A.*

Proof: It is easy to find a bounded sequence $\{f_j\}$ of continuous functions such that $f_j \to \chi_E$ pointwise. The result therefore follows from Propositions 1 and 2. ∎

Proposition 4. *Suppose A is bounded and B is a bounded operator such that $AB = BA$. Then $P(E)B = BP(E)$ for every interval $E \subset \mathbf{R}$.*

Proof: The set of operators that commute with B is clearly a strongly closed algebra. If it contains A, it therefore contains $P(E)$ by Proposition 3. ∎

Let π be a unitary representation of the group G on a Hilbert space $\mathcal{H}$. π is called *irreducible* if the only closed subspaces of $\mathcal{H}$ that are invariant under all the operators $\pi(g)$, $g \in G$, are $\{0\}$ and $\mathcal{H}$.

Suppose π_1 and π_2 are unitary representations of G on $\mathcal{H}_1$ and $\mathcal{H}_2$. An *intertwining* operator between π_1 and π_2 is a bounded operator $T : \mathcal{H}_1 \to \mathcal{H}_2$ such that $\pi_2(g)T = T\pi_1(g)$ for all $g \in G$. We denote the set of intertwining operators between π_1 and π_2 by $\mathcal{C}(\pi_1, \pi_2)$. If $\mathcal{C}(\pi_1, \pi_2)$ contains a unitary operator, π_1 and π_2 are said to be *unitarily equivalent*.

Schur's Lemma I. *Let π be a unitary representation of G on $\mathcal{H}$. Then π is irreducible if and only if $\mathcal{C}(\pi, \pi) = \{cI : c \in \mathbf{C}\}$.*

Proof: If π has a nontrivial closed invariant subspace $\mathcal{M}$, then $\mathcal{M}^\perp$ is also invariant since π is unitary, and it follows easily that the orthogonal projection onto $\mathcal{M}$ belongs to $\mathcal{C}(\pi, \pi)$. Conversely, suppose that $T \in \mathcal{C}(\pi, \pi)$ and $T \neq cI$. Since $\pi(g)^* = \pi(g^{-1})$, T^* also belongs to $\mathcal{C}(\pi, \pi)$, and hence so do $A_1 = \frac{1}{2}(T^* + T)$ and $A_2 = \frac{1}{2}i(T^* - T)$. A_1 and A_2 are self-adjoint, and at least one of them—say A_1—is not a multiple of I. Thus there is an interval $E \subset \mathbf{R}$ such that $P(E) = \chi_E(A_1)$ is neither 0 nor I. But $P(E) \in \mathcal{C}(\pi, \pi)$ by Proposition 4, so the range of $P(E)$ is invariant under π. Thus π is reducible. ∎

Schur's Lemma II. *Let π_1 and π_2 be irreducible unitary representations of G on $\mathcal{H}_1$ and $\mathcal{H}_2$. Then the space $\mathcal{C}(\pi_1, \pi_2)$ has dimension 1 or 0 according as π_1 and π_2 are unitarily equivalent or not.*

Proof: Suppose $T \in \mathcal{C}(\pi_1, \pi_2)$. Since π_1 and π_2 are unitary, we have $T^* \in \mathcal{C}(\pi_1, \pi_2)$ and hence $T^*T \in \mathcal{C}(\pi_1, \pi_1)$. By Schur's lemma I, $T^*T = cI$, so T is a multiple of a unitary operator. Thus, if π_1 and π_2 are inequivalent we must have $T = 0$, so that $\mathcal{C}(\pi_1, \pi_2) = \{0\}$. If π_1 and π_2 are equivalent and U is a unitary element of $\mathcal{C}(\pi_1, \pi_2)$, then $U^{-1}T \in \mathcal{C}(\pi_1, \pi_1)$, so $U^{-1}T = cI$ as above, and hence $\mathcal{C}(\pi_1, \pi_2) = \{cU : c \in \mathbf{C}\}$. ∎

The setting for Cotlar's lemma is as follows. We are given a σ-finite measure space (X, μ) and a family $\{A(x) : x \in X\}$ of bounded operators on a Hilbert space $\mathcal{H}$ such that the function $x \to \langle A(x)u, v \rangle$ is measurable for all $u, v \in \mathcal{H}$ and
$$\sup_{x \in X} \|A(x)\| = M < \infty.$$
Let
$$\mathcal{F} = \{ F \subset X : \mu(F) < \infty \}.$$
If $F \in \mathcal{F}$, we can form the operator
$$A(F) = \int_F A(x)\, d\mu(x)$$
defined by
$$\langle A(F)u, v \rangle = \int_F \langle A(x)u, v \rangle\, d\mu(x) \qquad (u, v \in \mathcal{H}).$$
Since $\|A(x)\| \leq M$, we have $\|A(F)\| \leq M\mu(F)$ for all $F \in \mathcal{F}$.

Cotlar's Lemma. *With notation as above, suppose there exists a measurable function $h : X \times X \to [0, \infty)$ such that*

$$\|A(x)A(y)^*\|^{1/2} \le h(x, y), \qquad \|A(x)^* A(y)\|^{1/2} \le h(x, y),$$

and

$$\sup\left\{ \int_X h(x, y) \, d\mu(y) : x \in X \right\} = C < \infty.$$

Then

$$\sup\left\{ \|A(F)\| : F \in \mathcal{F} \right\} \le C.$$

Moreover, the integral $\int_X A(x) d\mu(x)$ (or rather the net $\{A(F) : F \in \mathcal{F}\}$, where $\mathcal{F}$ is directed by inclusion) converges in the weak operator topology to an operator A such that $\|A\| \le C$.

Proof: If B is any bounded operator on $\mathcal{H}$, we have

(1) $$\|B\|^2 = \sup_{\|u\| \le 1} \langle B^* B u, u \rangle = \|B^* B\| = \|(B^* B)^n\|^{1/n}.$$

(The last two equations are easily derived by applying the spectral theorem to the self-adjoint operator $B^* B$.) Moreover, if $B_1, \ldots B_{2n}$ are bounded operators, we have

$$\|B_1 B_2 \cdots B_{2n}\| \le \|B_1 B_2\| \|B_3 B_4\| \cdots \|B_{2n-1} B_{2n}\|,$$
$$\|B_1 B_2 \cdots B_{2n}\| \le \|B_1\| \|B_2 B_3\| \cdots \|B_{2n-2} B_{2n-1}\| \|B_{2n}\|.$$

Taking the geometric mean of these inequalities, we obtain

(2) $$\|B_1 B_2 \cdots B_{2n}\| \le \left[\|B_1\| \|B_1 B_2\| \|B_2 B_3\| \cdots \|B_{2n-1} B_{2n}\| \|B_{2n}\| \right]^{1/2}.$$

Now, if $F \in \mathcal{F}$,

$$(A(F)^* A(F))^n$$
$$= \int_F \cdots \int_F A(x_1)^* A(x_2) A(x_3)^* \cdots A(x_{2n-1})^* A(x_{2n}) \, d\mu(x_{2n}) \cdots d\mu(x_1),$$

so by (2) and the hypotheses on $A(x)$, $\|(A(F)^* A(F))^n\|$ is bounded by

$$\int_F \cdots \int_F M^{1/2} h(x_1, x_2) h(x_2, x_3) \cdots h(x_{2n-1}, x_{2n}) M^{1/2} \, d\mu(x_{2n}) \cdots d\mu(x_1),$$

which by the hypothesis on h is at most $M \mu(F) C^{2n-1}$. Thus by (1),

$$\|A(F)\| = \|(A(F)^* A(F))^n\|^{1/2n} \le (M \mu(F))^{1/2n} C^{(2n-1)/2n},$$

and letting $n \to \infty$ we obtain $\|A(F)\| \leq C$.

This proves the first assertion. As for the second, what we must show is that the net $\{\langle A(F)u, v\rangle\}$ is Cauchy for all $u, v \in \mathcal{H}$. Once this is known, since $|\langle A(F)u, v\rangle| \leq C\|u\|\,\|v\|$ for all F we can define A by

$$\langle Au, v\rangle = \lim\langle A(F)u, v\rangle \qquad (u, v \in \mathcal{H}),$$

and we have $\|A\| \leq C$.

First suppose $v = A(x_0)w$: then for any $G \in \mathcal{F}$,

$$|\langle A(G)u, v\rangle| = \left|\int_G \langle A(x_0)^*A(x)u, w\rangle\, d\mu(x)\right| \leq \|u\|\,\|w\| \int_G h(x_0, x)^2 d\mu(x).$$

Since $\|A(x_0)^*A(x)\| \leq M^2$, we may assume that $h(x_0, \cdot) \leq M$, so that $h(x_0, \cdot)$ is in $L^2(X)$. Given $\epsilon > 0$, we can then choose $F_0 \in \mathcal{F}$ such that

$$\|u\|\,\|w\| \int_{X\backslash F_0} h(x_0, x)^2 d\mu(x) < \tfrac{1}{2}\epsilon.$$

But then if $F_1, F_2 \supset F_0$, we have

$$|\langle A(F_1)u, v\rangle - \langle A(F_2)u, v\rangle| \leq \|u\|\,\|w\| \left[\int_{F_1\backslash F_0} + \int_{F_2\backslash F_0}\right] h(x_0, x)^2 d\mu(x) < \epsilon,$$

so $\{\langle A(F)u, v\rangle\}$ is Cauchy.

Now let

$$\mathcal{M} = \left\{\sum_1^J A(x_j)w_j : x_j \in X,\ w_j \in \mathcal{H},\ J \in \mathbf{Z}^+\right\}.$$

From the preceding paragraph one easily deduces that $\{\langle A(F)u, v\rangle\}$ is Cauchy whenever $v \in \mathcal{M}$, and hence—thanks to the uniform boundedness of the operators $A(F)$—whenever v is in the closure of $\mathcal{M}$. On the other hand, if $v \in \mathcal{M}^\perp$ then $\langle A(x)u, v\rangle = 0$ for all x and hence $\langle A(F)u, v\rangle = 0$ for all F. It now follows easily that $\{\langle A(F)u, v\rangle\}$ is Cauchy for arbitrary u and v, and the proof is therefore complete. ∎

Remark. In many applications, X is $\mathbf{Z}$ or $\mathbf{Z}^+$ with counting measure. In this case the hypothesis is that

$$\sup_j \sum_k \|A(j)A(k)^*\|^{1/2} \leq C \quad \text{and} \quad \sup_j \sum_k \|A(j)^*A(k)\|^{1/2} \leq C,$$

and the conclusion is that the series $\sum A(j)$ converges in the weak topology to an operator of norm $\leq C$.

The proof that the operators $A(F)$ are uniformly bounded (which is the crucial part of Cotlar's lemma) first appeared in Knapp–Stein [91]. Slightly more general formulations of Cotlar's lemma can be found in Calderón–Vaillancourt [27], [28] and Unterberger [141].

BIBLIOGRAPHY

1. R. Abraham and J. E. Marsden, *Foundations of Mechanics*, 2nd ed., Benjamin–Cummings, Reading, Mass., 1978.

2. R. F. V. Anderson, The Weyl functional calculus, *J. Funct. Anal.* 4 (1969), 240–267.

3. R. F. V. Anderson, On the Weyl functional calculus, *J. Funct. Anal.* 6 (1970), 110–115.

4. R. F. V. Anderson, The multiplicative Weyl functional calculus, *J. Funct. Anal.* 9 (1972), 423–440.

5. M. A. Antonec, The algebra of Weyl symbols and the Cauchy problem for regular symbols, *Mat. Sb.* 107(199) (1978), 20–36; *Math. USSR Sb.* 35 (1979), 317–332.

6. V. I. Arnold, *Mathematical Methods of Classical Mechanics*, Springer-Verlag, New York, 1978.

7. L. Auslander, *Lectures on Nil-theta Functions*, CBMS Lectures no. 34, American Mathematical Society, Providence, R.I., 1977.

8. L. Auslander and R. Tolimieri, *Abelian Harmonic Analysis, Theta Functions, and Function Algebras on a Nilmanifold*, Lecture Notes in Math. no. 436, Springer-Verlag, New York, 1975.

9. L. Auslander and R. Tolimieri, Radar ambiguity functions and group theory, *SIAM J. Math. Anal.* 16 (1985), 577–601.

10. V. Bargmann, On unitary ray representations of continuous groups, *Ann. of Math.* 59 (1954), 1–46.

11. V. Bargmann, On a Hilbert space of analytic functions and an associated integral transform, Part I, *Comm. Pure Appl. Math.* 14 (1961), 187–214.

12. V. Bargmann, Note on Wigner's theorem on symmetry operations, *J. Math. Phys.* 5 (1964), 862–868.

13. V. Bargmann, Group representations on Hilbert spaces of analytic functions, pp. 27–63 in *Analytic Methods in Mathematical Physics* (R. P. Gilbert and R. G. Newton, eds.), Gordon and Breach, New York, 1970.

14. V. Bargmann, P. Butera, L. Girardello, and J. R. Klauder, On the completeness of the coherent states, *Rep. Math. Phys.* 2 (1971), 221–228.

15. M. Beals, C. Fefferman, and R. Grossman, Strictly pseudoconvex domains in C^n, *Bull. Amer. Math. Soc.* (N.S.) 8 (1983), 125–322.

16. R. Beals, A general calculus of pseudodifferential operators, *Duke Math. J.* **42** (1975), 1–42.

17. R. Beals, L^p and Hölder estimates for pseudodifferential operators: necessary conditions, pp. 153–157 in *Harmonic Analysis in Euclidean Spaces*, Proc. Symp. Pure Math. no. 35, vol. 2, American Mathematical Society, Providence, R.I., 1979.

18. R. Beals, L^p and Hölder estimates for pseudodifferential operators: sufficient conditions, *Ann. Inst. Fourier (Grenoble)* **29**(3) (1979), 239–260.

19. R. Beals and C. Fefferman, Spatially inhomogeneous pseudodifferential operators I, *Comm. Pure Appl. Math.* **27** (1974), 1–24.

20. R. Beals, P. C. Greiner, and J. Vauthier, The Laguerre calculus on the Heisenberg group, pp. 189–216 in *Special Functions: Group Theoretical Aspects and Applications* (R. Askey, T. Koornwinder, and W. Schempp, eds.), D. Reidel, Dordrecht, 1984.

21. F. A. Berezin, Wick and anti-Wick operator symbols, *Mat. Sb.* **86**(**128**) (1971), 578–610; *Math. USSR Sb.* **15** (1971), 577–606.

22. F. A. Berezin, Covariant and contravariant symbols of operators, *Izv. Akad. Nauk SSSR* **36** (1972), 1134–1167; *Math. USSR Izv.* **6** (1972), 1117–1151.

23. F. A. Berezin, Quantization, *Izv. Akad. Nauk SSSR* **38** (1974), 1116–1175; *Math. USSR Izv.* **8** (1974), 1109–1165.

24. J. M. Bony, Equivalence des diverses notions de spectre singulier analytique, *Séminaire Goulaouic-Schwartz* 1976-77, no. 3.

25. H. Brauer and H. Weyl, Spinors in n dimensions, *Amer. J. Math.* **57** (1935), 425–449; also pp. 418–442 in Brauer's *Collected Papers*, vol. III, MIT Press, Cambridge, Mass., 1980, and pp. 493–516 in Weyl's *Gesammelte Abhandlungen*, vol. III, Springer-Verlag, New York, 1968.

26. J. Brezin, Harmonic analysis on nilmanifolds, *Trans. Amer. Math. Soc.* **150** (1970), 611–618.

27. A. Calderón and R. Vaillancourt, On the boundedness of pseudodifferential operators, *J. Math. Soc. Japan* **23** (1971), 374–378.

28. A Calderón and R. Vaillancourt, A class of bounded pseudodifferential operators, *Proc. Nat. Acad. Sci. USA* **69** (1972), 1185–1187.

29. E. Cartan, Sur les domaines bornés homogènes de l'espace de n variables complexes, *Abh. Math. Sem. Hamburg* **11** (1935), 116–162; also pp. 1259–1305 in Cartan's *Oeuvres Complètes*, Partie I, tome 2, Gauthier-Villars, Paris, 1952.

30. R. R. Coifman and Y. Meyer, Au delà des opérateurs pseudodifférentiels, *Astérisque* **57** (1978), 1–185.

31. C. E. Cook and M. Bernfeld, *Radar Signals: an Introduction to Theory and Application*, Academic Press, New York, 1967.

32. H. O. Cordes, On compactness of commutators of multiplications and convolutions, and boundedness of pseudodifferential operators, *J. Funct. Anal.* **18** (1975), 115–131.

33. A. Córdoba and C. Fefferman, Wave packets and Fourier integral operators, *Comm. Partial Diff. Eq.* **3** (1978), 979–1005.

34. M. G. Cowling and J. F. Price, Bandwidth versus time concentration: the Heisenberg-Pauli-Weyl inequality, *SIAM J. Math. Anal.* **15** (1984), 151–165.

35. M. G. Cowling, J. F. Price, and A. Sitaram, A qualitative uncertainty principle for semisimple Lie groups, *J. Austral. Math. Soc.*, to appear.

36. M. G. Davidson, The harmonic representation of $U(p, q)$ and its connection with the generalized unit disc, *Pac. J. Math.* **129** (1987), 33–55.

37. N. G. de Bruijn, Uncertainty principles in Fourier analysis, pp. 55–71 in *Inequalities* (O. Shisha, ed.), Academic Press, New York, 1967.

38. N. G. de Bruijn, A theory of generalized functions, with applications to Wigner distributions and Weyl correspondence, *Nieuw Archief voor Wiskunde* (3)**21** (1973), 205–280.

39. L. de Michele and G. Mauceri, L^p multipliers on the Heisenberg group, *Michigan Math. J.* **26** (1979), 361–371.

40. A. S. Dynin, Pseudodifferential operators on the Heisenberg group, *Dokl. Akad. Nauk SSSR* **225** (1975), 1245–1248; *Soviet Math. Dokl.* **16** (1975), 1608–1612.

41. A. S. Dynin, An algebra of pseudodifferential operators on the Heisenberg group: symbolic calculus, *Dokl. Akad. Nauk SSSR* **227** (1976), 792–795; *Soviet Math. Dokl.* **17** (1976), 508–512.

42. Ju. V. Egorov, Canonical transformations and pseudodifferential operators, *Trudy Moskov. Mat. Obšč.* **24** (1971), 3–28; *Trans. Moscow Math. Soc.* **24** (1971), 1–28.

43. M. V. Fedoryuk, The stationary phase method and pseudodifferential operators, *Uspehi Mat. Nauk* **26** (1971), 67–112; *Russian Math. Surveys* **26** (1971), 65–115.

44. C. Fefferman, The uncertainty principle, *Bull. Amer. Math. Soc.* (N.S.) **9** (1983), 129–206.

45. C. Fefferman and D. H. Phong, On positivity of pseudodifferential operators, *Proc. Nat. Acad. Sci. USA* **75** (1978), 4673–4674.

46. C. Fefferman and D. H. Phong, The uncertainty principle and sharp Gårding inequalities, *Comm. Pure Appl. Math.* **34** (1981), 285–331.

47. V. Fock, Verallgemeinerung und Lösung der Diracschen statistischen Gleichung, *Zeit. für Phys.* **49** (1928), 339–357.

48. V. Fock, Konfigurationsraum und zweite Quantelung, *Zeit. für Phys.* **75** (1932), 622–647.

49. G. B. Folland, Applications of analysis on nilpotent groups to partial differential equations, *Bull. Amer. Math. Soc.* **83** (1977), 912–930.

50. G. B. Folland, *Real Analysis*, John Wiley, New York, 1984.

51. G. B. Folland and E. M. Stein, Estimates for the $\bar{\partial}_b$ complex and analysis on the Heisenberg group, *Comm. Pure Appl. Math.* **27** (1974), 429–522.

52. G. B. Folland and E. M. Stein, *Hardy Spaces on Homogeneous Groups*, Mathematical Notes no. 28, Princeton University Press, Princeton, N.J., 1982.

53. D. Gabor, Theory of communication, *J. Inst. Elec. Eng.* **93**(III) (1946), 429–457.

54. B. Gaveau, P. Greiner, and J. Vauthier, Intégrales de Fourier quadratiques et calcul symbolique exact sur le groupe d'Heisenberg, *J. Funct. Anal.* **68** (1986), 248–272.

55. D. Geller, Fourier analysis on the Heisenberg group, *Proc. Nat. Acad. Sci. USA* **74** (1977), 1328–1331.

56. D. Geller, Fourier analysis on the Heisenberg group I: Schwartz space, *J. Funct. Anal.* **36** (1980), 205–254.

57. D. Geller, Spherical harmonics, the Weyl transform and the Fourier transform on the Heisenberg group, *Canad. J. Math.* **36** (1984), 615–684.

58. H. Goldstein, *Classical Mechanics*, Addison-Wesley, Reading, Mass., 1950.

59. M. J. Gotay, Functorial geometric quantization and van Hove's theorem, *Internat. J. Theor. Phys.* **19** (1980), 139–161.

60. P. C. Greiner, J. J. Kohn, and E. M. Stein, Necessary and sufficient conditions for the solvability of the Lewy equation, *Proc. Nat. Acad. Sci. USA* **72** (1975), 3287–3289.

61. H. J. Groenewold, On the principles of elementary quantum mechanics, *Physica* **12** (1946), 405–460.

62. K. I. Gross and R. A. Kunze, Fourier-Bessel transforms and holomorphic discrete series, pp. 79–122 in *Conference in Harmonic Analysis* (D. Gulick and R. Lipsman, eds.), Lecture Notes in Math. no. 266, Springer-Verlag, New York, 1972.

63. K. I. Gross and R. A. Kunze, Bessel functions and representation theory II, *J. Funct. Anal.* **25** (1977), 1–49.

64. A. Grossman, G. Loupias, and E. M. Stein, An algebra of pseudodifferential operators and quantum mechanics in phase space, *Ann. Inst. Fourier (Grenoble)* **18**(2) (1968), 343–368.

65. V. Guillemin and S. Sternberg, *Geometric Asymptotics*, American Mathematical Society, Providence, R.I., 1977.

66. V. Guillemin and S. Sternberg, The metaplectic representation, Weyl operators, and spectral theory, *J. Funct. Anal.* **42** (1981), 128–225.

67. V. Guillemin and S. Sternberg, *Symplectic Techniques in Physics*, Cambridge University Press, Cambridge, U.K., 1984.

68. B. Helffer and J. Nourrigat, *Hypoellipticité Maximale pour des Opérateurs Polynomes de Champs de Vecteurs*, Birkhäuser, Boston, 1985.

69. L. Hörmander, Pseudodifferential operators and non-elliptic boundary value problems, *Ann. of Math.* **83** (1966), 129–209.

70. L. Hörmander, The Weyl calculus of pseudodifferential operators, *Comm. Pure Appl. Math.* **32** (1979), 359–443.

71. L. Hörmander, *The Analysis of Linear Partial Differential Operators*, vol. I, Springer-Verlag, New York, 1983.

72. L. Hörmander, *The Analysis of Linear Partial Differential Operators*, vol. III, Springer-Verlag, New York, 1985.

73. R. Howe, Remarks on classical invariant theory, unpublished manuscript, ∼1977.

74. R. Howe, Quantum mechanics and partial differential equations, *J. Funct. Anal.* **38** (1980), 188–254.

75. R. Howe, On the role of the Heisenberg group in harmonic analysis, *Bull. Amer. Math. Soc.* (N.S.) **3** (1980), 821–843.

76. R. Howe, The oscillator semigroup, to appear.

77. R. L. Hudson, When is the Wigner quasi-probability density non-negative? *Rep. Math. Phys.* **6** (1974), 249–252.

78. D. Iagolnitzer, Microlocal essential support of a distribution and decomposition theorems—an introduction, pp. 121–132 in *Hyperfunctions and Theoretical Physics* (F. Pham, ed.), Lecture Notes in Math. no. 449, Springer-Verlag, New York, 1975.

79. D. Iagolnitzer, Support essentiel et structure analytique des distributions, *Séminaire Goulaouic-Lions-Schwartz* 1974-75, no. 18.

80. J. Igusa, *Theta Functions*, Springer-Verlag, New York, 1972.

81. C. Itzykson, Remarks on boson commutation rules, *Comm. Math. Phys.* **4** (1967), 92–122.

82. C. G. J. Jacobi, *Vorlesungen über Dynamik*, 2nd ed., Supplementary volume in Jacobi's *Gesammelte Werke*, G. Reimer, Berlin, 1884.

83. A. J. E. M. Janssen, Weighted Wigner distributions vanishing on lattices, *J. Math. Anal. Appl.* **80** (1981), 156–167.

84. A. J. E. M. Janssen, Gabor representation of generalized functions, *J. Math. Anal. Appl.* **83** (1981), 377–394.

85. A. J. E. M. Janssen, Positivity of weighted Wigner distributions, *SIAM J. Math. Anal.* **12** (1981), 752–758.

86. A. J. E. M. Janssen, Bargmann transform, Zak transform, and coherent states, *J. Math. Phys.* **23** (1982), 720–731.

87. A. J. E. M. Janssen, A note on Hudson's theorem about functions with nonnegative Wigner distributions, *SIAM J. Math. Anal.* **15** (1984), 170–176.

88. D. S. Jerison, The Dirichlet problem for the Kohn Laplacian on the Heisenberg group, I and II, *J. Funct. Anal.* **43** (1981), 97–142 and 224–257.

89. M. Kashiwara and M. Vergne, on the Segal-Shale-Weil representations and harmonic polynomials, *Invent. Math.* **44** (1978), 1–47.

90. J. R. Klauder, The design of radar signals having both high range resolution and high velocity resolution, *Bell System Tech. J.* **39** (1960), 809–820.

91. A. W. Knapp and E. M. Stein, Intertwining operators for semisimple groups, *Ann. of Math.* **93** (1971), 489–578.

92. J. J. Kohn and L. Nirenberg, An algebra of pseudo-differential operators, *Comm. Pure Appl. Math.* **18** (1965), 269–305.

93. (a) P. Kramer, M. Moshinsky, and T. H. Seligman, Complex extensions of canonical transformations and quantum mechanics, pp. 250–332 in *Group Theory and Applications*, vol. III (E. M. Loebl, ed.), Academic Press, New York, 1975. (b) M. Brunet and P. Kramer, Complex extensions of the representation of the symplectic group associated with the canonical commutation relations, pp. 441–449 in *Group Theoretical Methods in Physics* (A. Janner, T. Janssen, and M. Boon, eds.), Lecture Notes in Physics no. 50, Springer-Verlag, New York, 1976.

94. L. D. Landau and E. M. Lifshitz, *Quantum Mechanics (Non-relativistic Theory)*, 3rd ed., Pergamon Press, Oxford, 1977.

95. P. D. Lax and L. Nirenberg, On stability for difference schemes: a sharp form of Gårding's inequality, *Comm. Pure Appl. Math.* **19** (1966), 473–492.

96. J. Leray, *Lagrangian Analysis and Quantum Mechanics*, MIT Press, Cambridge, Mass., 1981.

97. J. D. Louck, Extensions of the Kibble-Slepian formula for Hermite polynomials using Boson operator methods, *Adv. Appl. Math.* **2** (1981), 239–249.

98. G. W. Mackey, *Mathematical Foundations of Quantum Mechanics*, Benjamin, New York, 1963.

99. G. W. Mackey, Induced representations of locally compact groups and applications, pp. 132–166 in *Functional Analysis and Related Fields* (F. E. Browder, ed.), Springer-Verlag, New York, 1970.

100. G. W. Mackey, *The Theory of Unitary Group Representations*, University of Chicago Press, Chicago, 1976.

101. L. A. Mantini, An integral transform in L^2 cohomology for the ladder representations of $U(p, q)$, *J. Funct. Anal.* **60** (1985), 211–242.

102. G. Mauceri, The Weyl transform and bounded operators on $L^p(\mathbf{R}^n)$, *J. Funct. Anal.* **39** (1980), 408–429.

103. A. Messiah, *Quantum Mechanics* (2 vols.), John Wiley, New York, 1961 and 1962.

104. Y. Meyer, Principe d'incertitude, bases Hilbertiennes, et algèbres d'opérateurs, *Séminaire Bourbaki* 1985-86 = *Astérisque* **145-146** (1987), 209–223.

105. S. G. Mihlin, Singular integral operators, *Uspehi Mat. Nauk* **3**(3) (1948), 29–112; *Amer. Math. Soc. Transl.* no. 24, 1950.

106. W. Miller, *Lie Theory and Special Functions*, Academic Press, New York, 1968.

107. C. C. Moore, Representations of solvable and nilpotent groups and harmonic analysis on nil and solvmanifolds, pp. 3–44 in *Harmonic Analysis on Homogeneous Spaces*, Proc. Symp. Pure Math. no. 26, American Mathematical Society, Providence, R.I., 1973.

108. J. E. Moyal, Quantum mechanics as a statistical theory, *Proc. Camb. Phil. Soc.* **45** (1949), 99–124.

109. D. Mumford, *Tata Lectures on Theta* (2 vols.), Birkhauser, Boston, 1983 and 1984.

110. A. Nachman, The wave equation on the Heisenberg group, *Comm. Partial Diff. Eq.* **7** (1982), 675–714.

111. A. Nagel and E. M. Stein, *Lectures on Pseudo-differential Operators*, Mathematical Notes no. 24, Princeton University Press, Princeton, N.J., 1979.

112. E. Nelson, Analytic vectors, *Ann. of Math.* **70** (1959), 572–615.

113. E. Nelson, Operants: a functional calculus for non-commuting operators, pp. 172–187 in *Functional Analysis and Related Fields* (F. E. Browder, ed.), Springer-Verlag, New York, 1970.

114. E. Nelson, *Topics in Dynamics*, Mathematical Notes no. 9, Princeton University Press, Princeton, N.J., 1970.

115. G. I. Olshanskii, Invariant cones in Lie algebras, Lie semigroups, and holomorphic discrete series, *Funktsional. Anal. Prilozhen.* **15**(4) (1981), 53–66; *Functional Anal. Appl.* **15** (1981), 275–285.

116. J. Peetre, The Weyl transform and Laguerre polynomials, *Le Mathematiche (Catania)* **27** (1972), 301–323.

117. A. M. Perelomov, On the completeness of a system of coherent states, *Teor. Mat. Fiz.* **6** (1971), 213–224; *Theor. Math. Phys.* **6** (1971), 156–164.

118. J. C. T. Pool, Mathematical aspects of the Weyl correspondence, *J. Math. Phys.* **7** (1966), 66–76.

119. J. F. Price, Inequalities and local uncertainty principles, *J. Math. Phys.* **24** (1983), 1711–1714.

120. J. F. Price and A. Sitaram, Functions and their Fourier transforms with support of finite measure for certain locally compact groups, *J. Funct. Anal.*, to appear.

121. J. F. Price and A. Sitaram, Local uncertainty inequalities on groups, I and II, *Trans. Amer. Math. Soc.* and *Proc. Amer. Math. Soc.*, to appear.

122. M. Reed and B. Simon, *Methods of Modern Mathematical Physics I: Functional Analysis*, Academic Press, New York, 1972.

123. C. Rockland, Hypoellipticity on the Heisenberg group: representation-theoretic criteria, *Trans. Amer. Math. Soc.* **240** (1978), 1–52.

124. L. P. Rothschild and E. M. Stein, Hypoelliptic differential operators and nilpotent groups, *Acta Math.* **137** (1976), 247–320.

125. W. Schempp, Radar ambiguity functions, nilpotent harmonic analysis, and holomorphic theta series, pp. 217–260 in *Special Functions: Group Theoretical Aspects and Applications* (R. Askey, T. Koornwinder, and W. Schempp, eds.), D. Reidel, Dordrecht, 1984.

126. I. E. Segal, Foundations of the theory of dynamical systems of infinitely many degrees of freedom I, *Mat.-Fys. Medd. Dansk. Vid. Selsk.* **31** (1959), no. 12.

127. I. E. Segal, *Mathematical Problems of Relativistic Physics*, American Mathematical Society, Providence, R.I., 1963.

128. D. Shale, Linear symmetries of free Boson fields, *Trans. Amer. Math. Soc.* **103** (1962), 149–167.

129. J. Sjöstrand, Singularités analytiques microlocales, *Astérisque* **95** (1982), 1–166.

130. N. K. Stanton, The heat equation in several complex variables, *Bull. Amer. Math. Soc.* (N.S.) **11** (1984), 65–84.

131. S. Sternberg, Some recent results on the metaplectic representation, pp. 117–143 in *Group Theoretical Methods in Physics* (P. Kramer and A. Rieckers, eds.), Lecture Notes in Physics no. 79, Springer-Verlag, New York, 1978.

132. S. Sternberg and J. Wolf, Hermitian Lie algebras and metaplectic representations, *Trans. Amer. Math. Soc.* **238** (1978), 1–43.

133. M. H. Stone, Linear transformations in Hilbert space III: operational methods and group theory, *Proc. Nat. Acad. Sci. USA* **16** (1930), 172–175.

134. M. E. Taylor, Functions of several self-adjoint operators, *Proc. Amer. Math. Soc.* **19** (1968), 91–98.

135. M. E. Taylor, *Pseudodifferential Operators*, Princeton University Press, Princeton, N.J., 1981.

136. M. E. Taylor, Noncommutative microlocal analysis, part I, *Mem. Amer. Math. Soc.* no. 313, 1984.

137. M. E. Taylor, *Noncommutative Harmonic Analysis*, American Mathematical Society, Providence, R.I., 1986.

138. F. Treves, *Topological Vector Spaces, Distributions, and Kernels*, Academic Press, New York, 1967.

139. F. Treves, *Introduction to Pseudodifferential and Fourier Integral Operators*, vol. I, Plenum Press, New York, 1980.

140. A. Unterberger, Oscillateur harmonique et opérateurs pseudodifférentiels, *Ann. Inst. Fourier (Grenoble)* **29**(3) (1979), 201–221.

141. A. Unterberger, Les opérateurs métadifférentiels, pp. 205–241 in *Complex Analysis, Microlocal Calculus and Relativistic Quantum Theory* (D. Iagolnitzer, ed.), Lecture Notes in Physics no. 126, Springer-Verlag, New York, 1980.

142. A. Unterberger, Symbolic calculi and the duality of homogeneous spaces, pp. 237–252 in *Microlocal Analysis* (M. S. Baouendi, R. Beals, and L. P. Rothschild, eds.), Contemporary Mathematics no. 27, American Mathematical Society, Providence, R.I., 1984.

143. L. van Hove, Sur certaines représentations unitaires d'un groupe infini de transformations, *Mem. Acad. Roy. de Belgique, Classe des Sci.* **26** (1951), no. 6.

144. V. S. Varadarajan, *Lie Groups, Lie Algebras, and Their Representations*, Prentice-Hall, Englewood Cliffs, N.J., 1974; reprinted by Springer-Verlag, New York, 1984.

145. N. Ja. Vilenkin, Laguerre polynomials, Whittaker functions and the representations of groups of bordered matrices, *Mat. Sb.* **75**(**117**) (1968), 432–444; *Math. USSR Sb.* **4** (1968), 399–410.

146. J. von Neumann, Die Eindeutigkeit der Schrödingerschen Operatoren, *Math. Ann.* **104** (1931), 570–578; also pp. 221–229 in von Neumann's *Collected Works*, vol. II, Pergamon Press, New York, 1961.

147. J. von Neumann, *Mathematische Grundlagen der Quantenmechanik*, Springer-Verlag, Berlin, 1932; translated as *Mathematical Foundations of Quantum Mechanics*, Princeton University Press, Princeton, N.J., 1955.

148. A. Voros, Asymptotic h-expansions of stationary quantum states, *Ann. Inst. Henri Poincaré*, Sect. A, **26** (1977), 343–403.

149. A. Voros, An algebra of pseudodifferential operators and the asymptotics of quantum mechanics, *J. Funct. Anal.* **29** (1978), 104–132.

150. N. R. Wallach, *Symplectic Geometry and Fourier Analysis*, Math Sci Press, Brookline, Mass., 1977.

151. A. Weil, Sur certaines groupes d'opérateurs unitaires, *Acta Math.* **111** (1964), 143–211; also pp. 1–69 in Weil's *Oeuvres Scientifiques*, vol. III, Springer-Verlag, New York, 1980.

152. A. Weinstein, A symbol class for some Schrödinger equations on $\mathbf{R}^n$, *Amer. J. Math.* **107** (1985), 1–21.

153. H. Weyl, *The Theory of Groups and Quantum Mechanics*, Methuen, London, 1931; reprinted by Dover Publications, New York, 1950.

154. E. T. Whittaker and G. N. Watson, *A Course of Modern Analysis*, 4th ed., Cambridge University Press, Cambridge, U.K., 1927.

155. G. C. Wick, The evaluation of the collision matrix, *Phys. Rev.* **80** (1950), 268–272.

156. E. Wigner, On the quantum correction for thermodynamic equilibrium, *Phys. Rev.* **40** (1932), 749–759.

157. P. M. Woodward, *Probability and Information Theory with Applications to Radar*, Pergamon Press, London, 1953.

158. J. Zak, Dynamics of electrons in solids in external fields, *Phys. Rev.* **168** (1968), 686–695.

159. J. Zak, Angle and phase coordinates in quantum mechanics, *Phys. Rev.* **187** (1969), 1803–1810.

160. S. Zelditch, Reconstruction of symmetries for solutions of Schrödinger's equation, *Comm. Math. Phys.* **90** (1983), 1–26.

INDEX

www.ingramcontent.com/pod-product-compliance
Ingram Content Group UK Ltd.
Pitfield, Milton Keynes, MK11 3LW, UK
UKHW020034211224
452728UK00001B/74